CANCER FROM BEEF

CANCER FROM BEEF

DES, Federal Food Regulation,

and Consumer Confidence

Alan I Marcus

The Johns Hopkins University Press

Baltimore and London

© 1994 The Johns Hopkins University Press
All rights reserved. Published 1994
Printed in the United States of America on acid-free paper
03 02 01 00 99 98 97 96 95 94   5 4 3 2 1

The Johns Hopkins University Press
2715 North Charles Street
Baltimore, Maryland 21218-4319
The Johns Hopkins Press Ltd., London

Library of Congress Cataloging-in-Publication Data

Marcus, Alan I, 1949–
   Cancer from beef : DES, federal food regulation, and consumer
confidence / Alan I Marcus.
        p.   cm.
   Includes bibliographical references and index.
   ISBN 0-8018-4700-1 (acid-free paper)
   1. Feed additives—Toxicology. 2. Diethylstilbestrol—Carcinogenicity.
3. Meat—Contamination. 4. Meat inspection—United States—History.
I. Title.
RA1270.F4M37   1994
615.9′54—dc20   93-21505

A catalog record for this book is available from the British Library.

To Jocelyn Claire Marcus

for her keen mind and impish sense of humor

Contents

Acknowledgments   ix

Introduction   1

I / SCIENCE

1   Partners in Progress   11

2   Cracks in the Facade   26

II / GOVERNMENT

3   Regulation as Opposition   47

4   Monitors and Monitoring   68

III / SOCIETY

5   Hearts, not Minds   91

6   Entitle and Victimize   114

7   The New Synthesis   134

Epilogue   154

Notes   161

Bibliographical Essay   219

Index   229

Acknowledgments

Histories of the recent past, especially those dealing with issues of health or safety, often reveal and evoke intense emotion. The history of DES is no exception. Participants in the imbroglio over the feeding of DES to cattle remain passionate, even though the Food and Drug Administration (FDA) banned the use of the drug as a beef-cattle growth promoter more than a decade ago. In preparing this study, I wrote to DES manufacturers, feed producers named in the various FDA hearings, and many partisans and organizations outside of government asking for difficult-to-obtain documents. Almost all claimed that they had discarded the material I requested and could not provide the information I sought. In almost every case, however, respondents adamantly decried the FDA decision or the process that led to that decision as a travesty, no matter what side of the question they had been on. A handful graciously provided some additional information. My thanks to D. W. Simonsen, of Simonsen Mill, Quimby, Iowa; Charles A. Black, former executive director of the Council for Agricultural Science and Technology, Ames, Iowa; Kent S. Bernard, senior counsel at Pfizer, New York, New York; Peter Cleary, of the Toxic Chemicals Program, Environmental Defense Fund, Washington, D.C.; and Robin L. Rivett, attorney at the Pacific Legal Foundation, Sacramento, California.

DES played a large part at Iowa State University, and it remains controversial on campus today. I first became aware of this cattle feed additive some nine or so years ago when I was lecturing in Lush Auditorium here. A wall plaque read, "This auditorium-classroom with research facilities on the lower floor was financed in part by funds derived from an animal science stilbestrol patent." Current and former members of the university's administration—Daniel J. Zaffarano, former vice-provost for research, and dean of the Graduate College; Dave Topel, dean of the College of Agriculture; Patricia B. Swan, vice-provost for research, and dean of the Graduate College; and D. Michael Warren, former chair of the Committee on Agricultural Bioethics—have been supportive of my efforts. Individual scientists on campus have expressed less enthusiasm and been less forthcoming, while campus ethicists repeatedly revile the development of DES as the archetypal example of what was and is wrong with American universities. I thank them all for their perspectives.

Suzanne D. White and Wallace F. Janssen, FDA historians, helped me identify and gain access to FDA material on stilbestrol feeding. Their assistance was particularly crucial because the DES papers are as yet uncatalogued. The documents are controlled by the FDA, not the National Archives, and are housed in the Washington National Records Center in Suitland, Maryland. The mound of material I needed was taken from those vast warehouses and temporarily deposited in the FDA hearing clerk's office in Rockville, Maryland. I thank the staff for providing space and cheerfully copying the documents I requested. The Special Collections Department of the W. Robert and Ellen Sorge Parks Library at Iowa State University allowed me to peruse and to photocopy documents from the several relevant manuscript collections housed there. Scott Chaffin, a Ph.D. student in Iowa State University's Graduate Program in the History of Technology and Science, helped copy material I selected. He also brought agricultural periodicals from the University of Nebraska Library whenever he returned from his parents' home in Lincoln. Audrey Burton typed an early version of the manuscript, and Carole Kennedy helped with the voluminous correspondence.

Others also contributed. David Hamilton, R. Douglas Hurt, Deborah Fitzgerald, Joanne Goldman, Joyce Selzer, Richard Lowitt, Maureen Ogle, and Norman Jacobson read all or part of the manuscript and offered numerous helpful suggestions. The Vigilantes, an informal group of faculty members in the Iowa State University History Department, read a chapter and carried on a spirited discussion. I have benefited from many fruitful conversations with George McJimsey, Andrejs Plakans, Christie Pope, Zane Miller, and Hamilton Cravens about the recent past in America. Robert Brugger, of Johns Hopkins University Press, and my copyeditor Miriam Kleiger, patiently worked with me to craft this manuscript into its present form. James Harvey Young, the dean of FDA historians, provided encouragement and support, as well as gentle reminders about what I overlooked or ignored. His knowledge, experience, and kindness are truly impressive. Lastly, I thank Jean, Jocelyn, and Gregory Marcus for their interest, involvement, and confidence.

CANCER FROM BEEF

Introduction

Diethylstilbestrol, commonly known as stilbestrol, or DES, was a modern meat production milestone, perhaps the most important single occurrence in the chain of events that culminated in the current methods of production. Wise Burroughs, an Iowa State College ruminant nutritionist, discovered the hormone's cattle growth-promoting properties in 1954. That year the substance received Food and Drug Administration approval, and later a patent as the first artificial animal growth stimulant. DES speeded weight gain by more than 10 percent. DES-enhanced cattle reached market roughly thirty-five days sooner than untreated animals, and consumed about five hundred pounds less feed. Stilbestrol-generated economies tipped the balance from open-field grazing to confined feeding, and encouraged the creation of large commercial feedlots in the midwestern, western, and southern states. Cattlemen turned to DES feeds in droves. By the early 1960s, as many as 95 percent of the nation's cattle feeders used the drug to stimulate gains. Growth-enhancing hormones remain the centerpiece of modern beef production. In that sense, stilbestrol presaged practices associated with the current "biotechnology revolution." Its invention and application stand as radically important events.[1]

This drug did carry potential hazards. As an estrogen, DES was considered a potent carcinogenic agent. That aspect of stilbestrol feeds is the focus of this book. Indeed, no single chemical agent, except for DDT, holds as significant a place in recent regulatory history. The combination of DES's beef-producing powers and its possible harmful effects placed the hormone at the center of the considerable food safety discussion of the late 1950s and 1960s. This discussion occurred primarily within the government and focused on protecting the public from stilbestrol's possible carcinogenic effects while permitting consumers to benefit from its growth-promoting properties. It inspired additional federal food and drug legislation and regulations, as well as new interpretations of established laws and rules. Concern about stilbestrol proved to be a critical factor in the passage in 1958 of the Delaney Cancer Clause, which outlawed the sale of food containing substances that have been found to induce cancer in humans or animals; the DES Proviso, which, as part of the Kefauver-Humphrey Drug Amendment of 1962, modified the Delaney clause to permit stilbestrol use under specified conditions; and the 1968 Animal Drug Amendment, which reclassified and rationalized the federal government's animal growth promotant regulations.

By the 1970s, however, the character of the stilbestrol debate had changed. Discussion changed to agitation, marked by heated public controversy and repeated legal challenges. Congress held any number of hearings on the use of the drug and tried several times in the 1970s to enact special legislation ending its application in animal agriculture. The Senate, led by Edward M. Kennedy and William Proxmire, twice passed prohibitions, but the House did not assent. Scientific organizations became advocates for or against DES, as did the National Cancer Institute and the U.S. Department of Agriculture. Cattlemen, feed manufacturers, and pharmaceutical manufacturers reacted bitterly to any proposed ban. The Natural Resources Defense Council, the Environmental Defense Fund, and several organizations led by Ralph Nader scored the FDA for inaction and repeatedly pointed to the DES episode as proof that government often abdicated its regulatory responsibility. Several practitioners of the New Journalism, the journalism of advocacy, focused on diethylstilbestrol and exposed the cancer-causing drug's extensive integration into the human food supply. Stilbestrol's use in cattle feed became intertwined with the "DES daughters" tragedy. (Physicians had commonly prescribed diethylstilbestrol to pregnant women only to find, twenty years later, that some of these women's daughters suffered from a rare form of vaginal cancer.) The FDA probably spent more money regulating DES in beef cattle than it did on any other drug, and it underwent significant restructuring because of its stilbestrol-related regulatory efforts. In 1979 the FDA finally banned this cattle growth promoter, citing its potential threat to human health.

The response to DES spelled the beginning of the end of two interwoven historical themes. The first was a late nineteenth-century phenomenon; whereas the second had appeared in the 1920s. The identification of science as expertise, and the belief that, once applied, this expertise would revolutionize technique, provided the basis for both. The first theme was the definition of scientific expertise as the consequence of special knowledge and rigorous methodological training received by scientists, and possessed only by them. Beginning in the late nineteenth century, scientists claimed science as their own province, maintained that its pursuit marked them as virtuous and beyond partisanship, and worked to institutionalize themselves and this vision in virtually all spheres of American life. In the case of agriculture, this notion translated into a neat knowledge-application nexus. Agricultural colleges trained scientists and farmers; experiment stations employed scientists to produce new knowledge; extension agents and the agricultural press broadcast this new knowledge; and farmers applied it to their fields. In this framework, basic research was established as legitimate and desirable, even crucial, because it inevitably led to practical applications; and scientists were seen as the only parties capable of pursuing that research.[2]

Late nineteenth-century regulation also stemmed from a concomitant of the first theme: faith in science and, by extension, scientists. Governments adopted expert administration; scientific expertise seemed the sole means to guarantee quality and promote well-being. The new regulatory agencies took on a twofold task: to beat back what had become known as base self-interest, and to overcome ignorance. More precisely, they sought to end cheating in business and to educate the public to make "wise" choices. The passage of the 1906 Pure Food and Drugs Act provided a dramatic example of the new regulatory thrust. The scientific bureaucracy established by this legislation aimed to rid the nation of food adulterators and mislabelers, greedy businessmen who substituted unhealthy or impure foodstuffs for salubrious commodities; it also sought to develop guidelines to teach the public not to fall victim to unscrupulous vendors. Analyzing the chemical composition of marketed products and comparing these results to manufacturers' claims were at the heart of the effort, which promised to detect frauds and to expose scoundrels to an angry populace.[3]

The 1920s saw the beginning of the second theme, when the Progressive Era's dream of experts as perfectly objective decision makers in possession of unique abilities underwent subtle reconceptualization. For the first time, government acknowledged that business and industry had implemented the vision of some decades earlier. Many businesses and industries had hired scientists, each with his or her own specialty, and these scientists furthered corporate interests by translating nature's laws into profitable products. Government regulators and the persons they regulated had become virtually indistinguishable, a portentous omen for the regulatory mania of the 1920s and 1930s. The reliance of government and industry on scientists, who were considered to be impartial and above petty self-interest, fundamentally changed regulation's character. Manufacturers may not have become honest, but it was presumed that their scientists could be little else. Regulation was transformed in one stroke from opposition to partnership; it remained the province of scientific experts, but now these experts worked both in industry and elsewhere. Experts from many spheres of life lent their expertise to collectively solving society's problems, which were by definition resolvable by applying scientific initiatives.

Government worked as coordinator and convener. It needed only to put together multidisciplinary panels of qualified experts to deal with various regulatory problems, and then to execute these experts' recommendations.[4]

This later theme was again clear in agriculture. Manufacturers and business-people had employed countless agricultural college–trained scientists and provided direct research funding to colleges, which promoted numerous new scientific applications in the years after 1920. Hybrid corn, animal genetics, antibiotics, artificial fertilizers, and herbicides and pesticides, especially DDT, stemmed from this type of research. So too did flavor enhancers, emulsifiers, and spoilage retardants, as manufacturers increasingly not only acknowledged that agriculture and food processing depended on applied science, especially chemistry, but also promoted these new products and additives as panaceas that would raise farm incomes, lower consumer prices, and increase corporate profits.[5]

This "scientification" of industry—the vesting of scientists with unprecedented importance within manufacturing—also revolutionized regulation. Corporate scientists graduated from the same schools as did government regulators, and received the same degrees. More important, both groups shared the same professional orientation. A veneer of nonpartisanship and objectivity cloaked both, and ended the hierarchical regulatory relationship of the past. Government scientists had formerly checked the efforts of the self-interested. Now they were called upon to operate as peers with their new partners in industry; confrontation no longer seemed desirable, even necessary. This interactive "progressive partnership" between government scientists and their industrial and university counterparts was institutionalized in a wide variety of governmental initiatives, including the Food, Drug, and Cosmetic Act of 1938, which superseded the 1906 Pure Food and Drugs Act, expanded the government's regulatory power, and took for granted a close partnership between regulators and regulated. Industrial and university scientists, not bureaucratic regulators, provided the research protocols by which a substance was to be judged. These scientists were charged not with proving or guaranteeing safety but only with demonstrating that the product in question was not likely to be harmful. This streamlined bureaucratic procedure seemed to contemporaries to provide a substantial margin of safety because the studies showing that a substance apparently did not cause injury were generated by scientists, who were by definition impartial and objective. But the act did more. As it made regulation a scientific issue, it at once acknowledged the scientific transformation of industry and forced the few laggards, such as manufacturers of patent and over-the-counter medicines, who had cast their lot with advertisers (the new market-forming and market-manipulating experts), to embrace that transformation. FDA approval could rarely be achieved without scientists' participation; the act compelled these backward industries to hire dispassionate scientists to change or certify procedures so that the businesses would meet its provisions.[6]

DES as a cattle growth promoter appeared at a time when these assumptions about science and scientists held firm. The idea of partnership between government scientists and their industrial colleagues established as legitimate the legal rights and responsibilities of the various parties involved in DES regulation, and defined and constrained governmental action and intervention. That relative harmony marking stilbestrol's invention did not persist for long. By the 1960s, the drug became a persistent regulatory question. More significant, it emerged as an issue in a way that contradicted the premises of the progressive partnership.

Regulation under the terms of the partnership took scientific rationality for granted. It was assumed that the accumulation of scientific evidence weighs so heavily that if a substance is harmful first some scientists and then others speak out against and work to prohibit its use; scientific data gathered over time tips the balance, and the new scientific reality leads to calls for regulatory action. Prohibition itself may be delayed as vested interests and others selfishly fight to retain their prize, but the pattern is clear. Regulatory action is the consequence of the weight of scientific documentation.

That certainly was not the case with stilbestrol. Comparison with the history of DDT makes a striking parallel. Scientists in the 1940s, 1950s, and 1960s increasingly collected evidence that DDT harmed wildlife and made its way into the food chain. According to scholars, the evidence proved so overwhelming that by 1972 the Environmental Protection Agency had no recourse but to limit drastically the noxious insecticide's use. Mid-century scientists uncovered no similarly impressive scientific data about DES in beef cattle; the accumulation of scientific evidence played no role in the initial indictment. The DES controversy developed because of a breakdown of a consensus, a crisis of faith. Feeding DES to cattle became suspect and blossomed into a nationwide problem in the 1960s because ideas of a progressive partnership gave way to individuation, including individual perspective. Indeed, by the 1960s few believed that a basis for perennial partnerships remained, and this intellectual dissolution was quickly converted into a new social reality. A broad spectrum of opinions emerged, and groups such as consumer advocates, regulators, journalists, scientists, manufacturers, and cattlemen generally backed up the views they put forward by citing statements by scientists, or the results of "scientific" determinations. Although these new partisans retained the earlier era's form, they abandoned its substance. Ideas of scientific homogeneity and nonpartisanship had provided the foundation for earlier regulatory efforts. They had required nonscientists to suspend their own critical faculties, and to defer to scientists and depend on them exclusively. Acknowledgment by consumer advocates and journalists, however tacit, that scientific opinion could be marshaled on many sides of some regulatory questions shattered that traditional regulatory basis.[7]

Suspicion and contentiousness accompanied the new diversity. In this context, the situation that had marked stilbestrol since its beginnings—the fact that estrogens in quantities great enough to produce physiological and anatomical effects were carcinogenic—made the drug a regulatory "problem." Ironically, no more than a few dozen individuals, people such as Wise Burroughs, Edward Kennedy, Ralph Nader, and Thomas Jukes, an irascible industry scientist, were at the heart of the controversy, voicing their usually colorful opinions about stilbestrol usage and playing an active public role in a dispute that for two decades would captivate Congress, consume the media, and echo throughout the country. These persons attempted to manipulate the media, and this attempt was itself a telling reflection of the breakdown of established regulatory practice. The media matched these efforts with a propensity to fashion selected statements by DES partisans into dramatic verbal confrontations. Graphic declarations of doom or salvation made great theater.

A lack of definitive factual evidence also characterized the DES debates. Despite massive protestations to the contrary, no one—neither regulators, Naderites, toxicologists, journalists, nor anyone else—could prove that DES beef had harmed a single member of the populace. Conversely, no cattleman, agricultural scientist, feed producer, or pharmaceutical manufacturer could prove that it had not. That everyone lacked proof did not stop anyone from asserting that he or she possessed conclusive evidence. But without an agreed-upon frame of reference, what the factions uttered were opinions, not knowledge; hypotheses, not facts.

These forceful, essentially undocumented assertions are surprising. To be sure, persons with money invested in DES might argue volubly. And others whose credibility was at stake might protest just as strenuously. But what was really striking was the lack of effort on both sides to generate material that might help settle the question. The search for persuasive evidence was virtually abandoned, which perhaps implied that these parties understood that they lacked the ability to generate such compelling data, and that they considered stilbestrol's hazardousness at best a subordinate issue. What really seems to have mattered was carrying the day in the highly charged public political arena: winning or losing.

The events surrounding DES were in marked contrast to the regulatory policies and procedures of the preceding several decades. It was the breakdown of that traditional regulatory consensus, not the ability to monitor infinitesimal traces of the substance, that made DES regulation a public issue. Though important in its own right, the DES controversy was symptomatic of regulatory questions in late twentieth-century America. Atomic energy, insecticides, pesticides, genetically altered substances, and virtually everything that we eat, drink, and breathe have been subjected to the same sort of regulatory scrutiny as DES has been. In this partnerless world, a world of uncertainty and indeterminacy, a world predicated upon individuals' decision-making prerogatives, DES-style regulation has become typical rather than exceptional. The drug nonetheless holds a pivotal place in regulatory history. Stilbestrol was among the first food-production substances to suffer these new regulatory strictures; thus, much of the new regulatory apparatus was shaped around stilbestrol.[8]

No group in these debates had more to lose than did the scientists traditionally engaged in regulatory activities, whose authority was directly threatened. These scientists, working in government, business, industry, and academia, reluctantly relinquished their traditional claims to authority, only to adopt a new justification. They demanded continued regulatory preeminence because they were scientists, not because of their science. They conceded that they lacked conclusive evidence, but maintained that they had a special ability to interpret material, which would protect the public weal.

Ironically, the willingness of scientists in government, business, industry, and academia to abandon absolutism for indeterminacy enabled them to remain vital to the emerging regulatory nexus, although not as the primary force. Debates over DES and other drugs made the concept of cost-benefit analysis a regulatory touchstone in the 1970s. The cost-benefit approach established a power-sharing arrangement. With "cost" and "benefit" defined in a myriad of indeterminate ways—economic, environmental, social, or health-related—the approach substituted outrage for an objective determination of risk and preserved each individual's sense of power. It settled nothing in any absolute sense and suggested that a search for an absolute "truth" or self-evident adjudication is folly. Like beauty, cost-benefit analysis was in the eyes of the beholder. It depended entirely on what was chosen to be measured and whose measurements were accepted as valid; it was a completely relativistic intellectual construct. Although I doubt that it was the product of any conscious attempt to protect the prestige of scientists, government, consumer advocates, and the press, and to dramatize the fact that consumers have a significant stake and role in the regulatory decision-making process, it accomplished exactly those things. Indeed, its success in these areas was precisely the reason why the various contentious parties gravitated to and accepted the cost-benefit approach. It permitted each party, and each individual, to participate in regulation, but ensured that no decision ever vanquished anyone permanently. All retained their dignity and their pretense of authority, while living to fight another day. It was, in effect, no-fault regulation, and it offered a considerable degree of self-gratification, although often through self-delusion.

I / SCIENCE

1   Partners in Progress

The discovery of stilbestrol, as well as the drug's regulatory approval and application in feedlots, involved a partnership between state and federal government, industries, universities, producers, and consumers. Science and, by extension, scientists linked these diverse groups; each partner depended on science—seen as objective and nonpartisan, and personified in scientists—to ensure continued well-being. This idea of a partnership was well established at Iowa State College several decades before Wise Burrough's arrival at the college in 1950 and the discovery of DES's beef-promoting prowess. The college's agricultural experiment station was a leading food and fiber research institution, and by the 1940s it had assembled an impressive array of animal scientists. Its interests were market driven, for it concentrated on the state's four leading market crops—corn, soybeans, hogs, and beef cattle. Each of its scientists had an extensive reputation within his field and among farmers and manufacturers, and received crucial research support from the federal and state governments and from industry.[1]

Prior to Burroughs, research on beef cattle at Iowa State College lagged behind research on the state's other major market crops. The relative neglect of beef is striking because virtually every corn and soybean farmer fed fifty or so cattle, a practice that made Iowa the nation's cattle-feeding leader.[2] The omission was magnified further since feeding developments with other animals were often not transferable to cattle. Cattle and sheep are ruminants, cud-chewing animals with multichambered stomachs. Their microorganism-filled rumens digest and utilize cellulose and other substances undigestible by other animals, and thus they have different nutritional requirements.

Roswell Garst, promoter of agricultural science par excellence, helped draw Burroughs to the college. Proprietor of the Garst and Thomas Hi-Bred Corn Company, Garst grew hybrid seed for Pioneer Hi-Bred, the nation's largest hybrid corn producer. Loud, persistent, and forceful, Garst used his Pioneer connections and his clout with farmers to become a tremendous force in Iowa and in American agricultural politics: it was his farm that Nikita Krushchev would tour during the premier's 1959 visit to the United States. Garst was a relentless advocate of expanding production, using the newest scientific techniques, and opening new markets. His hybrid corn business, especially the great mounds of useless corncobs that seed production generated, piqued his interest in cattle feeding. In 1946 he learned of efforts to use finely ground corncobs as a partial replacement for hay or grass—cellulose—in daily cattle rations, and he started to experiment on his Coon Rapids farm. The initial results were promising, and Garst applied continual pressure on Iowa State College to adopt that investigative line. He even selected the serious, deeply religious Wise Burroughs, who had been pursuing similar experiments at Ohio State Experiment Station, as the appropriate investigator. The college ultimately arrived at the same conclusion, and when an offer was actually tendered in late 1950, Burroughs quickly accepted the opportunity to return to his native Iowa.[3]

Burroughs arrived at Iowa State and quickly secured a grant from the Waterloo-based Rath Packing Company to seek a substance—an unidentified nutritional requirement or hormone for rumen microorganisms—that would enable these microbes to digest poor-quality roughage, such as corn cobs, as efficiently as they converted good-quality roughage, such as clover hay, and with as much energy generation. Burroughs recognized only one significant difference, a definitional one, between these two roughage classes: the unidentified digestion-facilitating microbe nutrient existed in plentiful supplies within good roughage.[4]

Although relatively few scientists competed with Burroughs in his specific quest, the idea of finding "x factors" to rearrange the nutritional dynamics and to speed the growth of production animals was commonplace. More to the point, examples of the successful pursuit of that research strategy abounded. In the late 1940s, Karl Folkers and associates discovered Animal Protein Factor (later vitamin B-12), which greatly increased feed utilization in chickens and hogs. At about the same time, Thomas H. Jukes and others at Lederle Laboratories realized that the antibiotic Aureomycin enabled swine to metabolize their feed more effectively. And arsenic was revealed to hold important digestion-enhancing, growth-promoting properties, although it never gained general acceptance.[5]

That sort of analysis led the poultry industry to embrace diethylstilbestrol in the early 1940s. First synthesized by Sir Charles Dodds in 1938, the inexpensive drug almost immediately became the most commonly used estrogen.[6] Endocrinologists had known for some time that estrogens changed the lipid metabolism of birds. When stilbestrol appeared on the market, Fred W. Lorenz, professor of poultry husbandry at the University of California at Davis, subcutaneously injected or implanted large quantities of the drug in chickens. He found that DES radically changed male chickens. They developed female characteristics and, most important, a much more succulent meat; stilbestrol had chemically caponized them. Lorenz noted that when fed in rations, DES showed very little estrogenic activity—digestion seemed in chickens to render diethylstilbestrol virtually estrogenically inert—and he regarded feeding the chemical as useless to the poultry industry.[7] Others confirmed his conclusions,[8] while still others tried DES implants and injections on various animals. Beef cattle received special attention. The best, most sophisticated studies were performed by Frederick N. Andrews, William M. Beeson, and T. W. Perry at the Purdue University Agriculture Experiment Station.

In the 1940s and early 1950s, Purdue was the preeminent ruminant nutrition research center. Much of the early work there dealt with the effects of the endocrine system on meat production, but the Purdue group focused on thyroid hormones because of the thyroid's clear relationship to the basal metabolism rate. Not until Andrews attempted to extend Lorenz's poultry experiments did the group become interested in stilbestrol,[9] and it was not until 1947 that the Purdue contingent introduced DES pellets to lambs and beef cattle. Extrapolating from their poultry work and Lorenz's, they implanted massive doses of the drug in the ears of cattle and lambs. As in the case of chickens, these DES implants chemically castrated male animals. But in cattle and lambs, unlike chickens, drug-induced behavior clearly mitigated against estrogenic substance use: it resulted in numerous animal injuries and made handling stilbestrol-treated animals expensive and dangerous. Equally significant, diethylstilbestrol in doses large enough to simulate castration produced marked deterioration in the quality of meat. And since the prices producers received for animals depended on the grade of their meat, the lowered grade canceled any economic advantage caused by the growth-producing drug. The Purdue group tried to overcome the negative physiological effects of DES, but to no avail. Scientists could not create a dynamic hormonal situation in which the benefits of DES remained while its shortcomings were surmounted.[10]

Burroughs initially sought to confirm the Purdue DES studies but noticed after about seventy days that his controls grew about as fast as the chemically castrated cattle. Clover hay was the only ingredient fed to each group, and by early January 1952 Burroughs had devised a theory to account for this surprising occurrence. He hypothesized that there existed in this particular hay either a substance—an "x factor"—that enabled rumen microorganisms to convert the feed into energy with maximum efficiency, or a natural hormone that facilitated digestion and energy production. Burroughs gained additional data when he surveyed late-1940s Australian veterinary journals, which noted that some hays apparently contained estrogens sufficient to cause behavioral modifications in pastured animals.[11]

Burroughs supposed that the clover hay in his feeding experiment contained estrogen. Moreover, he recognized that if that were indeed true, and if the estrogen produced the growth-promoting effects he had noted, then he had uncovered something very important. His work would have taught him that at least some estrogens were orally active in cattle and that growth-enhancing quantities of these substances need not cause outward physiological changes. This apparent natural phenomenon seemed to Burroughs to hold great practical promise. He hoped to replicate this natural process by adding small amounts of an inexpensive estrogen to ruminants' rations. This idea was the essence of his cattle feed additive invention.[12]

Burroughs tested the hay for estrogenic activity by using the same method that others had used: the mouse bioassay, the standard way to measure the presence of estrogen in a substance and to determine comparative estrogenic strengths. First proposed in the 1920s, it remained an extremely sensitive, highly accurate test, but it took days and was expensive to perform. Investigators fed a group of immature female mice standardized units of a suspected estrogen-containing substance for up to ten days, and then slaughtered them to examine and weigh their uteri. The heavier the uteri, the greater the estrogenic effect of the substance, expressed in equivalent quantities of diethylstilbestrol. If the uteri of a particular lot were normal in size for mice of that age and weight, the sample contained no estrogenic material.[13]

Burroughs believed that he had "circumstantial evidence that the level of estrogens found in the hays fed contributed to the excellent gains made by the lambs." He knew that the estrogen received by his lambs constituted only one-fiftieth of the conventional minimal, or threshold, dosage, a dosage extrapolated from the oral administration of DES to chickens.[14] He also recognized that his stilbestrol studies required additional confirmation before he went public. Yet even as he planned the next set of investigations, he was so encouraged by his preliminary results that he reported them to Phineas Shearer, his department head, who in turn discussed the research with George Browning, the experiment station's associate director. As Burroughs's supervisors, both men had of course been aware of the nature of his work since early 1952, but now Browning recommended that Burroughs talk immediately with Quincy Ayres, secretary-manager of the Iowa State College Research Foundation (ISCRF).[15]

Established in 1938, ISCRF was a monument to the idea of a progressive partnership among industries, universities, and government. It took the Wisconsin Alumni Research Foundation as its model, seeking to control the patentable discoveries of Iowa State College professors and staff, to maximize advantages to the university and the public, and to speed the development of researchers' discoveries into commercial products. In late 1950, ISCRF began a concerted push to encourage the faculty to submit their inventions to the foundation, maintaining that discoveries "are seldom the results of the efforts of a single individual," that research at Iowa State College "usually employs funds that have been contributed by the state [i.e., state government]," and that it was "desirable to insure control for the benefit of the public of the inventions that grew out of the [research staff's] scheduled work." The foundation also promised inventors 15 percent of the foundation's net receipts from the patent, a rather generous amount given the administrative costs, and stipulated that all patent moneys after expenses would go "exclusively for the promotion of research at Iowa State College."[16]

Ayres convinced Burroughs in April 1953 of ISCRF's advantages and made the case for filing a patent application as soon as possible to protect Burroughs's and the college's interests. Burroughs agreed, and on June 3, 1953, he submitted an application assigning to ISCRF the rights to this and any subsequent stilbestrol patent.[17]

With this matter settled, Burroughs experimented to establish the parameters of effective stilbestrol use. These investigations yielded evidence that placing eleven milligrams of the hormone in the daily ration of each head of beef cattle "resulted in 37 percent faster gains, 15 percent greater corn consumption, 20 percent less feed required per pound of gain, and ⅙ less costs per pound." Equally important, cattle subjected to the regimen were, according to a blue-ribbon panel composed of representatives from the Rath Packing Company, the Wilson Packing Company, and the Sioux City livestock market, "better finished" than the controls.[18] Well before the cattle grading, however, Burroughs told his superiors that his research was likely to prove an unqualified success. They decided to end the secrecy surrounding Burroughs's work and to announce its results as soon as the meat packers completed grading, and scheduled a special Iowa Cattle Feeders' Day for February 18, 1954, expressly to announce Burroughs's triumph.[19]

Rumors of an extraordinary discovery at the college quickly circulated in beef-feeding circles, especially among manufacturers of feeds and feed supplements. Burroughs and the officers of ISCRF were responsible for these rumors, as they had dropped several hints about the college's revolutionary feeding research. The college printed five thousand copies of its feeders' day brochure, nearly two and one-half times the usual number, to handle the expected throng.[20] Even this number proved insufficient. The college quickly exhausted its supply of brochures as the crowd trooped through the mud on an unseasonably warm February day to learn about Burroughs's feeding research.[21] Iowa

State College's Information Service maintained that "animal science has achieved a new goal toward which it has been working for several years" and that diethylstilbestrol was "a new word in the cattle feeding industry." Adding this "hormone substance to the steers' feed" would result, it continued, in "speeding up gains and cutting [the] cost of producing beef," which in turn would culminate "in a revolution in the cattle industry." The service accurately reported Burroughs's investigations and data and took great pains to emphasize the preliminary nature of the conclusions. It also prominently stated that farmers could not lawfully use DES in cattle feed until the FDA gave its imprimatur, and that they then could use the drug only with the express permission of ISCRF, the patentee.[22]

As the college had planned, the popular press and the feed manufacturers' and cattlemen's presses began to spread the DES message as soon as the college and Burroughs made their initial announcement. Burroughs himself gave public addresses and wrote articles for feeders' and feed producers' periodicals, and personally answered numerous farmers' and manufacturers' queries. He made a special trip to Kansas City to speak to the annual meeting of the Midwest Feed Manufacturers' Association about stilbestrol the day after he announced his invention to the world. Not until several months later did information about oral stilbestrol's growth-promoting properties appear in conventional scientific or technical journals.[23] That hardly marked DES as peculiar, however. Agricultural science investigations and discoveries regularly were known to feeders and feed manufacturers well in advance of their announcement in traditional scientific periodicals, because agricultural scientists reported to the agricultural community as soon as they obtained the most preliminary findings.[24]

An increasingly formalized dissemination network augmented traditional mechanisms. Agricultural colleges, agricultural scientists, agriculturists, government agencies, and feed producers had long communicated through farmers' institutes, feeders' days, agricultural newspapers, experiment station bulletins, and extension agents, which guaranteed rapid dissemination of the latest relevant inventions and scientific theories.[25] By the 1930s, however, leaders of agribusiness and agri-industry had grown dissatisfied at receiving the newest knowledge in what they considered an irregular, haphazard manner. Equating financial well-being with possession of accurate, modern scientific information, these agribusiness and agri-industry groups established formal institutions by which agricultural college scientists routinely reported their work directly to the most powerful clients. They also sponsored a series of annual agricultural college–based nutrition conferences to bring the labors of college scientists to their attention; and virtually every meeting of these new agribusiness and agri-industry groups was marked by several invited addresses from leading animal nutritionists.[26]

A similar formalization of reporting procedures accompanied the entrance of scientists into agricultural industries in the second third of the twentieth century. These investigators formed technical committees within each agribusiness-dependent or agriculture-dependent group to survey and discuss the most recent developments and to announce new discoveries. The establishment of the Agricultural Research Institute in 1951 was an institutional extension of that approach. "Conceived by industrial scientists," the institute sought to promote "the kinds of research and policies needed to insure the best longtime utilization of agricultural resources for the national welfare." In effect, agri-industry scientists sought to set the agricultural research agenda and develop a firmer connection between the state and federal governments and the agricultural industry.[27] But they never wanted to disrupt extant information transmission pathways; instead, they wanted to expand these pathways and to assume a more vital role in decision making. For agri-industry scientists, as for their college-supported brethren, it made good business sense to keep consumers of science aware of recent investigative trends and familiar with the newest scientific breakthroughs. An informed, progressive public would readily and repeatedly accept new scientific initiatives and adopt new science-based practices.

Feed manufacturers and feeders, then, were well acquainted both with the newest animal nutrition research and with the concepts underpinning that research. Adding substances that do not occur in nature, such as diethylstilbestrol, to rations did not strike them as unusual. Those poised to employ the newest knowledge of nutrition had embraced de facto the plasticity of nutritional requirements, choosing to concentrate instead on results: successful feed rations were those substances that enhanced growth while reducing unit cost. The exact chemical compositions of the feeds were irrelevant. Harvey E. Yantla, editor of *Feedstuffs,* the weekly newspaper of feed manufacturers and dealers, expressed this sentiment simply. Each investigation, he remarked, "takes us farther away from a straight grain-and-hay feeding program. The profits in feeding will depend increasingly upon taking full advantage of the advanced knowledge of nutrition and the practical application of that knowledge."[28]

Federal law and regulation mandated that stilbestrol receive FDA approval as a cattle feed before feed manufacturers could offer the drug for commercial use. Technically, the FDA classified DES used in cattle feed as a new animal drug, reasoning that it was given to animals for a heretofore unapproved purpose. Whatever the drug's classification, the FDA recognized diethylstilbestrol as a potent substance, capable of inducing estrus in animals and menses in humans. Relatively small quantities, a milligram or so daily, could produce in humans enlarged mammary glands and other female secondary sexual characteristics. Even with these side effects, it had found extensive use in human medicine, paradoxically both as a "morning after" pill and as a means of warding off miscarriages. Yet almost from the moment of the drug's discovery, there surfaced evidence of a link between DES and cancer. As early as 1938, studies on cancer-susceptible laboratory rodents had demonstrated a statistically significant relationship between high doses of DES over extended periods and increased incidence of cancer. But scientists, including FDA officials, understood DES's carcinogenic properties as a by-product of the drug's estrogenic activity; they knew that any estrogen, even naturally occurring estrogens in plants and animals, was carcinogenic when ingested for long periods in doses large enough to produce anatomical or physiological changes. In that sense, the trouble was not with the drug but rather with the dosage and its duration.[29]

The need to secure FDA approval compelled Burroughs to do further experiments on stilbestrol's effectiveness and to set parameters for its application. Although FDA approval was an internal agency decision, the agency required patentees or their designates to submit significant scientific evidence testifying to each new drug's safety to help in that decision. And the rigor with which the FDA would scrutinize the DES application would have been unprecedented less than a decade earlier. The agency had considerably upgraded its safety standards after World War II as companies introduced remarkable numbers of new chemical agents.[30]

The dramatically increased volume of requests for FDA approval after the war swamped the already understaffed and underfinanced agency; meanwhile, other factors placed additional pressure on it. Perhaps the most important was the public hearings held in late 1950 and 1951 by the U.S. House of Representatives Select Committee to Investigate the Use of Chemicals in Food Production. Under the chairmanship of James Delaney (D-N.Y.), the committee stood as an acknowledgment of both the recent explosion of the use of chemicals in food and the fear that some substances could prove harmful. It was to be a reaffirmation that the public generally had more than adequate protection. Although stilbestrol played only a minor part in the committee's deliberations, the group did consider it. Because the drug had long ago been cleared by the FDA both as a chicken caponizer and for use in human medicine, the Delaney committee repeatedly questioned whether its use in fowl could constitute a human health hazard. Reports had surfaced that minks fed the heads and necks of implanted chickens grew sterile, and the committee apprehensively tried to determine whether a similar situation might arise among humans. Although the committee members recognized that there was little to worry about if poultry producers followed FDA guidelines, they were concerned that ill-intentioned, inept, or ignorant persons might circumvent proper procedures and cause contaminated material to enter the human food supply.[31]

The Delaney committee suggested that some regulatory law needed interpretation, clarification, and emendation. This was not the first such recommendation. In 1945 members of the New York State Bar Association had created a Food, Drug, and Cosmetic Law Section to pull together experts representing the FDA and industrial and consumer groups, as well as other lawyers and scientists, to discuss regulatory law and plot its course. The section's *Food Drug Cosmetic Law Quarterly,* first published in March 1946, became the medium through which the regulators, the regulated, and the beneficiaries of regulation formalized their progressive partnership and exchanged ideas. The new periodical supplemented the *Federal Register,* the government publication in which all new regulatory rules and procedures were announced, by providing interested parties, including FDA officials, the opportunity and space to explain or elaborate on various decisions or to give reasons for opposition. Two and one-half years later, in September 1948, this tripartite partnership was extended when the American Bar Association formed a 120-member Committee on Food, Drug, and Cosmetic Law to push for the establishment of postgraduate degree programs in that legal field at the nation's law schools. With the blessings of the American Bar Association and the FDA, food industry executives in 1949 formed the nonprofit Food Law Institute at New York University Law School to pursue research and instruction in the new area of food law.[32]

Food law did not long persist as the sole arena in which what today seem to be unlikely partners received a communal hearing. In September 1950, the chemical and food-processing industries provided funding to the Food and Nutrition Board of the National Research Council to create a Food Protection Committee. The new group was charged with making an "objective appraisal and evaluation of pertinent facts" about food chemicals "for the guidance of industry and Government," and by extension, consumers. It established numerous "liaison panels drawn from industry, governmental agencies and trade associations," each of which "integrates and promotes research and assembles and evaluates scientific information." The results of these panels' compilations and/or evaluations were published to inform the public. The committee distributed more than one hundred thousand copies of its 1951 publication *Use of Chemical Additives in Foods.*[33]

These new mechanisms had generated several new pieces of food legislation by the eve of Burroughs's DES discovery. Congress was also considering granting the FDA expanded powers. Burroughs proceeded cautiously in this uncertain milieu. John L. Harvey, associate FDA commissioner for food and drugs, warned him—as well as other interested investigators—that the agency would look favorably on the addition of stilbestrol to cattle feeds *only* if the edible meat would remain hormone free. Others made a similar point to Burroughs and urged him to comply patiently and completely with FDA directives at every turn.[34]

With this caveat in mind, Burroughs worked to determine the conditions under which stilbestrol worked most effectively, and to compile the carcass study information necessary for securing FDA approval. In this latter task, local beef-packing firms assisted him. Burroughs and the packers visually inspected the carcasses to confirm that DES did not reduce beef quality and then used the mouse bioassay to test the meat and internal organs for estrogenic activity, the presence of which would indicate stilbestrol residue in the tissues.[35]

Burroughs and the ISCRF also considered the important question of licensing. At least five chemical companies and twelve feed manufacturers were vying for the opportunity to sell DES-laced feeds. Two New Jersey–based chemical companies, White Laboratories and Merck, built their case on their long experience handling and manufacturing diethylstilbestrol for human medicine, and their familiarity with FDA guidelines and personnel. Several feed producers attempted to replicate Burroughs's early DES studies and to help the Iowa State contingent gather material for its case before the FDA. In mid-April Burroughs scheduled an eastern trip, first to Indianapolis to speak with the officials of Specifide, which made "farmaceuticals for poultry and animal feeds," and then to negotiate with White Laboratories.[36]

ISCRF initially spurned all suitors and planned to produce and market the DES premix (which was to be added to standard feeds) itself. The first trade name selected, "Stil-ISC-trol," was soon discarded for the awkward but more descriptive "Steerfatsrol." But Eli Lilly and Company convinced ISCRF to change its plans. Representatives of the giant pharmaceutical company, by far the nation's largest manufacturer of DES for human medicine, had met briefly with Burroughs during his April visit to Indianapolis. Lilly coupled its stilbestrol experience with a desire to enter the booming agrochemical market and recognized the Burroughs discovery as the perfect vehicle to achieve its ends. More specifically, it contemplated establishing an agricultural products division, the first product of which would be DES cattle premix. In return for generous terms, it sought a five-year guarantee of exclusivity.[37]

The college liked the Lilly offer but wondered whether exclusive licensing might create public antagonism. Leading feeders might suspect that monopolistic conditions were raising premix costs, while consumers might worry about the possibility of hidden human health hazards in stilbestrol-produced meats. To head off that threat, Iowa State president James Hilton called a special confidential meeting of parties likely to be involved in some aspect of DES feeding.

The college invited to the July 20 gathering the progressive partners: selected state agricultural and health officials, state legislators, meat packers, journalists, cattle feeders, feed manufacturers, veterinarians, and trade association representatives. Twenty-six individuals attended and voted unanimously to approve the college's granting Lilly a five-year exclusive DES contract. Bolstered by this show of support, ISCRF announced the agreement on July 29.[38]

About a month before the licensing question was finally resolved, Burroughs presented his initial case for DES feeds to the FDA. His material showed the cattle growth-promoting properties of DES-supplemented feeds, offered summaries of his experiments, and provided the results of tests for residual DES in the lean muscle, fat, liver, heart, kidney, and offal of cattle fed DES at Iowa State College. These tests bore most heavily on FDA approval. Burroughs had assayed the carcasses of forty-eight animals that had been fed various dosages of the drug for various periods of time, some up until slaughter. He used two types of assays, the modified mouse uterine bioassay and a new, slightly more sensitive chemical determination. Only one portion (the liver) of one carcass showed any estrogenic activity. The animal in question had been fed right up to slaughter, and its liver contained a minuscule 1.4 parts of stilbestrol per billion parts of liver. Burroughs calculated that a human would need to consume more than eleven hundred pounds of meat containing that quantity of the drug to receive a one-milligram dose, the normal human therapeutic dose. He also cited a radiotracer study indicating that animals excrete DES from their systems in forty-eight hours, and he suggested that feeders not give the drug to cattle within two days of slaughter.[39]

The FDA rejected Burroughs's evidence. Its statisticians found that the mouse uterine weight data did not always justify his contentions, that his method of establishing controls was faulty, and that tiny amounts of DES might well have been present in several other meat samples. Also, the one reported stilbestrol residue troubled John Harvey. An FDA regulation stated that the FDA "will regard the absence of satisfactory evidence showing that the meat . . . from animals fed the drug is entirely free of any poisonous or deleterious ingredient . . . as ground for refusal to make the application effective"; and Harvey interpreted that regulation as compelling him to reject the approval application. He notified Burroughs of his decision in late August.[40]

Burroughs's evidence had failed to convince the FDA, and the agency acted as a progressive partner should. But FDA rejection did not doom stilbestrol. The agency required its partner to provide further, more precise studies, and Iowa State enlisted Lilly to help make its case. The company enthusiastically complied, testimony to stilbestrol's money-making possibilities and the venture's partnership nature. After reviewing Burroughs's material, three senior

Lilly scientists joined Burroughs in Washington on September 16 to discuss the matter with seven high-ranking FDA officials. The Lilly representatives pressed the idea that the one DES residue uncovered by Burroughs "was so small as to be inconsequential and represent no health hazard," but the FDA demurred. Rather than argue that question explicitly, it maintained that Burroughs's analytical and statistical techniques were flawed. It disregarded his findings entirely, and pledged to work with him "to establish definitely whether there was an increase in estrogenic activity of the tissues of beef fed diethylstilbestrol."[41]

Burroughs brought FDA-amended procedures back to Ames and recruited Paul G. Homeyer, a statistician at the college's statistical laboratory, to assist him. In less than a month, Burroughs and Homeyer had completed their work and prepared reports for FDA scrutiny, and they chose to hand-deliver these materials at another Washington meeting. The Iowa State contingent was again joined by the Lilly people, but the latter group's assistance was not needed. The FDA concluded that the revamped "assay proved quite conclusively that the tissues examined contained no demonstrable diethyl stilbestrol [sic]." Even though the revised Burroughs-Homeyer technique gave evidence of being sensitive to DES at less than 2 parts per billion (ppb), the agency was in favor of "conservative handling," suggesting that it might "be safer to recommend a withholding period of at least 48 hours after the stilbestrol-containing diet has been withdrawn from these animals." This last stipulation became a requirement for use when the FDA officially approved stilbestrol on November 5, 1954.[42]

The newly approved drug took the feed industry by storm, seeming to herald a previously almost unimaginable era of plenty. Because of the drug's money-making potential, Lilly mounted a marketing blitz whose scope was unprecedented in animal agriculture. Not even the excitement attending the advent of antibiotics could approach it. The pharmaceutical giant named its DES premix "Stilbosol," and delivered the first shipment to the feed trade less than four weeks after FDA approval. A formal press conference attended by more than a hundred of the nation's leading farm editors marked the delivery. By that time several feed compounders had clearance to market stilbestrol-added feeds to farmers. Each feed manufacturer needed to file with the FDA a New Animal Drug Application (NADA) supplementary to Lilly's already effective NADA and receive the agency's approval before Lilly would ship the estrogenic premix. By early January, more than three hundred feed producers had filed the requisite forms. Two weeks earlier, Lilly had adopted round-the-clock production and shipment schedules to meet demand, which was heaviest in the Corn Belt, and had inaugurated a free assay service to help feed manufacturers incorporate the premix properly in their rations. It also planned to expand its new agricultural division, a reflection of Stilbosol's tremendous immediate success. The *Lilly Newsletter*, a new company-published information sheet targeted to feed producers, alerted prospective customers to the latest Stilbosol news. By the end of February 1955, nearly six million cattle were on feed, many because of stilbestrol's remarkable growth-promoting—and cost-cutting—powers. This record number was 8 percent higher than the previous year's figure and 19 percent higher than the five-year average. About one-third of these feeder cattle consumed stilbestrol-enhanced feeds daily. A month later, the total shot up an additional five percentage points.[43]

The number of cattle on DES was a tribute to Lilly's advertising. Some livestock producers abandoned grazing and adopted feeding just to benefit from diethylstilbestrol's growth-boosting effects. Others embraced the drug out of fear of falling behind in a highly competitive industry. Taking out full-page or longer ads in journal after journal, Lilly highlighted Stilbosol's revolutionary power to boost weight gains in beef cattle, and maintained that agrochemical firms, feeders, and feed producers had long been working toward and anticipating a breakthrough of this stripe. The ads accentuated the innovative and the traditional. To Lilly, Stilbosol was a partnership product; it stemmed from "research conducted by Iowa State College, Eli Lilly and Company, various feed manufacturers and experienced cattle feeders." Both "scientific experiments and on-the-farm feeding trials" indicated that Stilbosol was "the most important advance in animal nutrition since the introduction of antibiotics as growth stimulators." Economic incentives for both feed manufacturers and feeders to switch to Stilbosol were certainly not ignored. Arguing that "Stilbosol can carve more beef supplement business for you," Lilly's advertisements directed to feed producers championed the drug as "the key you've been waiting for to convert cattle feeders to a commercial fattening supplement." It also reminded cost-conscious producers that Lilly "is paving the way for you with a powerful advertising campaign to feeders." Those ads aimed specifically at feeders highlighted the drug's cost-cutting potential, proclaiming that Stilbosol "increased gains up to 37%," cut the "cost of gain . . . as much as 20%," boosted "daily gains . . . by ½ to ¾ pound . . . up to 3½ pounds," and produced "prime beef . . . at savings of 2¢ to 4¢ per pound."[44]

Agricultural journalists echoed the efforts of Lilly's public relations staff as they treated DES as a panacea. In all fairness, they had reason to exult, recognizing that "nothing has ever hit the meat-animal business with the impact of stilbestrol." To many writers, stilbestrol was "an exciting new example of how research can unlock the doors of science to help mankind." These journalists claimed that "one can easily realize how revolutionary" the feed additive seemed, and they sought to "credit the scientists for another accomplishment on behalf of agriculture." Indeed, the advent of DES appeared to be yet a further justification for actively embracing expert-based chemical agriculture.[45]

It should be noted that the acceptance of diethylstilbestrol in cattle rations occurred before many experiments had been done demonstrating the utility of feeding minuscule doses of stilbestrol daily. Yet that situation ought not to surprise. Decisions to implement the technique reaffirmed "the system"; they showed a blind faith in the wisdom of scientists, even those working in industry. A handful of scientists had begun to investigate the orally administered drug's gain-boosting qualities soon after Burroughs's announcement, of course, but most depended on Burroughs's results. Many more began the slow process of examining the drug after it received FDA clearance, but their motives seem suspect. DES was ready to be introduced into animal feeds, and college researchers in almost every state west of Pennsylvania and south of Maryland could not afford to ignore what their clientele viewed as a most promising substance. These scientists rushed to investigate the recently legalized feed additive, which enabled them to say that they were exhibiting their traditional commitment to their constituency's economic well-being. Their haste— and even apparent lack of seriousness—usually yielded slapdash, superficial research, often with duplicative and sometimes contradictory results. These confused studies proved to be self-generating. They did not inspire recriminations or condemnations, but ironically, they justified further investigation, because the poor research resolved few issues.

In the first year after Lilly offered the drug to feed manufacturers, scientists in no less than twenty American agricultural colleges presented the results of more than fifty separate DES experiments.[46] Agri-industrial scientists joined their experiment station counterparts. Those working for Charles Pfizer Company at the Pfizer Research Farm in Terre Haute, Indiana, conducted the largest experiment, using 192 steers. Quaker Oats investigators undertook a more modest quest. They labored at the company's research farm near Barrington, Illinois, with 32 steers, about the same number as were used by Purina researchers. Federal investigators also tested DES. A 10-steer experiment at the U.S. Department of Agriculture's Agricultural Research Center, in Beltsville, Maryland, confirmed that cattle on diethylstilbestrol-enhanced feeds grew more swiftly and more economically than those not given the estrogen.[47]

Within a year of FDA approval, DES had come to dominate the beef production industry. It captivated scientists, industry, and livestock producers. The remarkable speed with which farmers and manufacturers adopted stilbestrol feeding reflected the continued strength of the concept of the progressive partnership. Institutions erected on the premise of a progressive partnership, and to bolster that partnership, spread word of stilbestrol and conversely provided a potent justification for its use. Indeed, the case of DES seemed to be a model of the application of the partnership idea. A college scientist uncovered a new technique, pharmaceutical scientists produced the drug, feed-manufacturing scientists compounded the material as a premix, federal scientists approved its use, agricultural college scientists publicized it by demonstrating its utility, and farmers made use of it. That type of expert-based interaction had been the model for "progress" since the 1920s. With respect to stilbestrol, little in the mid-1950s seemed to undercut faith in that model.

2   Cracks in the Facade

A measure of corporate intrigue surrounded the marketing of diethylstilbestrol in the late 1950s, and some contentiousness surfaced among DES principals. In a few instances, scientists fought over even the most mundane issues. Money and pride were at the root of these problems. These skirmishes were among the first indications that the venerable façade of the progressive partnership had begun to develop miniature, almost imperceptible fissures. Beyond these petty disputes and jealousies, an occasional report surfaced claiming that DES in meat might cause human harm. But these public statements were never uttered by scientists well-regarded by their peers; the scientific community remained solidly behind stilbestrol feeding. Nor did any credible evidence appear suggesting that stilbestrol feeding might prove harmful. A more general concern about cancer and chemicals in the food supply spawned additional national regulatory legislation, but that legislation generally confirmed the partnership's virtues and flexibility, and served as public acknowledgment both of potential dangers and of confidence that the partnership remained capable of protecting the American people. DES certainly played an important part in that end-of-the-decade political process, albeit not because of its beef-promoting properties.

To be sure, not everyone outside the scientific community unconditionally hailed DES as a promoter of cattle growth. At least a handful of individuals questioned its use in beef production. This very tiny group—individuals really—explicitly rejected or ignored the scientific community's views and utterances, opting instead for a more personal perspective. Their objections to DES feeds reflected a suspicion of the virtues of the science-based partnership, a flat rejection of established authority. These minuscule cracks in the facade notwithstanding, the fact remained that there existed a staunch consensus about the drug's safety. The vast majority of Americans apparently felt secure as recipients of the benisons of scientific expertise.

Burroughs and his colleagues at Iowa State College had an intellectual, professional, and economic commitment to furthering stilbestrol, and they devoted considerable effort to defending the drug from allegations that it somehow lowered the quality of beef. These claims, which first surfaced at about the time that Lilly gained FDA clearance, persisted through the following year. Confusion between carcass studies of animals receiving the high-level DES implants used by Frederick Andrews and others and studies of animals being fed according to Burroughs's regimen generated misunderstandings, but Andrews explicitly claimed that his experiments with the feeding of minuscule amounts of stilbestrol daily greatly diminished meat quality.[1] Feeders not using the new product also spread rumors of its destructive impact on beef quality, and meat packers certainly recognized the economic advantages of paying reduced prices for stilbestrol-fed cattle and then selling the beef at premium prices. When the meat packers put these practices into action, it particularly frustrated feed manufacturers, who stood to lose the most and objected to "rumors running rampant . . . that the meat is soft and the carcasses are cutting dark." They complained that "the problem has reached the point where many of our feeders have gone back to a conventional protein supplement and others have returned feeds containing diethylstilbestrol," and called on Burroughs to employ his considerable scientific acumen to persuade packers and feeders of their errors. If a concerted, Burroughs-led scientific campaign failed to demonstrate that DES-fed cattle produced high-quality carcasses, manufacturers warned that stilbestrol could "be all wrapped up and put away in about 30 days time."[2]

The question of stilbestrol's effect on cattle grade was a matter for scientists to decide among themselves. Implicit in the debate was the idea that science would announce its decision and the feeders and packers would abide by it. Burroughs worked with the statistician Paul Homeyer in this professional dispute to ensure that statistical analysis would pass scrutiny. The several Iowa State carcass studies showed stilbestrol-produced beef as comparable in quality to meat from animals not given the drug, as did the occasional investigation at other experiment stations. Stilbestrol beef was sometimes slightly higher in protein and a bit less fatty, but it graded virtually identically. Burroughs publicized these results at every opportunity and met with packers on April 16, 1955, to "dispel disturbing rumors about stilbestrol." Packers' objections to stilbestrol-fed cattle virtually ceased following the conference and were further undercut in late September, when USDA scientists issued findings similar to those Burroughs had announced months earlier.[3]

At about the time that Burroughs's science mollified the meat packers, another threat to his invention surfaced. Perhaps because of his own long-standing investigations into the gain-boosting effects of stilbestrol and other hormones,

Andrews, who had received his Ph.D. degree at about the same time as Burroughs, had refused to recognize Burroughs's priority in the field and did not concede that the Iowa State scientist had developed a patentable process. In what can charitably be referred to as an excess of entrepreneurial zeal, Andrews, the 1949 recipient of the Sigma Xi research award, began soon after Burroughs's initial DES announcement to work secretly with Charles Pfizer Company to develop the safety data required to support a new basic New Animal Drug Application for a DES premix. Pfizer and other firms would use this FDA-approved NADA to compete with Iowa State and Lilly for the lucrative DES feeding market. Because the FDA had no organic affiliation with the Patent Office, its only concern was to determine the safety of processes outlined in applications, not to judge whether they infringed patents.[4]

Pfizer secured FDA approval in early June. By October, two other companies had gained FDA clearance, and four additional basic NADAs were in the pipeline. Yet each of these firms hestitated to market a stilbestrol premix and directly challenge ISCRF's patent, which was still pending. Each preferred to have some other firm enter into what promised to be an interminable and expensive legal proceeding. ISCRF and Lilly also wanted to keep the matter out of court. Despite the apparent infringement menace, sales of Stilbosol had achieved an all-time high. As of October, an estimated 50-plus percent of American feeder cattle were receiving the drug. More than five million Stilbosol-fed cattle had already reached market, and more than seven hundred feed manufacturers added the premix to their feeds. Lilly had purchased an additional 267 acres near Greenfield, Indiana, to expand its animal biologics investigations, and ISCRF had received nearly thirty thousand dollars in royalties. Neither the pharmaceutical giant nor ISCRF wanted to disrupt the status quo, and they chose to negotiate out-of-court settlements. Each of the offending parties ultimately applied for and received a Lilly license at generous terms to manufacture the premix under ISCRF and Lilly exclusivity.[5]

The various companies benefited from this propitious compromise. Lilly's sales soared 13 percent in 1955, to a record high of $141 million; rose an additional 10 percent the next year; and skyrocketed 26 percent more in 1957. Pfizer set a sales record in 1956 and increased that impressive total by more than 16 percent a year later. Other factors also helped these and other stilbestrol-dependent businesses boost sales and reap tremendous profits. United States beef consumption reached an all-time high in 1955 and zoomed about a billion pounds the following year, to an average of about 83.5 pounds per person. Consumers had demanded a leaner, less well marbled beef, and stilbestrol's slight carcass-modifying property had enabled feeders to produce that article without disrupting other facets of their operations. Rapid expansion of the feeder industry outside the Corn Belt, in parts of the West and the South, as well as the creation of huge feedlots in these fast-growing regions, provided additional stilbestrol markets. A revised market dynamic seemed to be the consequence of these changes. Increased beef production helped to satisfy demand and kept the market price of meat down—it even declined in the mid-1950s. These low beef prices spurred subsequent consumer demand, of course, but stilbestrol's cost-cutting propensity, coupled with other feed supplements and cheaper rations, permitted feeders to achieve record profits and inspired them to purchase and fatten additional feeder cattle. It appeared that a new era of production and consumption was at hand.[6]

Cattlemen and economists based their unparalleled optimism on more than stilbestrol's revolutionary growth-promoting powers. They expected scientists to continue regular production of what the editor of *Successful Farming* called new "test-tube feeds." These increasingly more potent feeds would be, according to *Flour and Feed*'s Theodore P. Thery, composed of "magic ingredients," incredible gain- or yield-boosting chemicals. Announcements of the wondrous properties of nitrofurans, pepsin, combinations of specific amino acids, estradiol, and progesterone—like similar announcements about arsenicals, stilbestrol, and antibiotics before them—appeared to prove that point. Scientists announced the discovery of each of these promising supplements before the end of 1957, but none had stilbestrol's psychological or physiological impact.[7]

Agricultural scientists maintained and even expanded their stilbestrol investigations during their search for other "magic ingredients." In 1956 no fewer than twenty-two state agricultural experiment stations published studies of diethylstilbestrol as a promoter of cattle growth, and in each of the next three years twenty-eight stations presented DES-related research results. Western and southern stations accounted for much of the new research.[8] Demand for the drug grew continually, in part because of a new, effective means of furnishing steers, even those on pasture, with regular daily stilbestrol doses. Investigators had devised a practical stilbestrol ear implant for cattle.

Andrews's megadosage studies had pioneered stilbestrol implants, and the utility of these new devices in a sense constituted a personal vindication. But other investigators, principally M. L. Clegg, at the California agricultural experiment station, and Earle W. Klosterman, at the Ohio station, had built upon Burroughs's work demonstrating the explosive growth-boosting potential of relatively small quantities of the drug; they had eliminated the disagreeable side effects that had plagued Andrews, and produced a commercially viable process. Effective beef cattle implants incorporated a dose of stilbestrol as low as twelve milligrams, significantly lower than that proposed by Andrews's group at Purdue.[9]

Confident that implant production in no way infringed Burroughs's patent, Pfizer became the first pharmaceutical firm to offer these implants for sale after FDA clearance in 1956. Hard feelings and the promise of corporate profits had fueled Pfizer's implant research. William H. Hale, a former Iowa State researcher who had participated in Iowa State's stilbestrol feeding experiments almost from the beginning, headed this corporate program. Hale had left Ames after his department voted against promoting him; this refusal of tenure had followed a falling out with Burroughs which involved the research of Charles D. Story, a graduate student and another principal in the development of DES feeding. Surely Hale relished the opportunity to reduce the value of ISCRF's diethylstilbestrol patent, which was finally awarded in early May. By then more than two-thirds of the nation's feeder cattle were receiving the drug.[10]

No matter what feelings of personal or professional animosity there may have been between Hale and Burroughs, or between Burroughs and Andrews, they would have agreed on at least one virtue of diethylstilbestrol implants. Implanting permitted virtually any livestock producer to use stilbestrol. Pastured or even range animals, not just cattle that were fed daily rations, could benefit from the drug. Just how much benefit pastured or range cattle derived from DES became the important research question, and scientists performed exhaustive experiments to determine its answer. Similarly, investigators compared the growth-promoting powers of oral versus implanted stilbestrol to help cattlemen make the rational economic choice.

The results of these two different groups of experiments were mixed.[11] Iowa State and Purdue animal scientists predictably took opposite sides. Never once did either group exhibit any public acrimony or explicitly criticize the work of the other, however. Such behavior would have violated the professional canon: scientific disputes never were argued in the public media. This sort of civilized disagreement was at the progressive partnership's core. Scientists might hold differing views, but patient consultation among scientists would necessarily result in consensus. The Purdue group and the Iowa State group each clearly framed its research to come to bear on questions the other group viewed as dear. The Purdue group maintained that implanted DES produced greater gains than fed stilbestrol, both in pastured cattle and among heifers and steers. They also found it more economical and claimed that estrogenic residues persisted in the muscle, kidney, and liver of stilbestrol-fed animals. The Iowa State group countered that no residues existed in carcasses of cattle that had received orally administered DES and that this beef was of higher quality than implant-produced meat. They also asserted that feeding stilbestrol resulted in larger gains because it enabled cattlemen to dispense precise, uniform daily doses of the drug, and that feeding was a safer and less time-consuming method of administration than implants.[12]

On the question of estrogenic residues, the federal government agreed with Iowa State. Extensive FDA-USDA experiments in which cattle were fed up to sixty milligrams of DES per day failed to detect estrogenic activity in the slaughtered carcasses. Conducted throughout 1956 and 1957, these joint studies offered some corroboration for tests done by Lilly in late 1955. Lilly researchers had given a handful of cattle two hundred milligrams of diethylstilbestrol per day in their feed for one hundred days, and reported that these animals' carcasses retained no detectable stilbestrol residue.[13]

The Purdue-Iowa State rivalry flared again when in late 1957 a Purdue-dominated committee of agricultural scientists issued a controversial report entitled *Hormonal Relationships and Applications in the Production of Meat, Milk, and Eggs*. Jointly sponsored by the prestigious National Research Council and the even more prestigious National Academy of Sciences, the five-member committee included Andrews and three of his former students. It surveyed the relevant literature published between 1952 and 1957 and summarized the state of the art. The committee warned in the preamble that "some of the uses which looked promising from a practical point of view at an earlier date now seem less promising" and gave DES cattle feeding short shrift. It acknowledged Burroughs's efforts, but dismissed feeding stilbestrol and concentrated instead on studies supporting the Purdue position.[14]

That intramural scientific and business squabbles sometimes spilled out into the public arena reflected the beginnings of a loss of a sense of cohesiveness, and the emergence of tensions within the progressive partnership. This apparent fragmentation actually involved relatively few individuals, and antagonisms remained superficial as established protocols remained firmly in place. Yet these events were significant. At the very least, they signaled a nascent lack of faith in established conventions, an incipient willingness to take it upon one's self to go outside traditional channels. But too much can be made of this split. Burroughs, for example, retained his honored position with feeders, feed manufacturers, and the public. Not only had he won the coveted American Feed Manufacturers' Association (AFMA) award and become the ruminant authority in farm and farm trade publications, but in 1958 the city of Philadelphia awarded him the John Scott Award for the Benefit of Man, an award established by the English chemist of that name in 1816 and presented annually by the city fathers. Previous recipients included Orville Wright, Thomas Edison, Madame Curie, and Guglielmo Marconi. Burroughs's invention of stilbestrol feeding merited the prize, the award citation read, because the technique provided "an almost free present to the farmer and to the meat-eating public."[15]

Not everyone thought Burroughs's invention worthy of celebration or without social costs. A very few questioned the wisdom of adding a substance as potent as diethylstilbestrol to animal rations and feeding the meat to humans. This sentiment emerged well before 1958, and Burroughs knew of it at least as early as February 1954. On February 19, the day after he had publicly announced his invention, Mrs. Robert Swenson, a River Falls, Wisconsin, homemaker, wrote to ask for a plausible account of his actions. She related that according to the "Science Report Tells" section of the February 1954 issues of the *National Police Gazette,* "75 percent of the rats tested" with DES "got cancer," and she wondered how it could be that "our public officials and educators are not aware of this harmful drug." Concluding that they must be cognizant of the hazard, she demanded that Burroughs answer the question "Why is our government letting . . . manufacturers manufacture the drug"? and invited him to explain why the government would "let the public suffer."[16]

Burroughs sent Mrs. Swenson a confident letter claiming that he knew of "no evidence that the hormone we are giving experimentally in cattle feeds causes cancer in man." He also told her to "rest assured that Iowa State College will not recommend the use of any material in cattle feeds which will in any way cause cancer in the human race." Both statements were technically correct. Published studies linked DES to cancer not in humans but only in animals; and no responsible scientist in the first half of the 1950s would have indicted Burroughs's invention as a possible human cancer threat. Only fringe publications bent on generating controversy and increasing their readership condemned DES use in animals feeds. The *National Police Gazette* was one such publication, as was *Organic Gardening and Farming,* published by Rodale Press. This latter periodical had been started in 1954 to capitalize on the interest the public had shown in the Delaney committee hearings of 1950. Under headlines such as "Poison by the Plateful," the journal offered its readers "an unmasked look at chemicals in foods" and various "health-robbing practices." Its very modest but surprising popularity marked a tiny but significant schism in the partnership between consumers, producers, scientists, and government that had been so much a part of the American socioeconomic and political system since the 1920s. Pitting consumers against scientists, *Organic Gardening and Farming* denied legitimacy to the coalition that had led America for three decades.[17]

Issues of stilbestrol licensing and implants had exposed a similar sort of crack in the facade of the partnership between scientists and pharmaceutical interests. No split developed over stilbestrol's merits or its safety, as the overwhelming majority of scientists unconditionally favored the drug. To scientists, their ability to decide any particular issue was crucial. They maintained that they played a central part in American life, serving to discover knowledge and to guard the public weal. C. N. Frey, writing a guest editorial in *Science,* the weekly magazine of the American Association for the Advancement of Science —the nation's largest, most influential scientific organization—summed up this position neatly. Faced by comments like those published in the *National Police Gazette,* he urged the public "to maintain a proper perspective." Acknowledging that "we are at present dependent upon the products of chemical technology in food production," he argued that "modern production cannot be maintained without the cooperation of the chemical industry, the food industry and the government." This partnership went much deeper. "Progress in the food industry is dependent upon the wise use of the discoveries of the agriculturist, the food technologist, the chemist, the biologist and the engineer," claimed Frey. To abandon this consensual arrangement would be disastrous, virtually spelling the end of American prosperity and consigning the nation to history's scrap heap. "We cannot turn backward. We must go forward with the help of increasing knowledge," Frey wrote, "guided by the best minds and best information that science has made available."[18]

Frey's comments demonstrated an unswerving commitment to a particular problem-solving approach. In that approach, practitioners of science and technology were uniquely qualified to define problems and develop solutions to them. Only those individuals expert and experienced in the scientific method, the method of science, possessed the detachment to analyze situations impartially, to determine appropriate actions, and to implement those actions successfully. Conversely, only evidence derived by individuals practiced in that method was worthy of consideration. Everything else was deemed spurious or speculative, not fit for examination.

Proponents of this formulation sometimes confused an ideal of science with what individual scientists did, but they brooked no outside interference. In the case of chemicals used as additives in food or animal feed, scientists in academia, industry, and government had in the early 1950s expressed a profound sense of concern in a number of forums. They wrote in the *Food Drug Cosmetic Law Journal* and the *Journal of Agricultural and Food Chemistry,* presented papers to the Agricultural Research Institute, undertook studies for the National Research Council's Food and Nutrition Board, testified before the Delaney committee, and helped draft national insecticide and food additive laws, the latter of which failed to gain congressional approval. But these scientists never equated concern with fear or panic, and like the vast majority of the interwar generation, they retained an unshakable faith that science would detect and deal with any problem before the public interest was seriously harmed.[19]

One concept pressed in public arenas raised scientists' ire: the idea that the

FDA should certify a substance as safe before it could be used in food production or preservation. To scientists, including those within the FDA, such an
objective seemed wrongheaded. Science could never prove a negative, and scientists could never guarantee safety—they could only state that no evidence
had been uncovered to indicate an additive's hazardousness. That some, such
as Mrs. Swenson, publicly argued for absolute safety confirmed to scientists
that many citizens were ignorant of fundamental scientific principles and even
lacked a basic understanding of the American judicial system. Proponents of
absolute safety assumed harmfulness, requiring others to prove safety, a situation that violated democratic precepts by presuming guilt and placing the onus
of proof on the innocent. More practically, such a position meant a virtual ban
on chemicals in food and in feeds, because governmental officials would have
"to play it safe and disapprove all . . . proposed additives." Scientists recognized a middle ground and claimed that "impossibility of absolute proof does
not mean good substantial evidence cannot be obtained that a substance is
harmless." But the question of who would decide what comprised good, substantial evidence of harmlessness divided even scientists. Those in government
generally wanted to place the burden of proof of relative harmlessness on manufacturing scientists, rendering each substance suspect until manufacturers presented persuasive evidence to the contrary. Manufacturing scientists and their
supporters usually reversed the equation. They demanded that any material be
considered safe until the FDA, identified by a Citizens' Advisory Committee in
1955 "as one of the weakest agencies in government," demonstrated its harmfulness. Academic scientists could be found in both camps, but animal scientists tended to side with manufacturing interests on this question.[20]

This cleavage among scientists did not involve the scientists' regulatory role.
All parties demanded that scientists retain their central place. It nonetheless
cast a pall over regulatory efforts, and the FDA's veterinary medical branch
moved to clear the air by scheduling a two-day scientific symposium on medicated animal feeds for January 1956. The agency's goal at this meeting was "to
clarify all positions and bring about an area of understanding whereby existing
problems can be eased or eliminated."[21]

In at least one regulatory area, scientists had long since reached agreement.
They claimed that it was their province to select the methods of analysis necessary to make food and feed additive determinations. The courts even recognized the Association of Official Analytical Chemists (AOAC), an organization
of chemists working for local, state, or federal government in a regulatory capacity, as the legitimate source of methods of analysis. The association began
the evaluation process when some chemist, generally a manufacturing chemist,
suggested a possible method of analysis to it. An organization member distin-

guished in the study of the particular chemical substance in question served as associate referee, designed a series of experiments to test the method, and selected other association members to perform these analyses. The associate referee finally surveyed the results and reported to the organization. If this expert felt "assured that the method is practicable, that results are reproducible, and that the determinations can be carried out routinely," he or she recommended that the association approve the method. If no opposition surfaced, the method became an AOAC "official method of analysis," acceptable in court. If the associate referee proposed that the investigation continue or that the method in question had failed to meet the challenge, the organization refused to sanction the method.[22]

This scientific tribunal reified science and scientists; and the FDA's medicated feeds symposium, which met in Washington, D.C., on January 23 and 24, 1956, continued in that tradition. More than 350 scientists registered for the conference. Twenty-nine of the participants presented technical papers, and more than one-third of these dealt with stilbestrol. A message from President Dwight D. Eisenhower set the tenor of this extraordinary meeting. Eisenhower called "the extensive and increasing use of medicated feeds supplied by a great and growing industry . . . a boon to small scale as well as large producers of livestock," a situation "which benefits all Americans." It remained for FDA commissioner George P. Larrick to remind the symposium of the grave issues involved, as well as the monumental nature of the meeting. "Never before has a group such as this had the opportunity to join together and discuss a national problem of mutual interest and concern," Larrick stated. Expecting sessions to be conducted "in an atmosphere of professional and business good-will," he felt it "increasingly apparent that meetings and joint educational activities" were the wave of the future. Only close, constant communication and interaction could provide "an understanding of the problems, their solutions, and a sound basis for voluntary cooperation and compliance."[23]

Most of the symposium was unexceptional. Scientists aired differences and sought common ground, for cooperation among the FDA and the feed and chemical industries remained a regulatory cornerstone. Only one participant volunteered a contrary view. William E. Smith, a New York physician, shocked the audience by contending that DES-supplemented feeds produced an immediate cancer threat. Smith was a graduate of Johns Hopkins University School of Medicine and had served as professor of industrial medicine at New York University until denied tenure. His controversial presentation stemmed from his research, which considered the consequences of subjecting strains of cancer-susceptible laboratory rodents to possible carcinogenic substances. He cited his studies and those of others to establish the "cancer inciting action of estrogens" and ominously claimed that "a variety of pathologic changes have been found to follow administration of estrogens." Maintaining that in some mice extremely small doses of DES yielded tumors, he argued that the drug's "cancer-producing dose . . . approaches the infinitesimal." The drug's use in beef cattle caused him consternation. The "sensitivity and accuracy" of the then-popular method of measuring DES residues in tissues, reliable to 2 ppb, provided consumers no cancer protection, because the "absence of detectable estrogen in the tissues" of treated animals "offer no assurance of the absence of a cancer hazard in such tissues." Smith also feared that stilbestrol might work as a co-carcinogen, inducing in young individuals cellular changes that agents later in life would convert into active tumors. He closed by condemning the nexus of food producers, regulators, feed manufacturers, and agricultural scientists for promoting stilbestrol as a feed additive, and asserted, "Addition to the cycle of food supply of any substance known to aggravate or incite cancer places a grave responsibility upon advocates of such a practice."[24]

A premise so far outside contemporary perspectives could not help but discredit Smith at the symposium. To the other participants, it simply demonstrated Smith's ignorance. Representatives of virtually every group in attendance—Lilly, Pfizer, the FDA, the U.S. Department of Agriculture (USDA), the National Cancer Institute, and the agricultural colleges—denounced Smith's claims as the product of flawed reasoning. These individuals stressed that no stilbestrol remained in meat, and that if any did it would constitute a dose so small as to be trivial. They also stressed that studies in one species cannot be used to predict cancer in another species (estrogenic tumor genesis was then unknown in monkeys, for example), accentuated that humans produced estrogens "as one of their bodily functions," and emphasized that natural estrogens were present in many common foods. Speaker after speaker rebutted Smith's arguments, and participants concluded "that all the scientific data on the subject positively assures that there is no danger from the consumption of meat from stilbestrol-fed animals."[25]

Smith fared better in the popular press, as the *New York Times* published the gist of his remarks without refutation. Opposition scientists were partly responsible for this unbalanced view. The *Times* focused on Smith's comments because he provided them with a copy. His opposition did not. Indeed, they could not, because taking a scientific dispute to the public virtually constituted an ethical violation. It might weaken the long-nurtured idea that scientists possessed the authority to resolve public issues. The *Times's* decision to publish Smith's criticisms also reflected a small but growing skepticism about the quality of the American food supply, as more and more individuals—the vast majority of whom were female, middle-class suburbanites—were complaining that the new chemical agriculture might be "an industry that's in too much of a rush." Stilbestrol seemed a particularly easy target, exceedingly potent and newly employed as a growth stimulant in beef cattle. The fear felt by those who were "just plain worried about the quick and widespread acceptance of stilbestrol" stemmed from a belief that "not enough time has elapsed to give researchers an [opportunity] to test fully the [long-term] consequences of stilbestrol feeding." Senators and representatives, led by James Delaney, of New York City, and Leonor K. Sullivan, of Saint Louis, certainly recognized the modest upswing in concern among middle-class women, and on the fiftieth anniversary of the first federal pure food and drug act, they forcefully lobbied Congress to enact new regulatory measures. There they met little opposition. Industry did not oppose new regulation per se but maintained that its own scientists and federal scientists already afforded consumers adequate protection. The National Research Council's Food Protection Committee counseled consumers to return to a state of calm and rationality and blamed the recent upsurge of consumer anxiety on "the press and other news media," which had reported "conjectures" about relationships between cancer and food additives. This reporting had "contributed to the present apprehension among consumers over the safety of the food supply, and to concern among food manufacturers over the possible loss of consumer confidence." The public would be better served, the committee remarked, if it left the issues of cancer and food additives to "governmental agencies, the food industry, and bodies such as the National Research Council." Consumers ought to offer these scientists their wholehearted support because "measures taken to safeguard the food supply can be only as effective as our state of knowledge permits."[26]

The FDA followed this general trend and pursued new legislation only to clarify its regulatory initiatives. Recent developments in medicated feeding had revolutionized that practice and made it a prime area of FDA concern. The agency now sought authority to treat medicated feeds exactly as it dealt with human food additives. DES had been subject to more stringent regulation than most feed additives because the FDA classified it as a potent drug, not an additive. But the agency's treatment of the drug apparently disturbed some agency personnel, who questioned whether a feed additive should be regulated as a drug and whether the bureau possessed the legal power to regulate DES-laced feeds if DES was not considered a drug. They suggested that animal feeds, no matter their composition, had long been and still remained the USDA's province, and they feared possible legal challenges by chemical and feed manufacturers unless Congress clarified matters by granting the FDA explicit power over medicated feeds in a new food additives bill. Although no challenges had in fact materialized on the stilbestrol issue, ambiguity within the regulatory agency persisted as manufacturers sought FDA approval of other medicated feeds almost daily. In brief, the FDA favored manufacturer-undertaken pre-clearance research in which applicants presented evidence of "harmlessness" to "a blue-ribbon advisory committee of nationally known scientists" who would determine whether the evidence met a scientific definition of safety, and would then recommend appropriate FDA action. The FDA would evaluate the merits of each individual case, with the courts restricted to determining whether the FDA had followed proper procedure, not whether the agency had ruled correctly.[27]

Feed and chemical manufacturers preferred to vest control of medicated feed in the more familiar USDA. But the FDA's claim to be the court of last resort would leave food and feed manufacturers without legal recourse, a potential threat that encouraged a few feed producers to voice their suspicions about agency motives. Even the recent medicated feed conference figured in the FDA's domination plan. It was "merely the opening in an effort by officials to gain control over all feeds which contain the new drugs which have been demonstrated to give great production increases."[28] The agency's desire to make food additives or medicated feeds achieve a "standard of usefulness" before they could receive legal sanction also troubled feed and food producers. The FDA sought to ban substances simply because they failed to meet arbitrary, agency-established criteria of utility. To manufacturers, this proposal smacked of an undemocratic power grab. The "FDA would be encroaching on the rights of individuals to exercise their own judgment on the relative merits of any particular product," argued W. E. Glennon, president of the American Feed Manufacturer's Association in 1956. "Freedom to make such decisions," he maintained adamantly, "for many good reasons should not be hampered by government dictate."[29]

This disagreement should be considered a weakening of the progressive partnership's bonds, not a defection from the partnership. Manufacturing interests criticized the agency's plans in surprisingly cordial terms. No doubt they realized that they were dependent on the FDA's goodwill, at least until Congress acted, and thus they downplayed their profound displeasure. And the situation was not quickly resolved. The Eighty-fourth Congress adjourned without passing additional legislation, and lack of consensus also plagued the initial session of the next Congress. Nearly a dozen bills reached the House or Senate floor as the Subcommittee on Health and Science of the House Committee on Interstate and Foreign Commerce held eleven days of hearings.[30] William Smith, who continued to seek DES's removal from animal agriculture, also contributed. But now he had a firmer, if not more prestigious, base from which to launch his assault. He had diligently resurrected a nearly defunct group, the

International Union against Cancer, a hodgepodge of European, Asian, and South American physicians woefully out of step with state-of-the-art medico-scientific research. As the first chairman of the Union's Cancer Prevention Committee, Smith had organized the group's São Paulo symposium in 1954 and orchestrated its 1956 gathering in Rome, which concentrated on feed additives. It was the latter meeting that proved to be a public relations coup. Smith wisely created the impression that the organization represented a substantial segment of expert opinion and therefore must be reckoned with. The Union fostered this illusion by acting decisively during the August conclave, as the Smith-inspired convention implored world governments to ban from use in food production and preservation any chemical "shown conclusively to be a carcinogen at any dosage level, for any species of animal, following administration by any route." Only a handful of Americans were at the Rome meeting, and only forty participants attended in all. Yet Smith parlayed this event and his role in it into column-inches in the *New York Times*. Not compelled to obey professional strictures, and now the Cancer Prevention Committee's executive secretary, he warned of the hazards of DES-implanted poultry, not cattle, and demanded that carcinogens "under no circumstances be introduced in food." Smith contended that in the United States "marketed poultry have contained, per bird, up to 324,000 times the amount of the drug sufficing as a daily dose to induce cancer in mice." He wondered aloud about the cumulative effect that regularly eating DES-implanted chicken would have on children "much later in life," and staunchly submitted that "speaking as a father," he would "react strongly if I knew that someone was feeding my children a drug that I knew would excite tumors in animals."[31]

Despite Smith's ability to turn a phrase, his lasting influence rested not with the public, but with Delaney. Smith's verve and pluck, as well as his arguments and sincerity, attracted the congressman, who met with him several times in early 1957. These conversations culminated in Delaney's revising his food additive bill to include "a carcinogen prohibition."[32] The only legislation of its kind submitted to Congress, the measure adhered to the principles of the International Union against Cancer. It banned from food and feeds any substance known to cause cancer in humans or in any animal species.

With the exception of James S. Adams, chairman of the American Cancer Society's legislative committee, Smith had been virtually the only public figure to press Delaney to include anticancer provisions. Lack of public demand for the measure troubled Delaney, and he read a letter from Smith into the *Congressional Record*. Dated January 28, 1957, Smith's letter, a polemic against American agri-industry, cogently summarized his reasons for pursuing a ban on cancer-causing substances. He stated unequivocally that "responsible bodies of

experts"—he cited his International Union against Cancer as proof of the contention—have demonstrated that "a few tiny molecules can start a body cell
propagating to form cancer."[33] Dosage or length of exposure to a carcinogen
appeared to be irrelevant. What truly mattered was the length of time an individual lived after receiving a dose, and that individual's condition.

This understanding, which was totally unsubstantiated and was supported
by no credible scientist, led Smith to demand that the federal government incorporate "a basic advance in philosophy" to protect Americans from these "more
subtle, delayed-action hazards." Congress needed to enact legislation outlawing all substances that caused cancer in any species, even if no evidence existed
that these materials could produce cancer in man. This prohibition must be
immediate and absolute, he claimed. If it was not, consumers would be required "to participate without consent, in the experiment of exposure to learn
whether man is or is not a susceptible species, or what percentage of human
beings respond to the dose involved."

Smith recognized that his proposal called for unprecedented regulatory
power, and he offered a metaphysical justification. He abhorred "the growing
custom of adding biologically foreign substances to food" because "a recurring
common denominator" among "cancer-inciting agents . . . is that they tend to
be biologically foreign substances that cells have not in the course of evolution
learned to handle." He congratulated food manufacturers for recognizing that
"the human digestive track provides an endless and understandably attractive
outlet for products of chemical ingenuity," but excoriated them as selfish. "It is
to our peril," he thundered, if we place "the human digestive tract . . . in the
role of a sewer for disposal of chemicals that afford only commercial advantages." Smith ended with a political reference, reminding his audience that in
1823 James Monroe had proclaimed his famous doctrine that the Americas
"were no longer open to exploitation by foreign powers. It seems time," he
declared solemnly, "to enunciate the doctrine that the American stomach is no
longer open to exploitation by biologically foreign food adulterants."

Smith had an interest as vital and as vested in his anticancer crusade as agri-
industry forces had in their food additive promotion. None of this is to suggest
that either side was less than sincere in its claims and efforts. Rather, it indi-
cates that the food additive question involved high stakes on both sides. The
industry would gain money. Smith and others in the business of cancer prevention, detection, or amelioration would gain professional and public standing—
rewards more psychological than material, perhaps, but every bit as dear. Nor
was this incident unique. When issues of science were considered in the public
sphere, they rarely remained free of outside pressures and benefits.

Delaney's use of Smith's rhetoric and arguments to gain backing for an anticancer clause proved a modest success. The support he received did not come from the medical or scientific communities. Only a handful of physicians, including a New Orleans cancer surgeon, a leader of the National Chiropractic Association, and a faculty member at the University of Florida medical school—the latter an attendee at the International Union against Cancer's Rome meeting— publicly offered their backing, while no significant scientific organization embraced the measure. Only one governmental employee of note, William C. Hueper, opposed his colleagues. A physician-participant at the Rome meeting, and Smith's hand-picked successor as chairman of the Union's Cancer Prevention Committee, he put himself at odds with his employer, the National Cancer Institute. Middle-class consumer groups, composed primarily of women not employed outside the home, were the bill's enthusiastic proponents. Indeed, attempts to defeat cancer through legislation struck a particular nerve in postwar America. The *Consumer Bulletin,* for example, lent its prestige to Delaney's initiative. Asserting that "we live in a sea of carcinogens," the *Bulletin* claimed that the "marked increase" in cancer deaths was "the result of the extreme activity and inventiveness of the manufacturers and the chemists in industry and government who are forever introducing new and untried substances into our working and living environments." Whenever "an adulterant is added to a food," it warned, "the balance of nature is disturbed." We can stop cancer by keeping these carcinogens out of the human body, the *Bulletin* concluded. To do otherwise would court disaster, because "the chemical and cellular processes within the body cells cannot react to the passing whims of chemists without disturbances in function."[34]

The Eisenhower administration also drafted food additive legislation—a measure that was similar to the FDA-sponsored bill, with one significant exception. The administration bill permitted government to establish tolerance limits, generally expressed in parts per million or parts per billion, for any known poison or carcinogen used as a food or feed substance or additive. Current regulations outlawed certain substances "per se," banning any additive that demonstrated in humans any degree of toxicity or carcinogenicity. More ignored than adhered to in practice, the "per se" rule long had proven unworkable. If interpreted strictly, the law compelled the FDA to exclude from foods numerous innocuous materials—including table salt, which is poisonous in huge doses. Moreover, many of these illicit chemicals possessed properties beneficial to consumers and manufacturers. Salt, for example, has potent bacteriostatic and bactericidal effects, which increase products' shelf life.

The Eisenhower bill quickly gained the support of the *New York Times* and the American Association for the Advancement of Science. These endorsements failed to galvanize legislators, and action during the Eighty-fifth Congress appeared most unlikely. One congressman even argued that "the job of producing a bill acceptable" to the parties involved "seemed insuperable."[35] This pessimism soon proved unwarranted. By mid-July 1958, Congress overcame its malaise. Smith helped provide the impetus.

New York City, home of Smith and Delaney, was pivotal, as was diethylstilbestrol. Jerome B. Trichter, the city's assistant commissioner of health for environmental sanitary services and a Smith partisan, drafted a city sanitary code provision banning the meat of DES-treated animals from the municipality.[36] As the city health board deliberated the provision, Utah and New York State began to consider similar measures. These local initiatives appalled the FDA and the various agricultural and chemical industries, and threatened the progressive partnership. The FDA feared losing its authority over all food and feed additives to states and localities. Maintaining that "restriction by a local government . . . is contrary to standards and permits granted and allowed by national and local governments," the agency threatened to precipitate a constitutional crisis if New York City and other localities passed regulations antagonistic to federal statute.[37] The agrochemical industry was even more anxious. The passage of individualized local and state regulations would give the trade the impossible task of conforming to each of several thousand potentially different codes, a Tower of Babel situation that would doom "nutritional progress." The industry countered that menace by reversing its stance, arguing that it was "in the public interest to clarify existing procedures at the earliest possible time," and calling for new national legislation. "Enactment of a Federal law at this time will set a pattern for state and municipal legislation," a consortium of industry officials maintained. Only then can "the scientific development of further improvements in the Nation's food . . . progress without uncertainty as to future legal requirements."[38]

Mississippi congressman John Bell Williams, chairman of the Subcommittee on Health and Science, seized on the industry's new flexibility to fashion a compromise food additive measure. It sailed unopposed through the House, encountered little resistance in the Senate, and was signed into law by President Eisenhower on September 6, 1958.[39] Cast as an amendment to the 1938 Food, Drug, and Cosmetics Act, the new law (known as the 1958 Food Additive Amendment) retained mandatory pretesting and utility clauses but abandoned scientific advisory committees. The FDA remained subject to judicial review, and the new statute specified procedures by which the agency would announce its actions and through which manufacturers could appeal. The law also contained a limited grandfather clause for any food additive in use on or prior to January 1, 1958; this was a modification of the "per se" rule. It permitted manufacturers to continue to add these substances for up to two and one-half years before having to conform to the new statute's proof-of-safety section. That dictum was not absolute, however. The secretary of health, education, and welfare could lengthen the grandfather period for cause if he considered that the substance in question posed "no undue risk to the public health." An anticancer clause also was included in the law. It stipulated "that no additive shall be deemed safe [and by extension a lawful food additive material] if it is found to induce cancer when ingested by man or animal." That prohibition was also in effect for any additive "found, after tests which are appropriate for the evaluation of the safety of food additives, to induce cancer in man or animal."[40]

The inclusion of this cancer clause was a tribute to Delaney's perseverance, as well as a recognition of middle-class, suburban concern. The House committee had felt that the measure's other articles adequately protected the nation from cancer, but they adopted the plank (commonly known as the Delaney Cancer Clause) out of respect for Delaney—then a congressman for more than a decade—because of "his deep and abiding interest in this subject." The Department of Health, Education, and Welfare (HEW), the executive department within which the FDA was located, also found the clause superfluous. Elliot L. Richardson, HEW's assistant secretary, worried that "to single out one class of diseases for special mention . . . could be misinterpreted." Nonetheless, he realized that "the widespread interest in cancer" dictated that the law "should mention the disease by name." His agency had not objected to this action since "a clear understanding" existed that the exceptional clause would not restrict the FDA's "freedom in guarding against other harmful effects from food additives."[41]

At least one congressman, Nebraska's Arthur L. Miller, supported the Delaney clause while admitting that it seemed unenforceable. The author in 1954 of pesticide legislation that placed the burden of proof of safety on manufacturers, Miller regarded as ludicrous the cancer clause's wording. Proving that something induced cancer would be nearly impossible, he reasoned. A statistical correlation between a particular substance and an increased incidence of cancer would be insufficient. Scientists would need to control and rule out virtually every other factor before they could possibly demonstrate conclusively that an additive induced cancer. Miller recognized that task as hopeless, and food, feed, and pharmaceutical manufacturers readily agreed. As a consequence, the manufacturers chose not to contest the addition of the clause to the food additive bill, although they remained wary about the "per se" argument. They had adamantly opposed any broad-based measure empowering the FDA to ban any food additive if the substance demonstrated any evidence of toxicity at any concentration. Dosage was paramount, they believed, since most chemicals exhibited toxicity at some level. The quantities in which many food additives were used were several orders of magnitude less than toxic doses; the additives were added to foods only at what industry and government scientists considered to be a no-effect level. By agreeing to the Delaney clause, manufacturers had abandoned their absolutist stance, perhaps creating a dangerous precedent.[42]

The new food additive legislation satisfied manufacturers in most other ways. They had eliminated the Tower of Babel problem and helped frame a law that they thought preserved their industry. The measure also buoyed the FDA, which maintained its nationwide preeminence while expanding its locus of concern. Scientists, too, felt vindicated, even though the law failed to mandate outside scientific panels. It forced the FDA to announce its regulatory proposals in the *Federal Register* several months prior to taking final action. That procedure gave "scientific experts . . . a large responsibility," noted Graham DuShane, editor of *Science*. If scientists objected to a particular proposal, the FDA was required to "hold hearings or otherwise gather additional information before making a ruling." This stipulation placed the regulatory burden where it belonged, on the scientists' shoulders, DuShane remarked; it empowered them "to see that no potentially dangerous additives escape their vigilance." DES as a cattle growth promoter also seemed to have successfully traversed the legislative morass. Despite the controversy surrounding the substance, and although it had figured prominently in the formulation of the food additive amendment, no clause apparently came directly to bear on cattle-fed diethylstilbestrol.[43]

Or so it seemed. In November 1958, at a Food Law Institute meeting, the FDA announced its first major interpretation of the new food additives law. The agency ruled that its mandate included feed additives, and gave manufacturers until January to comment on that proposed action. If no serious objections surfaced, the agency planned to implement its interpretation no later than the spring of 1959. Stilbestrol would be treated both as a new animal drug and as a food additive. Manufacturers would need to satisfy two sets of regulatory criteria, including the Delaney Cancer Clause.[44]

II / GOVERNMENT

3   Regulation as Opposition

The FDA's 1958 decision to treat medicated feeds simultaneously as animal drugs and as food additives subjected DES to new federal regulations and challenged the progressive partnership. DES's administrative defeat was not the first rift in that venerable relationship: fissures had begun to appear somewhat earlier. Nor did the defeat immediately terminate the partnership idea, which retained great strength for some time after that date. Yet by 1962 the progressive partnership concept was dead. A dramatic change in political leadership, from a staid Republican administration to a dynamic liberal one, paralleled the partnership's demise. So too did a dramatic upsurge in consumer interest, built on the already impressive gains of the early and mid-1950s.

Ironically, Congress enacted in 1962 a Delaney clause exemption for DES cattle feeds under specified conditions. This legislative victory for stilbestrol placed the drug on a secure footing, but it did not mark a return to the progressive partnership; the exemption came from premises radically different than those implicit within the partnership. Indeed, by the late 1950s the context in which regulation occurred started to change. Among the middle class first, perhaps, the consensual nature of the established order—the partnership wrought large—had begun to break down. Litterateurs and others attacked traditional bases of authority, the pillars of stability. Volumes such as *The Lonely Crowd,* by David Reisman; *The Organization Man,* by William H. Whyte, Jr.; and *The Man in The Gray Flannel Suit,* by Sloan Wilson, railed against conformity, corporatism, and scientism, and pleaded with readers to assert their individuality. Glorification of the disaffected, as manifested by the civil rights and antinuclear movements, but seen most clearly in the popularity of James Dean in *Rebel without a Cause,* suggested a movement away from conformity. Fears of a loss of conformity, as manifested by fears of rampant juvenile delinquency (allegedly furthered by individually held transistor radios playing rock and roll) and by fears of the subtle but no less insidious threat of beatniks, the ultimate noncomformists, showed a further breakdown of the established order. These various expressions had been transformed by the 1960s into a hardy and prized individualism in a hopeful new era. But individualism was a concept at odds with the idea of a progressive partnership. It cherished the one, casting the ideas of consensus and an agreed-upon public interest as problematic. More to the point, the individualism of the late 1950s and subsequent years developed as an explicit reaction to the idea of consensual arrangements; instead, it championed suspicion, disbelief, and opposition. Antagonisms and antagonists abounded, as each individual tended to identify vile self-interest as his or her opponent's primary motive for action.

This intellectual framework had profound implications for American society. In the case of regulation, former partners labeled each other as partisans, while regulation itself became like a wall, to be either erected unconditionally or beaten down and destroyed. The control of regulatory bodies emerged as a bone of contention, and since these bodies were part of the government, fights for regulatory control took place within the government. Congressional legislation, executive edicts, and administrative posturing became the standard forms through which newly acknowledged partisans engaged in these crucial battles. The debate moved from scientific societies to government, as fights within government for the control of government became common. In this milieu, the basis of the progressive partnership, an unfailing belief in the virtue of science and scientists, was untenable. Science had lost its cherished regulatory place, and scientists began to develop new justifications for their continued participation in regulatory action.

The enactment of the 1958 Food Additive Amendment had stood as an acknowledgment of the resurgence of consumer concern within the progressive partnership, not of the crystallization of this interest into a formal movement. Consumers had not sought to create for themselves a rigid organizational structure because it seemed unnecessary; the 1958 amendment marked the apex of the longstanding crusade to increase the FDA's food regulatory powers. They felt that the passage of the 1958 law had more than adequately protected the public—a sentiment based on a notion of the national government which had been articulated before the inception of Roosevelt's New Deal. The vast majority of Americans had explicit faith in the ability of government-created mechanisms to defend their interests. Government experts possessed the scientific-technical ability and objectivity, as well as the all-encompassing perspective, necessary to harmonize relations among American society's various segments. That the FDA had requested no additional authority suggested that the agency could effectively fulfill its time-honored responsibilities without more legislation. Put baldly, the task for consumers in the mid-1950s had been to urge legislators to modernize the FDA by clarifying its powers. In that way, government scientific experts could maintain their involvement in the food additive question. With that task apparently accomplished, nothing significant remained around which to organize a movement.[1]

That set of views also framed the FDA's consumer relations efforts after the passage of the 1958 law. The agency supposed that strident consumer demands stemmed from a lack of accurate knowledge among consumers, and it established a Consumer Consultant Program to rectify the situation. The program was staffed with seventeen homemakers—a tacit acknowledgment that middle-class homemakers were the persons most worried about adulterations of goods—working as part-time FDA representatives. These women spoke before community groups, gave radio and television interviews, created public FDA exhibits, and conducted informal consumer opinion surveys. These agency-tutored women explained federal laws and FDA policies and reported consumer concerns to the agency. Consultants confronted "ill-considered or nonrepresentative views" and gave "a positive explanation" for agency efforts by telling "the 'fair story,' not the SCARE story."[2]

The FDA justified consumers' confidence in the adequacy of the new regulatory law and in the willingness of government scientists to enforce it when in late May 1959 the agency decisively froze the use of DES and arsenic in cattle feed. Large doses of both substances over extended periods were associated with cancer in certain strains of laboratory mice, and the agency maintained that the Delaney clause forced it to move against the drugs. The new restrictions were not absolute. The FDA exempted DES implants on the grounds that they were different from food additives. More to the point, the agency remained unclear about exactly what the 1958 Food Additive Amendment had grandfathered. That uncertainty prohibited the creation of a single policy on DES- and arsenic-enhanced feeds, and agency officials established general principles to regulate these questionable substances. First, the FDA banned any new animal feed use for the drugs and declined to approve new products containing the substances. Second, it refused to allow feed manufacturers that had not added the chemicals prior to the freeze to employ them in the future. Third, and finally, it permitted feed producers already fortifying feeds with DES or arsenic to continue to use the drugs as long as they did not change their feed formulas.[3]

Feed manufacturers did not interpret the FDA action as irreversibly dissolving the progressive partnership or even as an unconscionable violation of the partnership's ideals. They expected careful scientific opinion to force the agency to revoke its dictate on DES and arsenicals. Manufacturers called on their trade associations, AFMA and the Animal Health Institute (AHI; a consortium of animal biologics and pharmaceutical producers), and over the next several months association representatives met repeatedly with high FDA officials. Both trade groups advised their members to be patient, discouraged law-suits, and recommended against seeking additional legislation. The associa-tions believed that the Delaney clause's lack of clarity and precision, as well as the absence of a definitive or restrictive court ruling, allowed considerable room to negotiate a more satisfactory regulatory outcome. FDA scientists and administrators fostered that impression when they commented in private that they generally supported continued use of DES and arsenic in animal feed and blamed the agency's legal department for ruling that the Delaney clause com-pelled the FDA to freeze the use of the drugs. The trade groups then worked to develop a legalistic, freeze-repealing interpretation to enable sympathetic FDA administrators to overrule their contentious legal advisors.[4]

AFMA hired a specialist in food and drug law, Bradshaw Mintener, to deal with FDA lawyers. A former general counsel for the Pillsbury Company, Min-tener had recently resigned as HEW assistant secretary. He had previously led Eisenhower's 1952 write-in campaign in Minnesota, which had helped secure the Republican nomination for the general, and was then currently on the same church board as Arthur Flemming, his good friend and former boss at HEW. The association openly acknowledged that Mintener's intimate acquaintance with the personalities within the agency and with the Delaney clause would assist the trade.[5] Whenever possible, Mintener brought representatives of Quaker Oats to meetings with the FDA. Quaker was the second largest distribu-tor of stilbestrol in cattle feed, and its executives had contributed heavily to Eisenhower's campaigns. The company's president had become the U.S. am-bassador to Canada, and another high-ranking executive had served as assistant secretary of state. Through Mintener, the AHI and AFMA focused on exposing the FDA's policy on DES and arsenic as discriminatory in two senses. First, the agency discriminated against feed manufacturers by singling out drugs in feeds while allowing those same drugs to be implanted or injected without stipula-tion; and second, it discriminated among feed producers by permitting some to manufacture feeds containing the targeted chemicals even as it prohibited oth-ers from doing so. This second matter seemed particularly silly, AFMA con-tended, because the total amount of DES- and arsenic-enhanced feed used remained about the same. Farmers merely bought enhanced feeds from pro-ducers that had long been licensed to sell them rather than from new suppliers.

But what really rankled AFMA was the FDA legal department's cavalier treatment of the historic partnership between agricultural science, the agri-cultural industry, and government. The agency's interpretation of the Delaney clause seemed a poignant demonstration of a new, offensive approach by the agency, and AFMA took exception to the de facto presumption of an adversarial relationship. Congress's intent in enacting the Delaney clause certainly had not been to destroy scientific progress, the trade maintained, but the FDA's DES and arsenic policy risked doing just that: it could "stifle the progress of animal agriculture, public health, and food production." AFMA urged the FDA to base regulatory decisions on those parameters it had used in the past. Foremost was "the judgment of well-trained capable scientists that have an appreciation of industry problems as well as a keen sense of responsibility and a genuine interest in protecting and promoting the welfare of the general public." In short, the trade association wanted FDA administrators to wrest control of the agency from the legal department by making that department subservient to agency and industry scientists.[6]

The AHI similarly called on "legislators, regulatory officials, manufacturers, farmers and consumers" to "withstand unrealistic and unjustified pressure of vocal minority groups and individuals" and to achieve "an acceptable balancing of safeguards . . . and enjoyment of scientific progress." The restoration of the progressive partnership was crucial. Congress had formulated the Food Additive Amendment, which established regulation "on a scientific basis," from that partnership principle, but the last-minute addition of the Delaney clause now seemed to negate the partnership. The clause "imposes an impossible situation upon both government and industry," removing regulation "from the realm of scientific judgment and placing it in the realm of conjecture." Effective social policy required scientific, not legalistic, decision making, and the AHI encouraged the FDA to circumvent the Delaney clause by setting a "zero tolerance" for DES in human food. Zero tolerance, a regulatory technique long employed for penicillin in milk and for pesticides, permitted agricultural utilization of a particular substance so long as no measurable residue remained in marketable commodities; zero was equated with the limit of sensitivity of the analytical method.[7]

Although FDA officials privately expressed some interest in zero tolerance, they now faced opposition within HEW. The National Institutes of Health (NIH), which favored strict adherence to the Delaney clause and lobbied HEW, asked the agency to vest it with complete authority to set the cabinet department's future cancer-related policies. To the feed and biologics trade, that would constitute an unmitigated disaster. The NIH was antagonistic to, not a member of, the longstanding agriculture-government partnership. The NIH, wrote John Cipperly, *Feedstuffs'* Washington correspondent, "can only be moved by proven data regarding total harmlessness of substances when used in human or animal foods." Placing the NIH in charge would end scientific agriculture, as few additives could meet the agency's exhaustive standards. For those that could, the approval process would be prohibitively expensive and agonizingly slow, comparable "to the erosion of Gibraltar by annual rainfall."[8]

HEW considered the interagency conflict during the fall of 1959, while the trade associations pressed for a relaxed interpretation of the Delaney clause. Control of regulatory mechanisms had become a bone of contention, and opposition, not partnership, began to emerge as a viable basis of regulation. Individual feed and biologics manufacturers generally had kept out of the imbroglio, deferring to their trade associations. A few independent souls had gone on record, but even they had chosen their words carefully so as not to offend. Eli Lilly and Company and Iowa State College, the two institutions most responsible for the discovery and distribution of DES-laced cattle feeds, had adopted a higher profile. Lilly had selected that route, whereas the choice had been thrust upon the college.

Lilly let the AHI make its case unimpeded until July. After that time, the giant pharmaceutical company became an active, even contentious, principal in the Delaney clause fight. Several salient points punctuated its arguments. Lilly maintained that the FDA had erred in declaring DES a carcinogen. The company acknowledged that in "some 200 cancer prone hybrid mice . . . tumors were produced in over half the mice fed stilbestrol," but denied that study's validity. None of the "429 rats, 149 fowls, 13 pigs, 9 dogs, 7 goats and 1 cow," as well as assorted guinea pigs, monkeys, rabbits, and "many other strains of mice" fed DES in tests had developed cancer, and human medicine had relied on massive doses of DES for two decades without apparent mishap or tumor genesis, a fact testified to by the National Cancer Institute. Lilly also contended that no estrogenic residue could be detected in the meat of cattle fed DES feeds. To those concerned that DES assays were reliable only to 2 ppb, the company responded that "several common foods . . . eaten daily contain much higher estrogenic residues . . . in their natural state."[9]

Lilly's complaint produced some internal FDA concern but no public response. In the case of Iowa State College, Clifford Strawman, a member of the Iowa State Board of Regents, created a public stir by submitting to the board several articles questioning DES's use in livestock feeding. Strawman feared that DES "produces residues in livestock that induce cancer in laboratory animals and might in humans," and asked the college to investigate the situation. If the investigation found stilbestrol to be "a danger to public health or [found] that it induces cancer in laboratory animals," he wanted the college to "request [that] the Food and Drug Administration . . . prohibit its use."[10]

The college's president, James Hilton, appointed Burroughs a committee of one to issue a report on the drug's safety. Burroughs conferred with D. C. Hines, director of Lilly's medical division, and within two weeks completed his mission. He maintained that "no evidence [existed] that stilbestrol in toto" induced cancer in humans or in animals, and he attributed Strawman's inquiry to unjustified fears engendered by the odious Delaney clause. To laypersons, such as the regent, the clause's meaning seemed crystal clear. But to scientists, the clause was anything but straightforward, implying that "we now know or can define what substances will and what substances will not induce or cause cancer in man and animals." These were phenomena that "unfortunately no one knows completely with certainty at the present time." Yet Burroughs argued that this uncertainty was no reason to ban the drug. "To deny on the basis of faulty or incomplete information the usage of a tool [stilbestrol] which has proven so beneficial during the past 5 years," he explained, would be "nothing short of a gross mistake," catastrophic "to progress in modern agricultural technology."[11]

Burroughs estimated that DES feeds had saved farmers more than $150 million and kept consumer beef prices down.[12] Regardless of Burroughs's economic calculations, his report contained an amazing admission: he conceded that scientists did not know that DES did not cause cancer. He contended, in effect, that there was no scientific proof to be had, but continued to call for the public to place its faith in scientists. He was asking citizens to trust scientists' judgment, not their concrete experimental results. The case for scientists had classically been based on their method, which produced verifiable results. But in the instance of DES, that was not so. Scientists could neither prove nor disprove that the drug induced cancer. This de factor admission might have been less than startling if scientists all judged the situation the same way, if their positions on the issue did not vary. But their positions did vary. With disagreements raging among scientists, and with experimental results seemingly no longer deterministic, how were consumers to know which party to believe and what action to take?

Burroughs's brief investigation into stilbestrol's financial benefits did not suggest that he felt that the matter's resolution rested in economics. He would never consent to ending scientists' decision-making perquisites. Yet Burroughs and his scientific contemporaries faced a new situation. Demystification of science and scientists, coupled with increasing dissatisfaction with the terms of the traditional partnership between agricultural science, agri-industry, and government, promised to undermine long-established adjudication principles even as those rejecting these principles searched for new ones. Scientists generally met the new challenge head-on, by adamantly defending their authority and methods. At the same time, they redoubled their efforts to uncover incontrovertible evidence—and, if that seemed distant, to use scientific techniques to generate new information—that might bear on the problem and enhance their credibility. For example, researchers in the late 1950s reported that stilbestrol and other endocrine hormones changed the rate at which already extant tumors grew in laboratory animals. Others identified a strain of mice that developed vaginal and cervical cancer at a statistically significant rate if huge doses of DES were implanted vaginally, while still others attempted to circumvent the bioassay problem by testing whether radioactive, tritium-labeled DES could be measured in cattle fed the drug. They used only one steer, feeding it ten milligrams of the chemical per day for eleven days, and then waited twenty-seven hours before slaughtering the animal. These scientists recovered almost all the radioactive material in the steer's urine and feces, but did find a 0.30-ppb concentration of DES in the lean meat. The research group noted that the residue constituted a mere one-seventh of the dose necessary to produce a physiological response in mice, and that an individual would need to eat 7,610 pounds of this meat to ingest one milligram of the drug.[13]

Hope that scientists might develop a practical means to adjudicate questions related to the Delaney clause evaporated in early November as the science-based progressive partnership received a devastating blow. Secretary of Health, Education, and Welfare Flemming announced that HEW would oppose efforts to remove the Delaney clause from the Food Additive Amendment, and that the FDA planned to extend the controversial clause to pesticides. As justification, he cited the inability of scientists to deliver concrete determinations about a carcinogenic chemical's no-effect dosage: "While in theory there may be a minute quantity of a carcinogen which is safe in foods," he wrote, "in actuality our scientists do not know whether this is true or how to establish a safe tolerance." Almost immediately, Flemming applied his new pesticide policy to a suspect carcinogen sprayed on much of the nation's cranberry crop. About two weeks before Thanksgiving, and despite the active resistance of Secretary of Agriculture Ezra Taft Benson, Flemming urged consumers to reject treated cranberries. His announcement produced an unprecedented public uproar, brought serious financial injury to cranberry growers, and drew attention to the Delaney clause and the Food Additive Amendment. The chemical industry deplored HEW's "government by publicity," and feed manufacturers saw the move as the triumph of the NIH over the FDA. Both industries wondered what useful agricultural material would become Flemming's next target. The national newspaper columnists Drew Pearson, Jack Anderson, and Helen Thomas identified stilbestrol as the next substance to go. Their prognostication was only partially correct. On December 10, HEW reported a ban on the use of DES as a chicken caponizer.[14]

Flemming did not arrive at this decision, which the USDA heatedly contested, arbitrarily or de novo. FDA analysis as early as September 1957 had found that caponized chickens contained DES residues on the order of 30 ppb in their livers and up to 100 ppb in subcutaneous fat, whereas agency attempts to find residues in treated steers repeatedly failed.[15] Though some claimed that Flemming "had castrated the poultry industry," his partial ban pleased attentive segments of the public as he defended the Delaney clause, pledging always to enforce its provisions. But he also reached out to cattle feeders and chemical manufacturers by arguing that the use of carcinogenic substances as animal drugs should be permitted so long as they left no residue in meat or milk. More important, he promised to ask Congress to modify the Delaney clause to reflect this position.[16]

Some attributed Flemming's seemingly abrupt about-face to his vice-presidential ambitions. Others saw it as the result of Mintener's incessant lobbying.[17] But Flemming's proposal was hardly different than that offered by those favoring zero tolerance, or than the status quo prior to the Food Additive Amendment's passage. A few physicians and others demanded an absolute ban, but the overwhelming sentiment among government scientists and within the Eisenhower administration was otherwise. Flemming's partial recantation of his commitment to the Delaney clause had not mollified the USDA, which now pressed in cabinet meetings for preeminence over HEW. The pretense of partnership vanished. Benson sought a direct White House repudiation of HEW or, failing that, of the Delaney clause. Numerous farm organizations joined the USDA's campaign against Flemming's "public scare tactics" and demanded that Eisenhower appoint a fact-finding committee to "recommend changes in existing laws and procedures" to "remove the daily threat of irresponsible public statements concerning agricultural chemicals and drugs." The cabinet as a whole gently upbraided Flemming by suggesting that the FDA "go slow" on food additives "pending the receipt of adequate scientific information." Benson also publicly criticized Flemming's action, declaring that the USDA "strongly endorses the safe use of carefully tested chemicals as required to maintain the excellence, variety and economy" of American foods. Arguing that farm chemicals, including stilbestrol, were as essential as tractors and other agricultural implements, he contended that the U.S. food supply was "the safest, cleanest, and most wholesome in the world." The use of chemicals created better, cheaper products, while extensive research prior to production ensured consumer safety.[18]

Lack of vocal opposition to Benson's position probably figured in Eisenhower's mid-February decision to try to recapture the spirit of the progressive partnership by forming a special food additives panel composed of scientists selected by the USDA, HEW, and the President's Science Advisory Committee. The White House instructed the panel "to find out all the facts from a scientific point of view," and promised to use the panel's determinations to set policy. The Eisenhower panel was neither the first nor the only scientific body established to investigate food chemicals. A few days before Eisenhower acted, Gaylord Nelson, the governor of Wisconsin, had bowed to farmer-generated anti-Flemming agitation and appointed a scientific panel to evaluate the situation. Selecting scientists from the University of Wisconsin College of Agriculture and from the state agriculture department, as well as a few physicians, Nelson authorized his panel to develop "a clear, consistent and decisive" public policy on agricultural chemicals. Flemming remained defiant in the face of these rebuffs, pledging to continue his rigorous enforcement of the Delaney clause. Yet as the panels deliberated, the administration admitted that Flemming's precipitous cranberry crackdown constituted "a multi-million dollar 'goof,' " and indemnified growers for the losses accrued.[19]

None of these initiatives forced Congress into dramatic action. As part of a more general consideration of a possible color additive amendment, the House Interstate and Foreign Commerce Committee debated the Flemming-backed clarification of the Delaney clause. In appearances before the committee, pharmaceutical and feed manufacturing concerns emphasized the importance of agricultural chemicals, stressed "elements of scientific judgment" over legal dictate, and argued that things as innocuous as sunlight caused cancer. They also questioned the objective of food law. Was its purpose "to protect our population against any conceivable hazard," or "to assure the greatest measure of good to the greatest possible number with a minimum of possible harm to the least number?" These industries adopted the latter interpretation and proposed modifying the Delaney clause to declare substances "unsafe [only] if the additive is found in amounts and under conditions reasonably related to the intended use, to induce cancer when ingested by man or animal." A few self-styled consumer representatives, such as W. Coda Martin, a physician and the president of the California-based American Academy of Nutrition, called for a total agricultural chemical ban. Martin insisted that Congress outlaw agricultural chemicals because "the human body can utilize only natural foods as nourishment and survive." Since agricultural chemicals were not "natural," Martin, a close ally of Smith, noted, they "can produce only negative or harmful results." It was "only a question of how much harm they will produce."[20]

Members of Congress spoke out on both sides of the Delaney clause issue,[21] while the affected trades mobilized to pressure Congress. AFMA and the AHI urged their members and their customers to demand that Congress adopt "a more realistic legislative approach" and pass the industry-sponsored Delaney clause modification. At its annual meeting, the National Institute on Animal Agriculture, a decade-old consortium of feed and biologics manufacturers, agricultural scientists, and farm organizations, scored Congress for enacting the Delaney clause and Flemming for enforcing it, ridiculing them with the slogan

"Fight cancer—avoid food." The tart-tongued Earl L. Butz, dean of agriculture at Purdue, spearheaded the assault by the Grain and Feed Dealers' National Association. In a widely covered, heavily promoted, and frequently reprinted address, Butz rejected "trial . . . by sensational news releases and conviction by carefully staged press conferences." Congress and the ignorant public received their share of abuse. Claiming that "consumers should be eternally grateful" for agricultural chemicals, Butz dismissed the idea of a risk-free environment as "inconsistent with generally accepted practice throughout nearly all phases of our way of living." The government does not outlaw "swimming pools or automobiles just because somebody might misuse them," he thundered; "if we eliminated from our lives everything that might cause cancer in animals, our twentieth century civilization . . . would be impossible." Instead of a blanket prohibition, he wanted a "common understanding among government regulators, research scientists and business executives." In this cooperative arena, "industry, government and colleges [would] undertake research" to set safe tolerances for every agricultural chemical, a policy that would negate the Delaney clause.[22]

Butz wanted to resurrect the progressive partnership on revised terms. He also argued that the Delaney clause throttled agricultural research, and he predicted disaster unless Congress overturned the anticancer stipulation. That analytical thread was picked up by Abbott Laboratories, which attacked the clause in its annual report, and by Charles Pfizer Company and Eli Lilly and Company, both of which claimed to have transferred scientific personnel from agricultural research to other divisions because of uncertainty over Delaney clause enforcement. Industry sources estimated that it cost more than one million dollars and took about five years to develop a new agricultural chemical. Faced with the odious clause, firms decided to spend their time and money in a less risky manner. This phenomenon was not lost on the editors of the *Chemical and Engineering News,* the weekly publication of the American Chemical Society, which stressed the anticipated decline in industrial agricultural research and issued a dire warning. The threat also provoked sustained comment in *Science.*[23]

The publication of William Longgood's *Poisons in Your Food* in early March provided a counterpoise to the trade's various anticlause efforts. Longgood, a former New York City journalist, presumed that industry and consumers had no common interests and suggested that industry had completely co-opted government for its own sullied purposes. *Poisons* assaulted the 1958 Food Additive Amendment for its authorization of the inclusion of minute quantities of potentially poisonous substances without any apparent long-term safeguards, and decried the continued legalization through grandfathering of suspect carcinogens, especially stilbestrol. In his DES section Longgood used almost exclusively material first presented by his long-time acquaintance William E. Smith several years earlier.[24]

Longgood's volume became a best seller but encountered scathing criticism from the groups he attacked, as well as other scientists. For instance, William Darby, a Vanderbilt University biochemist who was chairman of both Eisenhower's food additive panel and the Food Protection Committee of the Food and Nutrition Board, called it "an all-time high in bloodthirsty pen-pushing." Incorporating "all well-known methods of the irresponsible purveyors of the sensational," the book expressed only the views "of the nonscientific, natural food–organic gardening cult," citing as authorities "cult leaders, their gods." Darby lambasted Longgood for asserting that scientists who disagreed with him (Longgood) "have either been bought off by industry or by government." Only those sharing Longgood's paranoid delusions could welcome this text, Darby charged. Those sick individuals believed "that the public is the victim of a giant conspiracy joined in by the Food and Drug Administration, the American Medical Association, the 'big chemical companies' and, apparently, scientists in general."[25]

Longgood and his allies launched a vigorous defense.[26] But the publication of the report by Eisenhower's scientific panel in mid-May undercut Longgood's charges. The report rejected arguments that the nation's food supply faced a chemical risk, praised "the integrated contributions of the engineering, agriculture and chemical sciences," and recommended the creation of a special board of scientists to advise HEW on food additives. Composed of FDA, USDA, and National Cancer Institute scientists, as well as scientists nominated by the National Academy of Sciences, the board would substitute scientific judgment and "the rule of reason" for legislative decree and would evaluate on a case-by-case basis the safety of suspect "food additives under conditions of proposed use." The panel supported the principle that small amounts of many substances, including carcinogens, were harmless, and it urged Congress to amend food law to leave to biological and agricultural experts the question of what constituted an unnecessary risk. These recommendations were seconded by the Wisconsin group in its report two weeks later.[27]

Agri-industry sources hailed the panels' reports as a "turning point" in the quest to establish reasonable food additive regulations and a restoration of the progressive partnership.[28] But no transformation occurred within the Eisenhower administration. The USDA came out strongly for total abolition of the controversial clause, while HEW continued to back Flemming's proposed modifications. Eisenhower refused to settle the internecine squabble, set a single policy for his administration, or even incorporate the panels' recommendations into a congressional measure. When Eisenhower failed to take the initiative, the House wrote the Delaney clause into the color additive bill and left the Food Additive Amendment unchanged. Commentators asserted that a "new virus" had swept through Congress: "Delaney phobia." Legislators feared that if they tampered with the anticancer clause, Delaney would orchestrate a media confrontation, creating an election-year fiasco and dooming the modification. Not wanting to be portrayed as procancer, Congress passed the House's color additive measure containing the Delaney clause. It also refused to debate the 1958 Food Additive Amendment.[29]

Feed and pharmaceutical manufacturers greeted Congress's retention of the Delaney clause with the familiar refrain about how the clause hampered agricultural research. The facts suggest otherwise. Lilly continued to expand its farm chemical operations, and Hess and Clark, based in Ashland, Ohio, opened a huge new feed additive laboratory. Farm prognosticators claimed that during the next decade an unprecedented number of new chemicals would be added to cattle feed and would revolutionize the beef production business. In 1960, cattlemen raised a record number of cattle. An estimated 90–95 percent of these cattle received stilbestrol. Iowa State College's yearly share of DES royalties topped $175,000. Much of the DES-generated largess went to sponsor further animal nutrition work, and the number of faculty and graduate students doubled in the half-decade after Burroughs announced his discovery. But perhaps the biggest winner was the FDA. Concern over the nation's food supply led to expanded agency funding. Congress appropriated more than fifteen million dollars for a Washington, D.C., headquarters, complete with extensive laboratory space. It also allowed the FDA to open several new district offices and renovate at least a half-dozen more. The agency added more than seventy-five scientific and technical personnel to its payroll.[30]

The creation of the new FDA laboratory did not mark a fundamental cleavage in the progressive partnership. The laboratory served to beef up agency enforcement, not to move the FDA into the new area of prevention; its purpose was to detect and prosecute malefactors. Throughout the expansion, the FDA publicly supported Flemming. At the same time, however, feed manufacturers believed that the overwhelming majority of FDA functionaries sympathized with their anti–Delaney clause sentiments. Yet agency personnel, as part of their renewed enforcement campaign, were conducting investigations, gathering information, and offering opinions that would prove detrimental to the DES cause, especially the drug's anticipated chances under Flemming's proposed Delaney clause modification. For example, an agency study found that 8 percent of cattle feeders ignored the forty-eight-hour withdrawal precaution for DES prior to slaughter. This practice could result in stilbestrol residues in excess of two parts per billion in some edible beef, a quantity detectable by the mouse uterine test. Another FDA researcher, Ernest Umberger, understood that meat packers sometimes butchered steers with undissolved DES ear pellets, and maintained that in this situation "there is every reason to believe . . . [that] small residues of the drug are present in the edible tissues." He also wondered if some DES metabolites could be toxic or carcinogenic without being estrogenic. To settle these and other similar questions, and to protect the FDA from public ridicule, Umberger, who had extensive experience analyzing diethylstilbestrol in beef, wanted the new FDA laboratory facilities to conduct detailed DES research. "To let the facts develop from other sources," he worried, "might result in embarrassment to us [the agency]." If the agency refused to undertake this work, he feared that it might "be caught in a most indefensible position."[31]

Umberger's proposal was indeed at variance with the progressive partnership ideal. By declaring that the agency itself constituted a scientific interest group, different from and in opposition to manufacturing scientists, he rejected en passant the notion of a nonpartisan, truth-seeking scientific community. Umberger came out yet again for additional scientific research within the agency during early January 1961. The concepts of "no residue" and "no effect" made no sense, he asserted, because the "no effect level of today by our rather crude pharmacological methods may tomorrow have to be abandoned by more refined methods." In the case of stilbestrol, the scholarly Umberger envisioned a time when scientists would be able to determine estrogenic effects by examining the division of individual vaginal cells, a method one thousand times more sensitive than the mouse bioassay. Findings of that sort would undercut the Flemming position and require the FDA to institute a total stilbestrol ban. Umberger also expressed concern about what scientists did not know about potential carcinogenic substances, for if they remained unable to "sufficiently define the no effect level of a carcinogen," they could never be "in a position to say that any amount of a carcinogen is safe." He argued that uncertainty over a carcinogen's no-effect level ironically constituted "a valid scientific basis for the 'cancer clause,'" and he did not wish to see that situation persist. He pushed instead for scientific research into each carcinogen to produce "knowledge of the mechanism by which a chemical acts as a carcinogen and where a no effect level can be established." These facts would permit food and feed manufacturers to add carcinogenic substances in no-effect dosages with impunity.[32]

Umberger's critique resulted in no immediate substantive policy changes, as FDA administrators stressed the traditional issue of enforcement and limited the agency's experimentation to questions related to protection from food and drug adulterations. Agency researchers were particularly excited about gas chromatography, the use of specially constructed chromatographic columns to separate and identify minute components of complex mixtures. They left questions of the type raised by Umberger to researchers at other institutions, who by and large failed the challenge. Few studies published in the early 1960s explored mechanisms of carcinogenesis, determined DES metabolites, or even considered ways of estimating a no-effect level for a single carcinogen. One investigator attempted to determine a no-effect level for stilbestrol in cancer-susceptible mice, but his National Cancer Institute grant expired before he concluded his study and was not renewed.[33]

The inauguration of a new president and the installation of a new Congress in early 1961 did little to change the situation. Congress busied itself with the more popular, less controversial question of the skyrocketing cost of prescription drugs.[34] The FDA and agri-industry representatives sought a Delaney clause compromise sub rosa, but the FDA refused to accept the industry-sponsored modification. Trade associations acceded to the Flemming proposal but rejected agency attempts to revoke the grandfather clause that was then in place. Increasingly powerful consumer publications complained that the public was "dosed constantly with unwanted and poisonous chemicals." Maintaining that "every citizen has a right and a duty to criticize," they claimed that the scientists developing these invidious materials were creatures of the "industrial interests which pay their salaries or support their research." Industry representatives maintained that termination of food chemical usage would consign the American public to starvation—which, they reminded their readers, was "the natural 'organic' way of life." Writing off opponents as "alarmists, faddists and crackpots," they decried those writers and lecturers who "fan the fires of confusion to their advantage," thereby gaining "financially and politically from getting public attention by their untruthful propaganda." In addition to condemning "poison pen pushers with literary royalties in mind," they also lambasted "unscrupulous and cynically provocative . . . politicians who see . . . an almost perfect chance to appear in the noble role of protectors of the helpless and weak." But special scorn was reserved for the few scientists openly siding with these miscreants. Those alleged investigators emphasized the danger of substances that were all but innocuous because "the public fear" of cancer had led Congress to authorize "large sums . . . for research" into that disease; the glitter of gold had beguiled or seduced these men and women into making irresponsible statements.[35]

Recriminations on both sides reached a pitch not seen earlier. Both claimed to understand completely their opponents' motivation and purpose. To the combatants, their adversaries were not naive, misguided, or deluded. They were selfish, self-serving, or self-interested. That refrain became common during the very late 1950s and early 1960s, and would increase in popularity thereafter. It reflected a vision of reality vastly different than, and at odds with, the view that postulated cooperation and partnership as the natural state of affairs. Distance, antagonism, and opposition characterized the former view, as did the shattering of many of the established social units with which people identified. In the case of DES, the idea of a government partnership crumbled for both agri-industry and consumers. Albeit from different perspectives, both looked suspiciously at government as an agent of the opposition. This new vision of discord also put asunder any idea of a scientific community or a community of scientists. On the question of what constituted an unsafe, possibly cancer-causing DES dosage, scientists' opinions apparently formed a predictable, occupation-based or discipline-bound pattern, one that reflected their individual areas of concern. Cancer researchers tended to support claims that no safe amount of DES existed, whereas animal agriculture and pharmaceutical experts tied the drug's presumed carcinogenic properties to its estrogenic activity. This often vituperative disagreement exposed the foolishness of permitting (or even expecting) scientists to decide an issue when they clearly did not agree. But rather than fostering some other form of public adjudication, the scientific experts' inability to approach consensus merely engendered further antagonism. Each individual scurried to line up behind those scientists pushing that individual's preconceived point of view.

Although the particulars of the DES controversy were unique, its shape was not. Indeed, within the fractured and fracturing milieu of the late 1950s and thereafter, the individual repeatedly emerged as the crucial unit of concern. This relatively autonomous individual was free to seek participation in, or alliance with, virtually whatever collection of other individuals he or she desired, while those antagonistic were obliged to protest. Only those in agreement with an individual were identified by that individual as public spirited or acting in the public interest, or even as rational. An exacerbation of tensions almost always followed this relativistic assessment, as machinations for power, authority, prestige, and dominance seemed to supplant willingness to compromise.

In this context, President Kennedy's 1962 consumer program struck a resonant chord. It lent the prestige of the nation's highest office to the definition of each consumer as the ultimate arbiter, and thus gave validity to that definition, which encouraged others to assert this now-sanctioned prerogative. The program recognized four basic consumer rights—the right to safety, the right to be informed, the right to choose, and the right to be heard—rights that if realized would establish and legitimize each consumer as the regulatory sine qua non. Government would function as consumers' ally. It would exercise its traditional responsibilities to protect and provide information to the public, while breaking new ground by ensuring each individual the opportunity to offer his or her opinion in what could amount to a de facto national plebiscite on any single regulatory matter. Kennedy's FDA proposals reflected a more practical attitude. He called for a 25-percent increase in the agency's budget and sought improvement in food and drug law "to protect our consumers from the careless and unscrupulous." He incorporated these specifics into a draft of an omnibus FDA bill, but haggling within his administration prevented him from introducing the measure into Congress.[36]

Kennedy formulated his omnibus bill without waiting for his Citizens' Advisory Committee on the FDA to report. He had appointed this committee in late 1961 to recommend changes in the agency and in food and drug law. As the committee deliberated and as the administration remained circumspect, AFMA and the AHI framed a bill of their own. Brought to the House by Kenneth A. Roberts, an Alabama Democrat, the agri-industry bill dealt solely with feed additives. It permitted the use of any feed additive, even those known to cause cancer, so long as the additive did not harm the animal or leave a residue in edible products. It also did away with grandfathering for those feed additives that might violate the Delaney clause. This last stipulation initially put the industry measure at odds with FDA desires. The agency had wanted the power to end all grandfathering and to revoke its approval of New Animal Drug Applications in any instance in which there existed a reasonable doubt concerning safety, but it ultimately settled for the industry proposal.[37]

Hubert Humphrey introduced an identical measure in the Senate, and behind-the-scenes negotiations took place. W. E. Glennon, AFMA president, lobbied for the trades, while Winton Rankin, a top FDA official, offered the agency's complete support. Vital differences remained among scientists about what constituted a carcinogen and about the implications of that definition for public policy. Noted cancer researcher Robert E. Eckardt was an active combatant and launched perhaps the most effective salvo. In a guest editorial in *Cancer Research,* a leading periodical in the field, he took on Hueper, equating his high-profile, well-established foe's position on carcinogens with that of an ostrich burying his head in sand. Reflecting that being opposed to possible carcinogenic substances was "like being opposed to sin," Eckardt agreed with Hueper that it was "a matter of wise precaution to exclude as far as possible" these noxious materials. But he adamantly rejected his adversary's call for an absolute ban on all substances that in any form resulted in tumor genesis. Plastics, salt, sugar, and cobalt, an essential component of vitamin B-12, would fall victim to this absolutist prohibition, he maintained; the real question should be, "How will the results of such testing be interpreted?" Eckardt dismissed Hueper's apocalyptic vision of an "epidemic-like occurrence of cancers attributable to an indiscriminate decision made 10–30 years ago." He argued instead that problems of interpretation, problems of knowing which carcinogens were especially dangerous and at what dosages, can only be resolved through "vastly expanded research." They "cannot be avoided by simply avoiding discussion of them." Eckardt found himself agreeing with René J. Dubos, a scientist of the first order and later a winner of the Pulitzer Prize and the National Book Award for his environmental essays. "To demand a certified verdict of safety before accepting a new technological innovation would clearly result in paralysis of the economic and technological progress," Dubos wrote. A society refusing to take "educated and calculated risks inherent in technological civilization" is "not likely to survive long in the modern world."[38]

Eckardt's comments lost much of their force and credibility almost immediately, as word of thalidomide babies broke. Investigators discovered belatedly that thalidomide, a popular prescription tranquilizer in Europe and Canada, caused grotesque fetal deformities when taken early in pregnancy. America was spared a tragic epidemic of flipper-limbed children only because one FDA employee, Frances Kelsey, had steadfastly refused to approve the drug's commercial use in the United States. Citing only a vague feeling, she delayed deciding on the matter until the horrifying revelations, despite almost continuous pressure from the William S. Merrell Company, the pharmaceutical concern planning to bring thalidomide to the American market. As Americans saluted their newest hero, chemical manufacturers also braced themselves for a bombshell. Portions of best-selling author Rachel Carson's heavily promoted new book condemning agricultural chemicals as an unmitigated ecological disaster appeared in the *New Yorker.* The deftly named *Silent Spring* quickly became a book club selection, went into paperback almost simultaneously, and served as the basis of a television special.[39]

Both the thalidomide story and the attack on agricultural chemicals provided graphic visual imagery. They conveyed to a distraught public an additional sense of fragility, and seemed to confirm the increasingly popular perception that American consumers stood precipitously close to the abyss. Any further or collateral disclosure, no matter how small, merely reinforced this perception.[40] These two events together suggested to an ever-widening audience that consumers lacked the power to defend themselves and control their destiny. They constituted an interest group in opposition to other interest groups. The thalidomide and DDT catastrophes had only been prevented by a few public-spirited individuals nobly battling a corrupt juggernaut. But as in the story of David and Goliath, the underdog emerged triumphant as good vanquished evil. This was a metaphor of great portent, holding out hope that concerted action by individuals could make a tremendous difference—that consumers could seize and exercise control. What seemed imperative was to recapture, revamp, and revitalize government to serve the public's interest, with the public identified as those agitating for change. This hysteria for institutional change almost begged for political manipulation; and Kennedy, always the consummate politico, and established as the consumer's advocate, prevailed upon Oren Harris, an Arkansas Democrat, to introduce a version of his omnibus food and drug bill into the House. There it faced competition from a measure supported by Kefauver and Humphrey. The president repeatedly pressured Congress, and a measure very much like the Kennedy-introduced bill ultimately gained congressional approval. Ever the showman, Kennedy invited Kelsey to the White House for the October 10 signing ceremony.[41]

Embedded in the new law (known as the Kefauver-Humphrey Drug Amendment) was the feed additive bill advocated by AFMA and the AHI. Only Delaney and Leonor Sullivan had explicitly objected to its provisions. The overwhelming majority viewed it as a technical, DES-related amendment, necessary to redress inequities in the partial prohibition of DES cattle feed. Commonly known as the DES Proviso to the Delaney clause, it removed the distinction between food additives and drugs, treating substances that promoted cattle growth only as animal drugs and permitting the HEW secretary to revoke approval of any NADA whenever he detected "an imminent hazard to the public health." But the proviso also nullified the inflexible Delaney clause whenever a feed additive met two important criteria. First, "under conditions of use" that were "reasonably certain to be followed in practice," the additive must not harm the animal. Second, and in conjunction with the aforementioned conditions of use, it must not be possible to find any "residue of the additive . . . (by methods of examination prescribed or approved by the Secretary by regulations . . .) in any edible portion of such animal after slaughter."[42]

Congressmen hailed the DES Proviso as upholding the Delaney clause's safeguards, while agri-industry representatives recognized it as liberating, an overthrow of established precepts, because it permitted the use of feed additives previously banned per se. The trades' interpretation was much closer to the mark, as the proviso radically weakened the clause's consumer protection powers. It did not prohibit residues of the offending drug in meat, but outlawed only those residues found by methods of examination approved by the HEW secretary, a stipulation that made that official's decision on method pivotal. In the case of DES, cost and time considerations compelled the secretary to select the slow but reliable mouse uterine bioassay as the only conceivable method of evaluation. Designation of the mouse test as the official method of examination in effect legalized quantities of the drug in beef so long as they amounted to less than a 2-ppb concentration, the test's limit of sensitivity. Certainly all involved with the DES question understood the new law's implications. The proviso's passage codified as law the no-effect-level argument, equating the significant dose with the dose that could conveniently be measured. It dismissed out of hand the claims of those who contended that allowing any amount of a carcinogen courted disaster. Indeed, the sentiments expressed in the proviso neatly dovetailed with the overwhelming preponderance of scientific opinion about DES, which held that the drug's possible carcinogenic activity seemed to be a concomitant of its estrogenic effects.

Few persons other than feed and chemical manufacturers highlighted or even mentioned this aspect of the act, preferring to celebrate those sections that placed drug law more nearly in line with recent food and color additive enactments. Even Humphrey, the act's cosponsor, ignored the proviso and charged the increasingly visible FDA with lax administrative practices, incestuous fraternization with the industries it regulated, lack of scientific acumen, and failure to work closely with the NIH.[43] Kennedy's citizens' advisory panel offered a different critique when it finally reported in late October 1962. It, too, chided the FDA for staffing its top policy posts with lifelong administrators rather than scientists and for ignoring NIH personnel. But the panel also complained that the agency confined itself to "after-the-fact enforcement," which was "not always good consumer protection," and it encouraged the FDA to incorporate "approaches along preventive lines."[44]

A desire to increase the agency's scientific capabilities was behind the push for prevention. Industry scientists could not be expected to do anything more than represent the interests of their employers. But the citizens' panel, which included Bradshaw Mintener—who in the aftermath of the revelations about thalidomide represented Merrell in litigation—was hardly unanimous or consistent in its recommendations. It also complained of an appalling lack of cooperation between industry and the FDA. Contrary to popular assessment, the relationship seemed built "upon fear, questioning of basic motives and a lack of opportunity for discussion." Such a combative relationship was tragic. Only through a relationship forged "upon common understanding, trust and respect," a relationship in which industry identified itself as part of the regulatory team, could the agency effectively fulfill its regulatory mission. The panel based this last sentiment on recognition that voluntary compliance would likely continue as a major facet of the regulatory effort. The agency was unlikely to possess the resources and personnel necessary to monitor more than a small fraction of industry activity.[45]

The citizens' panel's call for an enhanced FDA-industry relationship soothed feed and chemical manufacturers, and the manufacturers quickly filed NADAs to add stilbestrol to cattle feeds.[46] But they recognized, despite the panel's positive words and the apparent victory regarding the Delaney clause, that regulation had become opposition. A new research- and prevention-oriented FDA, staffed by its own scientists, who were seeking to prove their own usefulness, promised to develop new practical tests to measure almost infinitesimally small quantities of suspect substances and to detect possible new hazards. In regulatory matters, the partnership idea was dead.

4   Monitors and Monitoring

The Kennedy panel, in its call for expansion of the FDA's research and prevention role, sought to renew the agency so it could move aggressively to protect the public. This act anticipated the tenor of the middle and late 1960s. Livestock producers proclaimed a "new beef industry." Journalists announced a "new consumer," assertive, unwilling to be duped. Consumers, livestock producers, and others agitated for a "new FDA." But the emphasis on newness was not limited to food and drug matters. It became a common cry of the 1960s, as many Americans expressed a conscious desire to change direction—to make the future different from, and presumably better than, the present. The present became the enemy, and established institutions—as creations of a flawed past—contributed to the unacceptable present. The church, the military, universities, and even marriage were seen as bulwarks of a failed status quo. Government appeared to be a particularly reactionary and reprehensible institution, guilty of perpetuating through its enforcement of laws an unfair, discredited social system.[1]

This notion that institutions, especially governmental institutions, were corrupt was not the only reason that Americans called for change. A significant group of individuals in the middle and late 1960s saw government not as corrupt but as corrupted, as having drifted away from traditional principles and values; and they called for a return to the way they wished to believe things had been before. Both this perspective (later memorialized in the concept of the "silent majority") and the diametrically opposed perspective of those who demanded newness were of particular regulatory importance. Proponents of both views turned to and depended on government, clamoring for it to act aggressively, not simply react. Activist government would seek and explore, not just respond on an ad hoc basis to complaints and situations. In this milieu, the FDA emerged as a central focus, remaining a bone of contention and a prize to be won.

Ironically, these notions of activist government resonated in an environment that prized individual autonomy. The individualism of the 1960s was no libertarianism. Individuals deemed it necessary and proper to move forcefully to secure their interests, an individualist perspective certainly congruent with a rejection of present practice and form. But although broad segments of the American population proclaimed a new age, or argued that the nation required new or restored institutional structures or arrangements, agreement ended there. The nature of this new or restored age and its new or restored institutional trappings remained a matter of significant and recurring dispute. Equally important, a sort of self-delusion accompanied announcements of newness during the middle and late 1960s. Americans simply proclaimed or described as new or restored many of the initiatives of the late 1950s and early 1960s. In that sense, the flurry of activity and exegesis of the mid-1960s and thereafter often constituted an unacknowledged acceptance and continuation of previously pursued actions, even as proponents lashed out at the present and the immediate past. A broad-based, virtually implacable stridency was among the few tangible indications of this activism.

FDA-consumer relations provided a striking demonstration of this new aggressiveness. A distinctly shrill response had greeted thalidomide, *Silent Spring*, and *Poisons in Your Food*, even the Kefauver hearings, as the middle-class public responded with expressions of outrage and shock. But except in the pages of the *National Police Gazette* and similar publications, these outrage-producing incidents seemed exceptional, unanticipated, outside the realm of normality. The agitation of the middle and late 1960s took on a different character. Antibiotics in milk, cyclamates, DDT in foods, red dye no. 2, nitrates, nitrites, and nitrosamines became subjects of neatly orchestrated campaigns. The campaigners clearly anticipated that such threats would recur: each crusade depended on institutionalized lobbying agents every bit as bureaucratic and persistent as those representing the various corporate entities and trades.[2] Within this institutionalized opposition, the *National Police Gazette* and similar fringe publications held no formal place: persons supporting the effort, as well as those leading it, worked through government, not outside it. Advocates of change across the political spectrum sought to restore or redirect the FDA by conventional political means; in that sense, they were anything but radical. They apparently retained a faith that traditional political action could force governmental reorganization. Indeed, it was as if the menacing revelations of the late 1950s and early 1960s, when interpreted through this individualistic filter, had inspired, terrified, and galvanized an ever-broadening segment of the American middle class—but in a most conventional, middle-class manner. These horrific disclosures seemed by the mid-1960s to have become commonplace, providing mandates for action and, in a perverse way, reasons for confidence that government could be fashioned to come to bear on these questions. At the heart of this faith-in-government formulation rested two critical questions: who would be the monitors, and what would they monitor? Vocal men and women overwhelmingly demanded a "new" FDA. It would operate according to "new" legislation, adopt "new" adjudication principles, and be staffed by "new" personnel who would employ "new" evaluation techniques. The idea of regulation as opposition would become firmly institutionalized in the FDA. Even as the agency embraced before-the-fact prevention, science and scientists paradoxically sought a legitimate role within that construct.

The recommendations of Kennedy's Citizens' Advisory Committee on the FDA encouraged the trades to introduce a measure they claimed would simplify the entire question of new animal drugs.[3] The industry bill would regulate each feed additive substance in only a single category—new animal drug, antibiotic, or food additive—and force the FDA to cease its regulation of the use of antibiotics in animal feeds. The FDA would replace its requirement that each feed manufacturer obtain agency clearance for each drug or drug combination in each feed with a master file provision requiring that each manufacturer describe its methods, facilities, and controls in order to receive permission to use all approved new animal drugs.

The FDA and HEW vehemently objected to the industry-sponsored bill on the grounds that it "repeal[ed] a number of important health protection provisions." Wise Burroughs spoke for the bill and claimed that "tighter government controls" had slowed feed additive research "almost to a snail's pace." His pessimism arose from what he recognized as the collapse of the progressive partnership, of the "teamwork of research scientists in both industry and university feedlots and laboratories with a minimum of governmental control." If present FDA policies had been in place in the early 1950s, Burroughs contended, no growth-promoting antibiotics or hormones would exist.[4]

Burroughs's statement struck a chord among scientists outside of government, whose regulatory connections had been sharply diminished. He further ingratiated himself with the beef industry and with industry scientists when he sought to reduce the cost of animal research further by permitting the sale of experimental animals as food. Tainted or dangerous meat would still be withheld from market, but the decision would be made by the animal scientists involved in each particular investigation. According to Burroughs, the then-current absolutist policy on such sales contributed mightily to outrageous animal experiment costs and "denied the [researcher] use of his scientific judgment." Animal scientists had exercised this regulatory perquisite for years before the FDA prohibition, and "essentially no difficulties [had] occurred." That led Burroughs to conclude that "the exaggerated cost of experimenting under present regulations" yielded no public health benison.[5]

In his statement, Burroughs urged the progressive partnership's restoration. His views received extensive trade paper coverage and Agricultural Research Institute support. The FDA, HEW, and Congress ignored his arguments.[6] Reorganization of the agency, championed by the Kennedy panel and by Hubert Humphrey, had been under way for some time and was virtually complete. The resurgent FDA stressed research-based prevention rather than after-the-fact enforcement.

The FDA's emphasis on prevention rather than enforcement reflected the regulation-as-opposition philosophy. In the context of the progressive partnership, an extensive FDA scientific research program would have seemed redundant. University and industrial scientists ensured that no danger reached the American people, whereas threats came from charlatans and malefactors— frauds. The redesigned FDA assumed that scientists outside of government could not be trusted. Agency changes included the creation of a new office, the office of the associate commissioner for science; a national scientific advisory council; and two new scientific bureaus.[7] In the case of potentially carcinogenic materials such as DES, the FDA's forceful thrust into research for regulatory purposes exposed it to questions—such as how to evaluate a possible carcinogenic hazard—that had long confounded those scientists working outside the agency. On that issue, research scientists had only the most superficial agreements. Few wished to conduct experimental studies on humans. All recognized the necessity of testing substances on laboratory animals. Most understood that the relationship between a human carcinogen and an animal carcinogen required interpretation.

Outside these few areas of agreement, chaos reigned. Since the mid-1950s, and probably earlier, scientists examining non-cancer-causing agents had viewed a safety factor of one hundred as sufficient to guard the public health. They determined the minimum dose of a substance necessary to produce toxic effects in laboratory animals and then urged government to permit quantities in human food no larger than one hundredth of that amount. Despite repeated nods to vaunted science, the reasons for selecting the one-hundred-to-one ratio were arbitrary and obscure. It seemed "high enough to reduce the hazard of food additives to a minimum and at the same time low enough to allow the use of some chemicals which are necessary in food production or processing."[8] Industry now rejected the one-hundred-to-one ration as inflexible and overly restrictive. Regulating carcinogens proved more controversial and complicated. Industrial researchers suggested that a one-hundred-to-one ratio provided more than adequate protection, but William Smith and like-minded men and women wanted an absolute carcinogen ban. In the case of every carcinogen, scientists lacked concrete evidence of what constituted the smallest disease-producing dosage. Investigators continually had failed to seek no-threshold or no-effect levels, focusing instead on the carcinogenic process, a focus that required the application of large doses to speed tumor development. Researchers also understood that cancers often had long incubation periods, and that different species, breeds, and individuals responded to carcinogenic agents differently. They hypothesized that cancer might result from possible "synergistic effects" between multiple carcinogenic agents present in dosages beneath the threshold level, and feared that tests on a limited number of experimental animals—generally for less than one thousand—might prove misleading, inadvertently establishing an artificially low, statistically meaningless no-effect level. That fallacious standard might hold catastrophic consequences for "a large population such as . . . the population potentially exposed to food additives in our culture."[9]

Perhaps an illustration of this last point will make this threat clear. Scientists might demonstrate that at a specific dosage of a particular carcinogen, not one out of a thousand laboratory animals developed cancer. Yet this dosage might not prove the substance's actual threshold amount. A strong statistical probability exists that a single incidence of cancer could be revealed if researchers similarly dosed a second thousand animals. If, on the basis of an experiment with one thousand animals, the decision was made to permit the use of a given dosage in a population of two hundred million (about the population of the United States), and if, in fact, the dosage produced cancer in one of every two thousand, an estimated one hundred thousand would contract the disease.

Despite this and other seemingly intractable questions, the Food and Nutrition Board's Food Protection Committee issued guidelines in 1959 to evaluate possible food-related carcinogenic hazards. Done in broad, general strokes, the guidelines included a disclaimer about transspecies comparability. The committee maintained that "using experimental animals cannot provide irrefutable proof of the safety or carcinogenicity of a substance for the human species." Having established as its premise science's inability to fix safety standards absolutely, the committee then outlined prudent procedures for what it deemed acceptable risk. Yet the committee felt compelled again to point out the folly of seeking a precise mathematical expression of riskiness in the absence of relevant data, concluding that no-effect levels in humans ultimately "must . . . rest largely on the evaluation of alternative risks and values" that "cannot be expressed in quantitative terms." These culture-based factors led the committee to recommend allowing the use of a cancer-causing food additive only when public "values" dictated such use or when no reasonable alternative existed.[10]

The committee's nod to acceptable risk and public values suggested that science at best could present evidentiary possibilities, not definite resolutions to questions of carcinogenic hazard. But the committee's apologia did not replace scientists as decision makers. Scientists alone would assess evidentiary possibilities and make final determinations. Since the late nineteenth century, scientists had received nearly absolute regulatory authority from government. They refused to relinquish arbitrarily this awesome power. Nor was that all. Many apparently truly believed in the myth their predecessors had created, the myth that they and they alone were indispensable to the public weal. These men and women retained an abiding sense that only they stood between the public and some horrific cataclysm. Expecting them to abandon that post would have been naive.

None of this meant that American scientists were monolithic. They generally agreed that they and their studies must play a crucial role in the determination of carcinogenic hazard, but disputes centered on what constituted a hazard, and on what was unacceptable risk. Scientists tackled this disagreement directly. Investigators on all sides of the issue, both inside and outside government, attempted to convince their colleagues of their sagacity; peer review, the essence of expertise within the progressive partnership, continued to have scientific currency. Their willingness to broach the issue in a straightforward manner stemmed from their manner of collective identification, and characteristically, they employed the tools of their trade to make their cases. Put simply, scientists confronted the issue head on precisely because they were scientists and sought redemption in the techniques and the language of the scientific method. But since the dispute-producing question proved not to be reducible to simple mathematical calculations or models—these calculations and models were the consequence of individual assumptions and definitions—researchers generally failed to sway each other. No investigator could offer the type of evidence others would accept unconditionally. This fruitless exercise merely exacerbated tensions among scientists and exhibited their uncertainty to an increasingly skeptical public.

A paper published in 1961 by Nathan Mantel and W. Ray Bryan, two National Cancer Institute biometricians, neatly demonstrated the intellectual lengths to which scientists went to obtain a prominent regulatory role in the new prevention-oriented milieu. Mantel and Bryan's plan tacitly disavowed the basis upon which scientists traditionally received regulatory authority and reflected the great disparity of views among scientists about what comprised a carcinogenic hazard. The authors made no attempt to disguise their arbitrariness, setting as matters of definition "what we mean by safety and what kind of feasible results we are willing to accept as proof of safety." At risk of one in one hundred million satisfied their "virtual safety" requirement, which they decided could be achieved "only by indirect methods extrapolating" from already extant data. Mantel and Bryan saw a carcinogenic dose-response graph's slope as essential to their extrapolation, but not in raw form. They cautioned investigators to substitute "an arbitrarily low slope" to obtain "a conservative result" and "to avoid the risk of overestimation." Only then were statistical methods applied—only after "arbitrary selection of the assurance level, the level of safety desired, the slope value used for extrapolation, and even the extrapolation curve." Not surprisingly, perhaps, they assumed that the extrapolation curve possessed a constant slope rather than approaching an axis asymptotically.[11]

A more thoroughly grounded study a few years later approached the issue of the threshold level directly. Three Southern Illinois University researchers— George H. Gass, Don Coats, and Nora Graham—plotted the carcinogenic dose-response curve for oral stilbestrol administration in hybrid mice highly susceptible to mammary cancer. A desire to estimate the highest oral dosage of diethylstilbestrol that did not cause cancer when administered continuously prompted the investigation, which the *Journal of the National Cancer Institute* published in 1964. The researchers divided the mice into male, male castrate, and female groups, and fed each member of each group food containing a specified concentration of DES daily throughout its life span. The concentrations ranged from 6.25 ppb to 1,000 ppb. The scientists anticipated finding no statistically significant cancer occurrences at the lowest dosages. At higher dosages, they expected a linear relationship between the size of each group's daily stilbestrol intake and the percentage of that group developing tumors. The second of these predictions proved out more nearly than the first. Except among the intact males, the results for which could not be fashioned into a linear plot, any increase in DES dosage over 25 ppb yielded a corresponding rise in cancer genesis. Conversely, the two male groups required daily stilbestrol amounts of at least 25 ppb to show a significantly higher proportion of tumors than controls. Interpreting the pattern seen in female mice became more problematic. Daily DES quantities in excess of 25 ppb resulted in increased tumor formation, but so too did 12.5 ppb and 6.25 ppb, the smallest administered amounts. Equally puzzling, a concentration of 6.25 ppb produced more cancerous neoplasms than did either 12.5 ppb or 25 ppb.

The Southern Illinois University contingent did not attempt to resolve the anomalous situation in female mice, and few other researchers cited the study. Little reason existed to note the conclusions, because they seemed unexceptional. True, the carcinogenic dose-response results in female hybrid mice were somewhat surprising, but they hardly seemed menacing. The vast majority of scientists recognized DES's carcinogenic properties as a concomitant of its estrogenic effect. A 6.25-ppb concentration of DES in a set daily ration provided more than three times the dosage necessary to produce uterine changes in mice. Thus, the fact that experimenters detected increased cancer incidence was not surprising, although the lack of linearity proved difficult to rationalize. More important, no scientist openly questioned whether dosages below 6.25 ppb stimulated cancer. The critical relation between estrogenic effects and carcinogenic properties blinded medical, scientific, and regulatory personnel to the potential implications of the oral DES investigation. The Southern Illinois researchers had sought the initial point at which increased carcinogenesis resulted from persistent daily diethylstilbestrol feeding. They proved unable to determine that value, and in female hybrid mice they found cancer caused by dosages smaller than anticipated.[12]

Legislation concerning new animal drugs remained stalled throughout the mid-1960s. AFMA and the AHI failed to find language acceptable to the FDA, and Congress refused to act without FDA concordance.[13] The FDA had little reason to compromise. An increasingly large segment of the middle class and Congress questioned industry's motivation, and the agency gained prestige simply by refusing the AHI's overture. Equally important, the FDA, especially its scientists, welcomed agency autonomy. Freedom from the progressive partnership was not to be sacrificed for another, possibly flawed, arrangement. But while FDA scientists clung to the myth of autonomy, practical considerations demanded that the agency capitalize on contemporary regulation-as-opposition sentiments. It needed to act decisively to placate its often-boisterous foes outside of industry, many of whom advocated a "new" FDA. Practitioners of the period's New Journalism proved particularly meddlesome, in part because of their zeal and in part because they had nothing to lose. The investigative reporters of the later 1960s, like scientists both within and outside of the FDA, and like the agency itself, sought a legitimate regulatory role; they abandoned any pretense of journalistic detachment to become active participants in the fray. Proponents of the new advocacy journalism often displayed a blatant disregard, even contempt, for tradition, "the establishment," convention, precedent, process, and law, bowing instead to an individually determined higher calling. Morton Mintz's inflammatorily-titled *The Therapeutic Nightmare* captured the New Journalism's essence. An undergraduate degree in economics from Michigan helped Mintz secure positions at the *St. Louis Star-Times* and *Globe-Democrat* before he came to the *Washington Post* in 1958. He covered the National Institutes of Health and the FDA for the *Post* and won the Heywood Broun Award for Journalism in 1962. These two institutions oversaw American prescription drug policy, the subject of his study, and Mintz carefully selected incendiary snippets from congressional hearings to make his case. Including in his book thalidomide, birth control pills, excesses of pharmaceutical promotion, and other subjects chosen for shock value, he contended that the book stood as an unbiased, factual "report on prescription drugs, the men who make them and the agency [FDA] that controls them." John Kenneth Galbraith, former ambassador to India, ventured on the dust jacket that "avaricious drug manufacturers, the A.M.A.," and the FDA would rate Mintz's book as "the most unpopular book of the decade."[14]

Mintz's volume provides insight into the nature of the crusade in the middle and late 1960s to reshape the FDA and regulatory policy. Prior to the mid-1960s, FDA critics had either attacked the agency sporadically or, in the wake of some cataclysm, questioned the relevant regulatory action. As part of their regular reporting of alleged industry-government conspiracies (which was a perverse reflection of the centrality of the progressive partnership idea even to those on society's margins), fringe periodicals had often lambasted the agency for being a creature of the industries it regulated. In contrast, some middle-class women living in suburban areas and not employed outside the home had engaged in a genteel, after-the-fact consumerism. In the spirit of community service or volunteer work, they reacted to particularly heinous revelations to prevent the problem-engendering situations from recurring. No formal connection existed between these two groups of FDA critics, although some women might have browsed scurrilous periodicals as a guilty pleasure. But the *Washington Post* was not the *National Police Gazette*. It was the largest daily paper in the nation's capital. Nor was Mintz a hack journalist. He had been the *Post*'s chief reporter on consumer health affairs, and would continue for many years in that capacity. And Galbraith certainly was no fringe character; rather, he was a pillar of the liberal intelligentsia. What had earlier been scandal mongering now seemed respectable, as the public's perception of food- and drug-related threats, once unusual, became constant, even perpetual. With that change came a sharp broadening of interest and involvement. Criticism of the FDA had become a mainstream activity. Persons identified as respectable felt free to enlist in that cause.

Ironically, the FDA's food regulation under President Lyndon Johnson had proven aggressively independent, even sympathetic to consumer concerns.[15] The agency repeatedly had warned livestock producers against feed additive misuse, and it had established an advisory committee to discover whether antibiotic residues could be found in animals slaughtered for food and to investigate whether any such residues it might detect created human sensitivity or drug resistance problems. Intensified scrutiny of animal agriculture also produced a new Bureau of Veterinary Medicine to handle animal drugs. Not all Great Society measures let immediately to new programs. The gentle removal of George Larrick from the FDA commissionership positioned the agency to "enter a new era," a fact the outgoing commissioner had recognized in his retirement speech. Larrick's retirement, coupled with vacancies in the associate commissioner of science post and in the newly formed veterinary medical bureau, meant that Johnson's FDA could begin to accentuate research-based prevention rather than after-the-fact enforcement.[16]

The Johnson-era USDA also reached out to the masses, a move that led the National Livestock Feeders' Association to dismiss it "as a department of consumers." Long the food and fiber producers' staunch ally, the USDA now recognized that it had a significant obligation to those who purchased agricultural products. The agency's sudden embrace of the zero-tolerance concept, even in the face of distinguished scientific opinion, reflected the new consumerist slant. A joint National Research Council–National Academy of Sciences committee investigating pesticide residues in foods had declared the reliance on zero tolerance "scientifically and administratively untenable." Maintaining that it was "illogical to associate a tolerance value with the ability of chemists to detect smaller and smaller amounts," the committee demanded that Congress abandon the concept. The USDA actively opposed the report. The creation of a Consumer and Marketing Service (CMS) within the USDA to separate meat and poultry inspection duties from scientific research functions enabled the department to target a constituency different from that which immediately benefited from department-sponsored research laboratories; the new service more nearly presented a consumer viewpoint.[17]

DES-fed cattle were an early focus of CMS activities. As part of its new duties, the service investigated ways to measure stilbestrol residues in beef carcasses. It initially explored paper chromatography, testing the technique on animals butchered in violation of the FDA's forty-eight-hour withdrawal period. The CMS increased the likelihood of securing positive specimens by checking only each animal's liver, the organ that metabolizes DES. Fifteen out of 558 animals tested DES-positive, generally in concentrations less than 6 ppb. The CMS released these preliminary results to Congressman Thomas M. Pelly, a Seattle representative, when he requested information about stilbestrol-treated meat. That event coincided with a reemergence of concern about diethylstilbestrol-fed cattle in fringe publications. The *National Police Gazette* and *Natural Food and Farming*, a small magazine based in Atlanta, Texas, both published short essays implying that DES feeding caused many of the nation's yearly 570,000 cancer cases. Readers wrote their congressional representatives, who in turn contacted the FDA for clarification. John N. S. White's campaign was particularly noteworthy. A former Washington State meat inspector, White wrote in *Natural Food and Farming* outlining stilbestrol's dangers. He cited Longgood's *Poison in Your Food*, studies of the deleterious effects of huge doses of DES on cattle (the early Purdue research, discussed above), and reports from Brazil and Italy describing mammary gland development among males eating large quantities of stilbestrol-implanted chicken necks and heads; and he demanded that the drug "not be used in food animals." White learned from Pelly of the ongoing CMS experiment, interpreted it to serve his own purposes, and brought the matter in early 1966 to the Federation of Women's Clubs, the Farm Bureau Federation, and the Meat Cutters' Union. What White claimed in no way matched the USDA investigation. He assumed that the CMS had randomly chosen stockyard cattle and that DES quantities found in livers were indicative of those that would have been found in flesh. White also maintained that the ongoing USDA studies grossly underplayed stilbestrol contamination of the nation's beef supply, and cited a University of Massachusetts chemist's unpublished gas chromatography investigation that found at least ten parts of DES per *million* in "2 out of 10 livers purchased at a slaughterhouse"—a determination White deemed most menacing. He extrapolated from USDA experiments and claimed that about 672,000 cattle carcasses yearly must contain discernible amounts of stilbestrol. The shocking Massachusetts results suggested that more than 4.5 million beef carcasses carried physiologically active quantities of stilbestrol to American consumers.[18]

White's charges concerning stilbestrol concentrations quickly collapsed as the Massachusetts research failed to survive peer review and was eventually retracted. The FDA labeled White's entire argument "propaganda," and the recent appointment of James Goddard to head the agency helped make that designation stick. A physician formerly in charge of what now has become the Centers for Disease Control in Atlanta, Goddard had worked his way up through the ranks of the U.S. Public Health Service to lead a premier government scientific institution. He was the prototypical scientific research administrator, and even persistent FDA critics hailed his appointment as promising to transform the agency into a crack investigative unit. Like his predecessors, Goddard understood the FDA's task as law enforcement. But unlike them, he was strongly committed to the idea that only through first-class scientific research could the agency develop the capabilities necessary to make sound law enforcement judgments. The agency would become the "new FDA."

Goddard had resources that Larrick had lacked when he became FDA commissioner. In 1954, the FDA had employed 800 people and operated on a $5.5-million annual budget. Goddard commanded a 4,441-person agency with a $52-million budget. But the Goddard-led agency's responsibilities kept pace with its personnel and finances. Congress had enacted legislation assigning the FDA additional duties, and the public clamored for food, drugs, and cosmetics that were nearly pristine. Goddard's attack on pharmaceutical manufacturers provided a tangible demonstration early in his administration that he recognized and sympathized with the public mood and that he understood the politics of regulation as opposition. He excoriated the pharmaceutical industry for its "irresponsibility" and "dishonesty," arguing that "too many drug manufacturers . . . [have] obscured the prime mission of their industry: To help people get well." Goddard identified as particularly unconscionable "withholding unfavorable animal or clinical data," relying on test results only from those physicians beholden to industry, and resorting to "emotional appeals" in advertising.[19]

Goddard further showed his mettle by creating a committee of prominent university-affiliated veterinarians to advise him on "means of enhancing competence and stature" among FDA veterinary scientists. The committee helped select Cornelius Van Houweling to direct the FDA's Bureau of Veterinary Medicine. An Iowa State University veterinary medicine graduate who had headed several Agricultural Research Service science units, Van Houweling appeared to be Goddard's veterinary counterpart, providing the FDA with a second experienced research administrator.[20] But revitalizing the FDA's antibiotic residue advisory committee proved to be Goddard's crowning feat. The committee's August 1966 report noted that "unauthorized and undesirable" antibiotic residues had been found in human foods, and warned the agency to tighten the regulations controlling antibiotic use or "serious hazards to man may arise." The committee recommended mandatory withdrawal periods prior to slaughter as a possible solution but expressed frustration that the "paucity of data" prohibited the establishment of a rationally determined withholding policy, thus making any action not scientific but arbitrary.[21]

The FDA immediately announced a unilateral ban on the use of antibiotics in food preservation, but it made no similar moves against antibiotics in feed additives, preferring to sponsor a National Academy of Sciences symposium on medicated feeds to amass the opinions of key university and industry research personnel.[22] The NAS-organized symposium met in early June 1967. Thirty-seven prominent researchers presented papers at the symposium, and more than 750 interested individuals attended. Virtually all agreed that medicated feed issues required additional research. But then matters got murky. The speakers offered no hard evidence linking antibiotics or other commonly used drugs to human health problems, but most refused unequivocally to rule out the possibility of some such relationship. Partisans interpreted the speakers' statements as justifying their previously held positions. AFMA, the AHI, and other animal agriculture groups pointed to the lack of concrete data as vindication for their operations and for the status quo before the recent upsurge in consumer interest. Their opponents seized on the speakers' general refusal to state categorically that antibiotics in agriculture could cause no human harm to proclaim antibiotic use a "potential hazard." Much to industry's consternation, Goddard's FDA adopted the latter perspective.[23]

Drug and medicated feed manufacturers wondered why the FDA commissioner placed antibiotics so perilously close to the regulatory crossroads when "no case [had been] made" against them. "Considerable evidence that low level antibiotics are of value" existed, but no evidence demonstrated these drugs' harmfulness. "We demand facts," not the opinions of "individuals in the FDA," insisted producers. Feed and drug manufacturers were not the "new FDA's" only critics. James Appel, American Medical Association president, scored the agency for creating "an atmosphere of hysteria" among the public, while at least one high-ranking FDA official resigned to protest Goddard's refusal to regulate drugs "in a scientific and flexible manner."[24]

Goddard and the agency concentrated on antibiotics, but when they found themselves unable to generate quick answers to questions of transspecies sensitivity or resistance and the like, they narrowed their investigation to residues. The FDA sought to determine what percentage of the nation's meat and milk contained residues, and in what quantities, reasoning that extremely small amounts of contamination provided lesser grounds for alarm. The agency did not perform residue analyses itself but gave the questionable material to the USDA's Consumer and Marketing Service, which had become the federal meat inspection entity.[25]

The CMS tested meats for all residues, not just antibiotics. Its inspectors expected to find occasional residues because they believed that farmers sometimes ignored withdrawal periods and rushed to sell animals as soon as the market peaked. In the case of DES, they noted in late 1967 an increased number of carcasses bearing minute quantities of stilbestrol, an apparent sign of noncompliance. Inspectors confiscated contaminated flesh but neither punished violators nor attempted to ban DES feeds. The reason was simple. The CMS had used an experimental fluorometric method, a technique not approved by the HEW secretary, to determine DES tissue residues. And only the approved method, the mouse uterine bioassay, could constitute the basis for regulatory proceedings.[26]

Suasion was the FDA's only legal option.[27] But even if the CMS had employed the mouse test, the FDA would have had trouble penalizing the feeders responsible for stilbestrol-contaminated carcasses. Difficulties inherent in the bioassay made punitive enforcement measures nearly impossible. Officials would have had to use that method on each of the estimated five thousand livers examined yearly. An ironclad legal position required duplicate results, moreover, which meant that technicians actually would have needed to feed and slaughter two groups of mice for each liver tested. These exorbitant regulatory costs led the agency to seek an alternative that was quicker, simpler, and cheaper than the mouse bioassay but as sensitive and precise; fluorometric methods and paper chromatography had proven inadequate. The Association of Official Analytical Chemists joined the FDA and the CMS in this quest, but to no avail. Throughout the 1960s the AOAC and the agency failed to devise a suitable substitute.[28]

Several major FDA-related events in the late spring and early summer of 1968 diverted attention from stilbestrol. The resignation of the flamboyant, glib Goddard drew substantial press coverage. His replacement, his former assistant Herman Ley, was dramatically different from him in style but not in substance. A former microbiologist at the Harvard School of Public Health, Ley had helped Goddard formulate FDA policy. Unlike Goddard, however, he shunned public confrontations whenever possible, preferring to work behind the scenes rather than in full view of the press. Goddard dropped a final bombshell before he left. He announced that the agency would restrict some high-level oral and injectable antibiotics in food-producing animals, an action he hoped would spur "the animal drug and feed industries" to "pool their resources and do some research" on alternatives to the banned substances. Feeders and the drug industry immediately recognized that capitulation to the proposed limitations would open the way for the agency to outlaw all antibiotic substances and remove many animal agriculture drugs.[29]

Goddard lacked the power to effect an instantaneous ban because FDA regulations provided disgruntled individuals or corporations sixty days to register complaints. If matters of substance divided the two sides (a requirement intended to block frivolous objections) and the FDA chose to pursue the issue further, regulations required the agency to hold a full-scale administrative hearing. The commissioner could override these stipulations and act unilaterally only when he declared an imminent health hazard, an action legally difficult to justify. The affected parties retained the right to appeal and to secure damages if victorious.

The FDA clearly had not designated high-level antibiotics an imminent health hazard. To industry, the lack of palpable danger mattered. The FDA recognized no threat in antibiotics other than a hypothetical "potential public health hazard." Even Van Houweling noted that "no evidence" existed "linking use of antibiotics in animals to human disease." The agency had acted preemptively, maintaining that high-level antibiotics had "not as yet been fully evaluated" and might prove harmful. The FDA's reasoning contradicted established legal premise, as the agency was planning to restrict an activity simply because some individuals thought it might ultimately constitute a menace; industry maintained that the agency's capricious behavior would undermine the economic viability of an entire social group. Animal agriculturists argued for the insignificance of minuscule antibiotic residues and called for additional scientific research on the subject, not outright prohibition. After Goddard's departure, the FDA extended the comment period to 120 days and then held its original proposal in abeyance for two years. The AHI's offer to fund an antibiotic residue research program probably contributed to the change, as the institute sought to "stimulate industry integrity from an inert ingredient to an infectious ingredient" and to secure the data necessary to make an informed regulatory choice.[30]

The passage of the 1968 Animal Drug Act, a version of the animal drug bill long advocated by the AHI, probably influenced the trade association to support antibiotic residue studies. The new law streamlined and consolidated procedures and also reduced the time between the submission and approval of a New Animal Drug Application by requiring the FDA to issue a final decision within ninety days of receiving a completed NADA. Consumer groups and physicians favored this latter plank on the grounds that it made beneficial drugs available significantly more quickly. The industry, for its part, dropped its master file and antibiotics demands.[31]

The FDA presented the new animal drug law as enhancing consumer protection. In 1967, at least twenty-seven states had standing consumer organizations, and consumer groups existed in numerous cities. Placing the FDA within HEW's new Consumer Protection and Environmental Health Service in mid-1968 also reflected the "new consumerism." The FDA, the National Air Pollution Control Administration, the Environmental Control Administration, and other, lesser agencies were joined together to take into "account . . . the interdependence and interrelatedness of all environmental factors as they affect man." This new superbureau stressed research-based prevention, not after-the-fact enforcement. "Before we could prove health hazards" in the past, the new bureau's partisans maintained, "we had to be able to count the corpses." Now the situation was different. We (i.e., government scientists) "apply the scientific knowledge we have—and it will always be incomplete—to the problems" we face. Just how the new service proposed to implement this program was by no means clear.[32]

Agency reorganization, the 1968 Animal Drug Act, and the attack on animal-fed antibiotics made the FDA's future course seemed obvious. The agency stood poised to enforce law aggressively, using the latest scientific techniques to uncover potential sources of harm and to convince the electorate and the judicial system of the virtue of its determinations; it would regulate by opposition. Discrepancies between legal niceties and agency activities could be hammered out after the fact. The public good, as identified by the increasingly sovereign FDA, took precedence.

Richard M. Nixon's 1968 election cheered the pharmaceutical industry and the medicated feed industry considerably. Nixon's campaign had reached out for the "silent majority," those individuals desperately seeking stability in a new world. It had suggested that he understood and sympathized with industry's problems and would probably work with its captains to restore order rather than treating industry as an implacable foe. Executives and lawyers of the feed and drug industries formed themselves into a "Committee of Sixteen" to offer the incoming administration recommendations on the FDA's proper scope. The committee wanted to specify who would be the monitors, and what they would monitor. Its members argued that they wished to strengthen the FDA and "free [it] from bureaucratic or political control" by separating it from the new Consumer Protection and Environmental Health Service. Restoring the FDA to agency status would enable it to withstand the persistent onslaught of consumer-generated emotionalism. Independence would position the agency to fulfill its "congressional mandate to protect consumers through enforcement," which the committee asserted could best be accomplished by career FDA officials stressing voluntary compliance. Scientists would "be relieved of any responsibility for enforcement action" so that they could focus exclusively on analyzing new drug applications. Making new drug approvals a purely scientific process in which FDA scientists would examine data presented by companies would bring new substances to market more quickly. Removing scientists from research-oriented regulatory activities ensured that they would not overstep the neatly circumscribed bounds laid out by congressional legislation. More specifically, the Committee of Sixteen recommended that the Nixon administration compel the FDA to accept and not expand congressional dictates; accept and not just evade judicial interpretations; terminate enforcement by news release; stop publicizing voluntary recalls; consult with consumer and industry groups prior to issuing proposed regulations; reevaluate "non-significant indirect food additive regulations," such as those for antibiotics; and establish a long-range planning entity to plot the agency's agenda.[33]

Neither the FDA nor nonindustry claimants to agency policy-setting preserves would concede anything to the industrial opposition. In particular, Ralph Nader labored to keep the FDA on its present course. Nader was consumerism's true organizational genius. He parlayed past successes into future initiatives, plowing psychological profits back into his consumer business. Through interlocking directorates and hard work, Nader began to form a horizontally integrated consumerist engine. The Center for the Study of Responsive Law functioned as his organizational cogwheel. Chartered on June 28, 1968, as a charitable trust, with Nader its only trustee, the center claimed as its mission research "relating to the general question of the responsiveness of law and legal institutions to the needs of the public."[34] It trained and housed "Nader's Raiders," who in a few brief months completed investigations of complex federal agencies and published book-length studies and articles. The raiders characteristically determined that federal agencies failed to meet "the needs of the public" and condemned them in these exposés, which mimicked the New Journalism. Nader now accused the food production and preservation "industry and the Federal Government of laxity in protecting the public against impure products and potentially dangerous chemical additives," and he predicted that food would become "the next big consumer issue." He chided the FDA for being "unduly meek" and for carrying out "a passive enforcement policy"; and he established a task force of raiders to slice through the agency's "bureaucratic secrecy" to reveal the real, seamy story.[35]

As the Naderites began their investigation, Senator Gaylord Nelson examined manufacturers' new-drug testing programs—the basis for FDA approvals—and found them "of poor quality," "abbreviated," "ineptly designed," "deficient," and "bluntly unacceptable." Drug companies' scientists had a "special interest" in gaining agency sanction for their products, and Nelson wanted only scientists with "no direct relationship with manufacturers, who cannot benefit financially . . . and who are not motivated even subconsciously by the desire to get anything but the truth" to conduct these crucial tests. That meant government-employed scientists, and Nelson called for the creation of a national drug-testing and drug-evaluation center to handle new drug applications. Run entirely by FDA scientists but paid for by industry levy, the center would scrutinize and test at manufacturers' expense all new substances tendered by drug companies.[36]

An internal FDA report leaked to the press overshadowed Nelson's indictment. A Ley-appointed committee formed shortly after Nader announced plans to investigate FDA practices had produced the report, which took a startling direction. The recommendation that the agency admit its relative impotence to the American public was the report's most daring feature, but the sections on medicated animal feed also proved controversial. The committee asserted the FDA's independence from both consumer and industry advocates by acknowledging that "maintaining a readily available food product" made "risk-benefit relationships" the unacknowledged regulatory basis. It then questioned whether the FDA could continue to "justify a conservative [feed additive] risk or safety philosophy" as the world's food supply "becomes critical." To prepare for the eventual shortage, the committee suggested that the agency "reevaluate its conservative drug residue policy and develop objective criteria for determining risk-benefit ratios." It also called for "user and consumer education" programs. The former would familiarize producers with potential hazards and urge them to take care, while the latter would reassure consumers that risks were minimal.[37]

Several industry pundits took the Ley committee report as a sign of a thaw in relations between government and pharmaceutical manufacturers, but this assumption was fundamentally incorrect: the report was an FDA declaration of autonomy. This fact was supported by the agency's mid-October cyclamate ban. The FDA interpreted a series of tests undertaken at Abbott Laboratories subjecting rats to high cyclamate doses for extended periods as showing an increased incidence of bladder cancer. These results placed cyclamates in violation of the Delaney Cancer Clause and compelled HEW to announce the ban.[38]

The removal of the most common artificial sweetener from the market brought additional attention to food additives and provided focus for manufacturers' and consumers' ire.[39] The creation of a cause célèbre apparently disturbed the Nixon White House. The ban found little favor in the administration's upper echelons or even in HEW, which technically had outlawed the substance. It seemed to be a remnant of the high-handed Goddard regulatory approach, of which Ley was the most tangible legacy. The public first learned that the administration disapproved of the ban when Nixon announced in his consumer message to Congress that he had charged HEW secretary Robert Finch with undertaking a "thorough examination" of the FDA. Finch went on record almost immediately favoring a "softening" of the Delaney clause to give the government the "flexibility to set maximum or tolerance levels for cancer-causing additives." He quickly gained the AMA's support and separated the FDA from the Consumer Protection and Environmental Health Service, a key industry aim. Finch also removed from their positions Ley and two other high-level Johnson-era administrators. He appointed as FDA commissioner Charles C. Edwards, formerly a vice-president and managing officer of a Chicago management consulting firm and an AMA staffer. The distinguished Edwards looked every bit the businessman. His forte was business administration, not research administration, and Finch challenged the new commissioner to give the FDA aggressive leadership, to increase the agency's efficiency, and to give industry clear regulatory guidelines. Expeditiously determined, judicious rules painstakingly presented to the relevant industrial interests would reverse the FDA's then-current "orientation of [acting] after the fact, hitting them over the head."[40]

Ley initially said little about his dismissal and the implicit criticism that accompanied it. But persistent unflattering news leaks, presumably by HEW insiders, led him to strike back. He accused the drug industry of applying "constant, tremendous, sometimes unmerciful pressure" on the FDA. The agency lacked "motivation" to beat back pharmaceutical companies because its "professional staff was inadequate and filled with retreads." Ley claimed to have experienced "a total lack of top-side support from the current administration." But he insisted that "everybody realizes that this is a business-oriented Administration." He only wished that the American people understood the implications that this allegiance had for FDA activities. It "bugs me . . . that the people think the FDA is protecting them—it isn't," he stated.[41]

The widespread belief that the Nixon administration had a probusiness bias strengthened the consumer movement. For individuals who were already concerned that government administrators had failed to provide adequate protection against the seemingly never-ending menace of self-interested businesses and industries, this perceived bias reenforced the conviction that consumers needed to organize to force governmental redirection. It also encouraged Nelson to introduce into the Senate a measure that would extend Delaney-like coverage to substances (e.g., teratogens and mutagens) "found to induce chronic biological injury or damage when ingested by man or animal." Nor did industry stand idle. Pharmaceutical companies manufacturing DES for feed responded to the administration's assumed support by repeatedly urging the FDA to permit greater DES doses in animal feeds. Research results demonstrated conclusively that cattle fed ten milligrams of stilbestrol daily and implanted with the drug grew more quickly and inexpensively than those treated with either method alone. The FDA had ruled that simultaneous stilbestrol feeding and implanting violated regulations, but the agency now winked at transgressions. It could do little else. The FDA determined violations only by finding residues of the substance in carcasses. Throughout the year, CMS personnel noted a decrease in DES-contaminated beef, a situation FDA officials attributed to the AHI's and AFMA's educational campaign to get livestock producers to observe the forty-eight-hour withdrawal period.[42]

This no-measurable-residue, no-harm philosophy led Elanco, the agricultural chemical subsidiary of Eli Lilly and Company, to develop the dosage-doubling data necessary to pass FDA muster.[43] The agency's approval of Elanco's petition drew little comment, for the FDA's future course, not stilbestrol, remained the central question. Who would monitor and what would be monitored in this regulation-through-opposition milieu took precedence over all other issues. In the seven years after 1962, contemporaries had repeatedly redesigned the FDA to meet challenges of newness. Battle lines had rigidified as consumer advocates firmly ensconced themselves both within and outside of government, developing new bureaucratic arrangements and full institutional networks. These consumer "trade" associations proved every bit as determined as their commercial and industrial counterparts.

Science was becoming increasingly irrelevant to this debate. It stood like a dinosaur, a relic of the past, unable to provide the definitive answers individuals sought. Scientists expressed conclusions in terms of probabilities and possibilities. They lacked certainty, but not conviction. They continually stressed their regulatory importance and claimed for themselves an inordinate share of regulatory responsibility. That proved equally true inside and outside government service. Interested scientists rarely failed to make their opinions known.

III / SOCIETY

5   Hearts, Not Minds

"The consumer is without armor," wrote Erma Angevine, the Consumer Federation of America's executive director, "in a battle with an adversary equipped with computers, a giant corporate structure largely concerned with making profits, far-reaching influence on the Congress and Administration, and almost unlimited access to financing. This Goliath can ignore the law," she continued, "and even if proved guilty can deduct from its income taxes the cost of repaying the consumer."[1] Angevine sounded a refrain familiar during the late 1960s and early 1970s. The consumer, armed only with personal outrage, challenged a nearly omnipotent amoral foe. Contemporary variations of this good-versus-evil motif included the theme of concerned citizens against unfeeling bureaucracies and bureaucrats, and the theme of those with the public interest at heart against those with private interests at heart. In this politics of personalization, consumer advocates seized the high moral ground in each case by relentlessly denigrating their adversaries, a tactic calculated to place their targets on the defensive.

The battle for public opinion intensified in the late 1960s and early 1970s as Americans virtually abandoned the pretense of working through established government mechanisms to reorient public policy. Partisans sought victory in the public arena, where they attempted to win the hearts, not necessarily the minds, of the public. Regulatory disputes were no longer played out in scientific papers or government documents—traditional rationalist sources—but in the press or in mass market books. Posturing became the predominant form of debate, and government hearings emerged as an important locus of that dramatized activity. Both sides talked emotionally of the wisdom of the people in this predatory, accusatory democracy, and both sides recognized the people as victims. Those without regulatory power demanded that the people be heard, and cast their pleas in terms of a burgeoning middle-class social movement that would sweep aside the status quo. Those in power maintained that the people had already been heard: voters had elected officials, who now found themselves constantly abused by a self-righteous minority seeking to seize power for its own personal ends.

Supporters of DES feeds and, to a lesser extent supporters of the FDA found themselves in a politically untenable position. Their opponents' rhetoric suggested recurring regulatory abdication, an unspeakable crime. That bodies did not litter the streets because of the presumably unconscionable actions permitted by the FDA mattered little. Some hypothetical day it could happen, and the American public would suffer. Thus, DES in food was a matter for concern even though the evidence that DES was a human health hazard seemed slim. Stilbestrol was a known carcinogen, but only in doses large enough to cause physioanatomical changes. The amounts used to boost cattle growth were comparatively less than the quantities approved for use in human medicine. Physicians had prescribed the substance for nearly three decades with little apparent incident. Only rarely did stilbestrol-produced beef contain hormonal residues great enough to register on the mouse bioassay. Even in cases in which residues exceeded 2 ppb, contamination was restricted to the liver, the drug-metabolizing organ. The long-forgotten 1964 Southern Illinois University dose-response study was the sole scientific study that might have given pause. That investigation had failed to determine a threshold DES dosage in mice. Yet the lack of scientific evidence pointing to a threat due to stilbestrol in beef made little difference. The terms of the debate had changed. Scientific opinion, itself increasingly fragmented, seemed irrelevant; and reference to science and scientists seemed to be a mere formality, a style of discourse the substance of which appeared no longer germane. Consumer advocates, New Journalists, and industrialists all called upon scientifically certified partisans of their own positions in a pro forma credentialing demonstration. Scientists now rarely represented their studies as conclusive, but called on the public to honor their judgments, their opinions. In a milieu marred by suspicion and individuation, the fact that scientific assessments contradicted each other further undermined scientists' historical basis for authority. The most dire predictions and threats of a pending apocalypse became the most actionable. This demagoguery preyed on the safe-rather-than-sorry mentality of a frightened middle class that, although sometimes unconvinced by a specific claim, dared not risk inaction. The hypothetical consequences seemed far too great.

The publication of *The Chemical Feast* in April 1970 reflected the new accusatory politics of the visceral. Written by James S. Turner, a twenty-nine-year-old Ohio State University law school graduate, this long-awaited Nader's Raiders study of the Food and Drug Administration employed innuendo and unsubstantiated rumor and capitalized on unparalleled access to agency documents and personnel to produce an unequalled attack on the FDA. Turner chose phrases for their attention-grabbing value as he reported as factual or as probable virtually every conceivable harmful occurrence or situation. Industry exhibited "corporate greed and irresponsibility." Many animal scientists "routinely produce scientific studies that support the most recent industry marketing decision." The FDA's food regulations "read like a catalogue of favors to special industrial interests." Agency officials "regularly tailored scientific activities to support already arrived at administrative positions," while "scientific opinions and memoranda [within the FDA] have been distorted, altered, misrepresented and ignored." Turner pointedly contrasted the agency's methods to the raiders' virtuous inductive approach. "None of the students expected to find in the FDA the shocking disarray and appalling failure of responsibility that their investigations revealed almost daily." Only when "the number of altered documents, misrepresented facts and suppressed studies began to mount" did "the conviction that most agency efforts were a failure" surface.[2]

Turner aimed to tap emotions, not to inspire reasoned dialogue. That his book, so obviously hostile and biased, received a serious, measured hearing is fascinating.[3] Turner and his helpers seemed young and inexperienced, unprepared and unfit for their task. They claimed for themselves almost no scientific training. Yet they acted as if they were superbly qualified to evaluate the scientific activities of one of the foremost federal science-based regulatory agencies, and were comfortable in doing so. Their admitted lack of formal scientific education provided them a perverse sort of credential, that of disinterested outsider. They touted their lack of scientific training not to glorify ignorance or antiintellectualism, but to signal unsulliedness and the belief that anyone could participate in effecting change. Indeed, a romantic belief in the possibilities of everyman and in the power of concerted action characterized and unified the consumer movement. The movement promised a future different from the present. The "outs" would replace the "ins" and set things right, precisely because they felt—although they did not necessarily know at the present instant—what was right.

The Chemical Feast offered several prescriptions to remedy FDA ills. The least surprising of these prescriptions was the demand that consumer representatives be given a vital place in agency decision making, but the others followed a similar line. Turner urged the FDA to grant its personal "scientific independence" by establishing a well-equipped investigative research laboratory; to restrict "the discretionary authority of the politically appointed [HEW] secretary"; and to strictly enforce the letter of the law rather than working with industry to encourage voluntary compliance. In short, he echoed recommendations that others before him had advocated and that, to a certain extent, the

Johnson administration had implemented: he urged that the agency be research based and prevention oriented, and emphasize regulation as opposition. Turner also declared medicated feed residue monitoring programs a sham, demanded that legislation modeled on the Delaney clause protect the public from mutagens and teratogens, and identified antibiotic use in food-producing animals as a cause of the American diet's deterioration. In sum, Turner charged that "the FDA effectively obscured for Congress and the public that food protection and drug control are inextricably bound together."[4]

The medicated feed industry's response to Turner's indictment proved surprisingly mild and somewhat unexpected. Perhaps Nader's announced intention to send his raiders next to investigate pharmaceutically enhanced feeds and drug residues in meats clouded the industry associations' judgment. More likely, the industry felt frustrated by its inability to counter effectively assertions piled upon assertions. It certainly rejected the raiders' critique of industry-FDA relations. It felt itself constantly "burned by improper or inaccurate use of information" on the part of a group that "establishes positions and then tailors its activities to support already arrived at positions." But it elected to fund for consumer advocates and their staffs a recurring "symposium" on the use of animal drugs in food production, to show these naive youngsters their errors.

The industry's recognition that it could no longer ignore Naderism and the increasingly broad-based consumer movement rested at the initiative's heart. So too did the belief that the data clearly sustained the industry position. In a charmingly antiquated way, the industry expected to persuade consumer advocates and their supporters simply by laying out its version of the scientific facts; the industry assumed that its opponents operated from the same rationalist assessments as it did, and therefore that the only logical explanation for their actions was that they had failed to confront the evidence. Since "agribusiness and agricultural science positions are sound, why not expose them to more people," reasoned Roger Berglund, editor of *Feedstuffs*.[5]

The FDA's Charles Edwards brought the agency back into the public eye as he faced the issues of antibiotics in feeds, and cyclamates. He agreed with consumers by acknowledging "protection of public health [as the agency's] top priority," but dismissed the consumer movement's leaders, calling for "the Ralph Naders, our congressional critics and even some of the 'experts' within the government structure to leave us alone." Edwards wanted an independent FDA, but on terms more nearly consonant with industry objectives. The agency was "constantly confronted with balancing the benefit vs. risk ratio," he argued, and "our decisions must be based on scientific facts and our best judgment of what is in the public interest." Absolutism had no place in regulatory policy.

"We must be careful how" we use animal drug data when we "speculate on the possibility of [drugs'] effect in humans." Only "expert judgment" can "evaluate the scientific evidence" to yield an informed decision. Slavish application of the Delaney clause would soon reduce us, Edwards exaggerated, "to a nation of vegetarians, and even some of the vegetables would have to be banned."[6]

Edwards's actions spoke louder than his words, however, and the independent-minded agency's initiatives displeased the medicated feed and pharmaceutical trades. Edwards established a scientific task force on antibiotics in feeds and allowed industry experts no representation. In spite of trade objections, he began to institute a total cyclamate ban. But nothing that Edwards did in 1970 maddened, threatened, or disrupted the medicated feed trades as much as a single Associated Press wire story. John S. Lang wrote the accusatory piece, which began with the charge that FDA officials were permitting DES residues "known to incite cancer" in the nation's beef supply, "in violation of federal law." Agency personnel compounded this criminal culpability with the sin of lack of concern for the public weal. Lang reported that Van Houweling had stated, "Most of us [at the FDA] can't get too excited about the occasional carcass tinged with minuscule DES amounts." Lang then sketched a broad beef conspiracy emanating from Iowa State University. He noted that Van Houweling had graduated from the veterinary school there and had worked in Ames at the National Animal Disease Laboratory before coming to Washington. And it was at Iowa State that Wise Burroughs had discovered DES's growth-promoting properties, which had netted the university nearly three million dollars in royalties. Cattlemen reaped a "$300 million annual bonanza" because of Van Houweling's agency's illegal permissiveness. They routinely ignored the mandatory forty-eight-hour withdrawal period to achieve top market price, and they added stilbestrol in quantities well in excess of the approved dose to get "faster gains and greater feed savings." Feed and pharmaceutical companies, particularly Elanco, encouraged beef producers to ignore the legal limit by advertising that researchers had secured greater economies at a higher daily dose. Agricultural colleges produced most of those studies. Iowa State remained the leader in the investigation of stilbestrol feeding.[7]

Lang mentioned that DES caused impotence in male animals and nymphomania in females, and that the drug had been banned from poultry production more than a decade earlier. The crux of his complaint, however, was the FDA's refusal to prohibit stilbestrol feeding. He had learned that during 1968–69, out of nearly thirteen hundred animals tested, the USDA's Consumer and Marketing Service had found seven with DES traces in excess of 2 ppb in their livers. According to Lang's story, under the Delaney clause and the DES Proviso those results meant that the FDA was required to outlaw the drug. "The law explicitly prohibits all residues," he wrote, and he selectively quoted those sections that seemed to sustain his point.[8]

Lang's article received a wide hearing precisely because of its emotional appeal. One clipping service compiled a file of more than 180 reprints. But although the story made excellent reading, its author's central point could hardly bear legal scrutiny. The DES Proviso empowered the agency to stop diethylstilbestrol feeding only in specific circumstances. The absence of one of these circumstances in particular blocked FDA action in 1970. Drug residues needed to appear in beef "under the conditions of use and feeding specified . . . and reasonably certain to be followed in practice." Thus, to make a legal case, the FDA had to demonstrate either that residues showed up when producers conformed to FDA-prescribed procedures (e.g., the forty-eight-hour withdrawal period) or that a substantial portion of the target population could not or would not adopt those procedures. The evidence available to the agency suggested that the current situation met neither condition. A study undertaken some years earlier indicated that cattle fed DES metabolized and excreted the substance completely within twelve hours. That investigation proved to the agency that no residue could remain when cattlemen conscientiously adhered to FDA withdrawal requirements. Similarly, the relatively small proportion of DES-contaminated carcasses—about 0.5 percent—seemed to demonstrate that the great majority of livestock producers followed regulations.[9]

The agency responded to what Van Houweling termed "biased reporting" and promised to bring legal proceedings against those apparently violating the forty-eight-hour withdrawal period. Lang's article led Allen H. Trenkle, a former student and then colleague of Burroughs at Iowa State, to call on his animal science peers to lessen the "fears of anxious consumers who don't know who or what to believe when they learn about things like this." Trenkle claimed that FDA-approved DES testing often produced false positives because it measured total estrogenic activity, including the activity of estrogens "normally produced by the cow," and he professed indifference about occasional DES-contaminated beef escaping detection and reaching private homes. "There are no known cases of humans ever developing cancer" from diethylstilbestrol, he remarked.[10]

Trenkle's public utterance reflected a position that scientists, both those who supported DES and those who opposed it, increasingly adopted. His statement presumed that his judgment on an issue carried considerable weight simply because he was a scientist. But traditionally, scientists had gained credence and public sanction for a very different reason: their results, which could be confirmed or verified by their peers and therefore could be agreed to as correct by their peers. Trenkle reversed that time-honored equation, arguing that his views were right because of who he was, not because of what he did, "knew," or demonstrated. His statement was a shrill, emotional assertion, requiring no confirmation and providing no incentive to conduct potentially conclusive experiments. Moveover, experiments resolving complex regulatory issues may not have been possible in this individualistic milieu. Individual perspective, the "right" of individuals to come to their own decision from their own unique personal conditions, virtually precluded the possibility of a single event furnishing the decisive justification for a solid, broad-based regulatory consensus. Fear proved to be the compelling political motivator.

Pressured by the adverse press coverage, the publicity-conscious trade associations of the animal pharmaceutical manufacturers and the cattlemen started a joint education program to teach producers the importance of withdrawing cattle from DES-enhanced feeds at least two days prior to slaughter.[11] Advance notification of the long-anticipated FDA decision to approve formally a doubling of the daily stilbestrol cattle-feeding dosage no doubt contributed to the program's formation.[12] In October, the nation's two largest cattlemen's organizations drew up a certificate for livestock producers to sign testifying that they observed the FDA's stilbestrol withdrawal regulations, and prevailed upon meat packers not to slaughter animals lacking these documents.

Leading farm and feeder periodicals applauded the proposition and warned that a frightened public would be increasingly dissatisfied with the FDA and the cattle industry unless producers eliminated residues.[13] Even an appearance of insensitivity could cause intense public reactions. The response to the Associated Press's late-November revelation that the CMS had reduced by half the number of carcasses tested for DES in 1970 shocked that agency. The CMS had reallocated its resources to monitor salmonella and heavy metal contamination, which had "been hot." Officials justified their move by arguing "that benefits . . . must be weighed against . . . risks," stating that they judged the risks of cancer from DES as "slight."[14] The CMS had turned up only a solitary stilbestrol-contaminated carcass in 1970, but a maelstrom of protest greeted the Associated Press story. In the face of these emotional outbursts, the bureau announced a grand diethylstilbestrol inspection plan to examine six thousand carcasses during 1971 "to satisfy consumers [that] no problem exists."[15]

The CMS's optimism stemmed from initial successes in applying a new chemical DES assay to replace the mouse uterine bioassay. The new test, gas-liquid chromatography (GLC), developed primarily by Elanco, had proved much quicker and less expensive than the mouse test, and seemed nearly as reliable and almost as sensitive. Faith in the new technique had encouraged the

FDA some months earlier to raise the daily stilbestrol feeding limit.[16] The agency, like its agricultural affiliate, apparently felt that it would soon possess a tool superbly suited to regulatory tasks.

Although the CMS reversed its decision on stilbestrol testing reduction,[17] two separate congressional hearings examined DES beef as part of a broader exploration of chemicals in the nation's food supply. L. H. Fountain, a North Carolina Democrat and longtime FDA sparring partner, chaired the House Intergovernmental Relations Subcommittee, which met to investigate food additive and medicated animal feed regulation. The Senate Subcommittee on Executive Reorganization and Government Research, headed by former HEW secretary Abraham Ribicoff, a Connecticut Democrat, convened to consider "chemicals and the future of man."[18]

The Fountain hearing immediately placed the FDA and the CMS on the defensive. It pilloried Edwards, Van Houweling, and other FDA officials for approving the doubling of the maximum daily DES dose in the absence of a residue test suitable for regulatory purposes. The committee questioned Richard Lyng, assistant secretary of the USDA; Clayton Yeutter, CMS administrator; and Yeutter's assistants because the CMS tested so few beef livers for stilbestrol residues. Members of both agencies talked optimistically of current enforcement plans, the cattlemen's drug certification program, and the GLC test, which so far seemed reliable to 2 ppb. These men could not know that GLC would never pass AOAC or FDA muster for stilbestrol residues. For their part, several members of Congress had difficulty grasping regulatory concepts. Infinitesimally small quantities proved particularly slippery and threw more than one legislator into a tizzy. So too did dosage. That physicians regularly prescribed comparatively huge diethylstilbestrol doses for women while the Delaney clause outlawed even parts per billion in meat struck some as intellectually inconsistent. Fountain adjourned the proceedings with the moderate assessment that "the public is getting less than the maximum protection possible in this vital area."[19]

The Ribicoff committee, on the other hand, called only a single witness to testify on DES. Harvard Medical School's Samuel S. Epstein implied that beef from stilbestrol-treated cattle was of poor flavor and texture, claimed that farmers failed to follow withdrawal rules, and maintained that the CMS had proven incapable of commanding compliance. He noted that Sweden had banned stilbestrol feeding and demanded that Congress give Americans identical protection. When asked to name individuals on whom the committee should call to develop the full DES story, Epstein came up with a single name, that of Harrison Wellford, the Nader's Raider heading the Center for the Study of Responsive Law's USDA-FDA meat investigation.[20]

Beef producers predictably leapt to stilbestrol's defense,[21] but the situation abruptly changed on April 22–23, 1971, when Arthur L. Herbst, Howard Ul-felder, and David C. Poskanzer published in the *New England Journal of Med-icine* their classic study of DES daughters. In 1970 Herbst had identified in seven women aged fifteen to twenty-two a form of vaginal cancer rarely found in the young. He and his colleagues then set out to determine the outbreak's epidemiology and ultimately focused on the almost indiscriminate use since World War II of DES, both to prevent miscarriages and, paradoxically, as a morning-after pill to induce spontaneous abortions. They found that these young women's mothers probably had been given doses of DES as large as 125 milligrams daily during the first trimester of pregnancy. The group hypoth-esized transplacental transmission of the drug and suspected a causal relation-ship between diethylstilbestrol and the tumors.[22]

The journal hailed the study as being "of great scientific importance and serious social implications," and as adding "a new dimension" to the drug safety question. Reporters recognized DES as the substance known for promot-ing cattle growth, and immediately quizzed the FDA about a possible human health hazard. Van Houweling denied any such possibility because "no [DES] residue is allowed in foods," and noted that the agency had given "no approval" to "add any stilbestrol to food intake."[23] The press also contacted Wise Bur-roughs, who dismissed as "conjecture" Herbst's tentative conclusions, recog-nized "no evidence whatsoever" that DES caused cancer, and claimed that the mass "alarm" over stilbestrol in cattle feed has "become ridiculous." Even if livestock producers ignored the forty-eight-hour withdrawal period, Burroughs claimed, a human must ingest "220 to 2200 pounds of liver to get a therapeutic dose of stilbestrol."[24]

Senator Ribicoff and Senator William Proxmire, a Wisconsin Democrat, demanded federal action. Ribicoff addressed open letters to USDA secretary Clifford Hardin, FDA head Charles Edwards, and Environmental Protection Agency administrator William D. Ruckelshaus decrying the current policy on the use of DES as a feed additive. Proxmire urged an immediate stilbestrol-feeding ban, identifying DES as a "cancer-causing" agent and complaining that the FDA "is allowing meat containing DES residues to be sold to consumers" in "clear violation of the Delaney Amendment." Addressing Edwards personally, Proxmire reminded the commissioner that it remained his "responsibility to protect the consumer from unsafe foods." In the case of DES, "the only totally effective method" for Edwards to adopt was "to prevent DES from being used to promote the artificial growth of meat cattle." Proxmire thought the situation especially tragic because DES use was not "essential," for "cattlemen have other growth-promoting drugs at their disposal." Twenty-one nations apparently had outlawed the substance in animal agriculture, and he called on the United States to become the twenty-second.[25]

The FDA persisted in warning beef producers to follow mandatory withdrawal regulations. The USDA aggressively rejected Proxmire's contention, as Lyng claimed flatly that agency "tests show that no red meat containing detectable levels of DES residues has reached the consumer." Lyng also disputed those who assumed that CMS-tested livers came from "a cross section" of meat animals rather than only those "suspect to start with"; congratulated cattlemen for their certification program; and hailed GLC testing as providing consumers even greater protection. The USDA also calculated the possible costs of instituting a ban on DES feeding. It figured that the removal of DES from cattle feed could run cost-conscious consumers as much as $460 million yearly.[26]

Livestock producers and their supporters greeted with derision the statement that twenty-one countries had banned DES. None of these countries possessed an agricultural system structured to benefit from stilbestrol feeding. American beef producers knew that refusals to import DES-fed beef (a policy adopted only by Sweden and Italy) simply constituted trade protectionism, a non-tariff trade barrier to keep a cheaper commodity out of a country to enable local production to survive. Cattlemen claimed that the public received only one side of the DES story, and urged their trade organizations to alert consumers to the virtues of medicated feeds.[27]

Livestock producers, pharmaceutical companies, and feed manufacturers apparently based their argument on two major points: first, the contention that no residue remained in edible tissues when cattlemen withdrew animals from DES feed forty-eight hours prior to slaughter; and second, the belief that DES's carcinogenic properties stemmed directly from its estrogenic effects. Together, these points suggested that there was no hazard from stilbestrol-produced beef, but the partisans of DES certainly lacked sensitivity to public fears. Moreover, their explanations often skirted the line between truth and falsehood. Van Houweling's claim that the FDA permitted no DES residue in food did not mean that beef was residue-free. It indicated only that the FDA was obligated to condemn any carcass in which it found measurable stilbestrol. Lyng's contention that no red meat with detectable DES levels reached consumers was especially disingenuous: it presumed that USDA investigators condemned all contaminated meat, which almost surely was not the case. And the USDA's multimillion-dollar estimate of the costs of banning DES centered on the fact that nothing exactly duplicated stilbestrol's growth-promoting properties. In fact, however, animal agriculture certainly possessed several substances nearly as effective and inexpensive as diethylstilbestrol.

The great disparity among rivals' claims created dissonance, further undermined public confidence, and raised the larger question of what ailed federal regulatory agencies. Scientists contended that the FDA suffered from "poorly managed laboratories, a lack of sound scientific facts [and] a curious aura of secrecy." They noted that "food and drug technology have [*sic*] outstripped the FDA's abilities"—for example, the "public health significance of drug residues in meat . . . is not well understood"—and that "conflicting pressures" come to bear on the "overworked and undermanned" agency. Journalists often reduced FDA problems to a simplistic central tension: whether the agency should move toward "greater recognition and sympathy for the pharmaceutical industry's economic problems, or else towards a more unyielding pro-consumer line focusing only on potential public health problems." Lawyers sometimes lamented that "food and drug law establishes an unwieldy regulatory system," while FDA old-timers deplored the transformation of their bailiwick from a "police agency" to an "education agency." Each of these self-serving critiques singled out a particular arena in which the FDA had failed, and held out hope of a swift resolution through the transfer of regulatory authority. A cynical antiestablishment voice maintained that the FDA was impotent because all parties (except consumers) wanted it that way. It served their special, selfish interests. No one in authority wanted a "strong FDA" because "it would interfere with the medical profession's freedom to prescribe; challenge the scientific predominance of the drug industry; deprive Congressmen of a tempting political target, and involve HEW and White House superiors in constant controversy."[28]

Even within this contentious milieu, the publication of Harrison Wellford's *Sowing the Wind* drew attention. This long-awaited follow-up to *The Chemical Feast* included extensive coverage of DES feeds. Wellford demanded that the FDA improve the sensitivity of its stilbestrol residue monitoring and ban the drug should residues be detected by these more discriminating methods. He argued that "it is possible that DES is present [in meat] at levels potentially hazardous to man, but too small to be detected" by current USDA assays, which he believed to be accurate only to 10 ppb. The 1964 Southern Illinois University dose-response study by George Gass and colleagues fueled Wellford's fears. It indicated that as little as 6.25 ppb of stilbestrol "produced cancer in 50 percent of mice tested." He contrasted the gross amount of the drug fed these mice daily with the amount in one pound of beef containing just under 10 ppb and concluded that a USDA no-residue result could mask "up to 14 times" the quantity of DES found to cause cancer in the SIU investigation.[29]

Wellford based his comparison—whether accurate or not—on total number of atoms of stilbestrol. But just what was he comparing? Was he comparing one pound of liver contaminated with 10 ppb of DES to the total daily diet of a stilbestrol-fed mouse or to the amount a mouse would consume over its lifetime? or was he comparing some other figure? Nor was the total number of atoms of DES the only possible comparison. Gass's 6.25 ppb was based on the entire daily mouse diet; everything the animal ingested contained precisely 6.25 ppb of stilbestrol. If that concentration of diethylstilbestrol was compared to the percentage of DES in the human diet—even figuring ingestion of one pound of liver daily at a concentration of 10 ppb—a quite different result would be reached. Correcting for relative body weight would produce yet a different comparison, as would the consideration of the relationship in terms of a single exposure to one pound of 10-ppb liver, or a similar exposure daily or at some other frequency. Over the course of the stilbestrol controversy, many comparisons would be offered. No two commentators compared the same things. Each partisan used a comparison that showed his or her argument off to good advantage.

Wellford saw the stilbestrol situation as symptomatic of American agribusiness practices. Once "a chemical has become entrenched in the marketplace," the "definition of safety often takes on political implications." Wellford explained that "corporate agriculture fears that its huge investments . . . would be threatened" should the FDA or the USDA discover a human health hazard. This "anxiety leads to tremendous pressure" on regulatory scientists, many of whom "have promoted the chemicalization of meat production." Safety concerns originate almost exclusively among scientists with "no direct voice in [regulatory] decisions." Wellford thought it absurd to presume that "objective science will prevail," for "chemical industry and the producers and processors of meat" lobbied against "objective science"—the science of consumerism—"to protect their investment." They foisted their definition of "acceptable risk" on "society and its political leaders," generally through the "use of code words such as 'toxicological insignificance' and 'no-effects levels' and 'limits of analytical sensitivity.'" Without "full and timely disclosure of risks and benefits to the consumer," consumers remained at agribusiness's mercy. Only if there were an absolute and complete public investigation into each substance's potential repercussions before it received government approval could consumers gain the power necessary to protect themselves.[30]

Wellford's licensing proposal received little notice as individuals and institutions focused on the specific stilbestrol allegations, including the resurrected Gass experiment. This study, in the wake of the Herbst correlations, influenced subsequent consideration of DES. Seemingly incontrovertible evidence existed associating stilbestrol with cancer in humans. This data suggested a degree of certainty and a previously unrecognized risk, and tipped the balance against DES cattle feeding.[31] Stilbestrol received another blow when a consumer advocate caught the USDA in an apparent lie.

Throughout 1971, the CMS claimed that it had detected no new incidents of DES contamination. Yet David Hawkins, a lawyer for the Natural Resources Defense Council, a small group of young professionals "dedicated to the preservation and defense of the human and natural resources of the United States," learned that low-level CMS personnel had found fourteen DES contamination cases. These workers had received orders not to notify their superiors until a second assay provided confirmation, but Hawkins argued that no second assay was to be performed: the whole matter was merely an elaborate cover-up to make "it administratively impossible" to find DES residues. He demanded that it "be admitted publicly" that American beef remained contaminated with DES, "the very same chemical . . . which caused a previously rare form of cancer in daughters of women treated . . . during pregnancy."[32]

Hawkins leveled these charges in a letter sent to senior USDA, FDA, and HEW officials and simultaneously released to the newspapers. USDA superiors admitted that some violations had occurred, but pleaded ignorance and promised a speedy investigation. The FDA simply expressed anger at being misled. Four days later, the two agencies tightened DES feeding controls. The CMS admitted to ten positives in twenty-five hundred carcasses tested, asserted that it had withheld announcement until the bioassay validated the GLC test, and reported that all of the residue remained in the liver. It also revealed a mandatory DES certification program to replace the cattlemen's voluntary effort. The FDA extended the withdrawal period to seven full days—"a more practical period," which was "more certain to be followed" and offered "an additional margin of safety"—and warned that it was "prepared to take whatever further action may be required" to end stilbestrol residues.[33]

In the emotionally charged atmosphere, the CMS's apparent ethical failure proved more meaningful than new regulations. It dramatically confirmed the visceral presumptions of those who were certain that the federal government protected special interests at the expense of consumers. Only concerted public action could redress their grievances. Fountain announced plans to hold hearings on the USDA's residue-reporting fiasco, and Hawkins sued for an immediate ban on the use of DES as a cattle growth promoter. The suit was filed on behalf of his group, as well as the Environmental Defense Fund, perhaps the most aggressive and litigious activist entity; the staid Federation of Homemakers; the National Welfare Rights Organization, an association of welfare recipients and other low-income persons; and Mary Ann Stein, pregnant, "who desires to resume" eating liver and "seeks protection for herself and her unborn children" against DES's "demonstrated carcinogenic effects."[34]

Hawkins's carefully constructed coalition included virtually every element of the long-active New Deal Democratic congressional consensus. That it boldly accentuated women, the poor, environmentalists, consumers, and the like demonstrated that Hawkins himself targeted specific interest groups in his emotional quest. His brief advocating an immediate stilbestrol ban proved to be nearly as broadly crafted. In addition to presenting stock arguments about DES's carcinogenicity in animals and its potential human carcinogenicity, and recounting the legislative history of the Delaney clause, Hawkins neatly expanded FDA regulatory responsibility to include the drug's metabolites. He contended that the agency had never possessed an analytical method sensitive enough to protect the public, and he noted that a pound of meat contaminated by stilbestrol in a concentration of 2 ppb still contained approximately two trillion DES molecules, which meant that the absence of a positive assay provided "no basis for assumption" that a sample contained no DES residue. Hawkins labeled DES inspection procedures a sham because of the small number of cattle sampled and the USDA's apparent duplicity; indicated that livestock producers widely misused DES; and described the recent regulatory upgrading as laughably inadequate. He roundly condemned agricultural journalists and agribusiness for promoting improper use of DES in feed; the FDA for a lack of clarity in issuing its regulations; and the USDA for publishing no manual showing proper stilbestrol feeding techniques.[35]

The court refused to grant Hawkins injunctive relief,[36] and the matter began its slow legal march. Proxmire introduced a Senate measure to outlaw the drug as a cattle growth promoter. Cosponsored by George McGovern (D-S.D.), Birch Bayh (D-Ind.), Clifford Case (R-N.J.), Frank Moss (D-Utah), and Abraham Ribicoff, the bill called USDA surveillance and FDA regulation "one failure after another," and declared that an absolute diethylstilbestrol ban was the only certain way to remove residues from beef. Proxmire resurrected long-repudiated but serious charges against the drug: that some residues appeared in beef muscle, not just in the liver; that the USDA knew that the vast majority of cattlemen ignored the voluntary certification program; and that the meat of stilbestrol-treated cattle proved tougher and less tasty. Proxmire found DES "a very bad bargain indeed [especially] if it brings with it the threat of poor health." Ogden Reid (D-N.Y.) submitted similar legislation to the House a day later.[37]

Proxmire's indictment gained credibility when on November 9, the same day that Reid's measure was submitted to the House, the FDA sent a special mailing to the nation's physicians advising them that Herbst's revelations had made DES treatment contraindicated during pregnancy. Together the FDA warning and the Proxmire bill cast a formidable shadow over the renewed Fountain committee hearings, which convened two days later. To set the hearing's tone and to serve as a counterpoise to animal scientists' opinions, Fountain arranged for prominent physicians to begin the proceedings. Yet this tactic failed to provide the stark contrast and visceral appeal that Fountain had anticipated. The physicians refused to state categorically that stilbestrol constituted an actual menace. The testimony of Umberto Saffiotti, the National Cancer Institute's associate scientific director for carcinogenesis, and Mortimer B. Lipsett, the National Institute of Child Health and Human Development's associate scientific director for reproductive biology, typified this circumspect approach. Both found DES feeding a "hazard, potentially, theoretically . . . a problem to be investigated." And they advocated "a cautious attitude" and were willing to "recommend any measure that could reduce or eliminate" any possibility of danger, even when no evidence of danger had surfaced. But they did not think that the then-current DES residue rate and level made a stilbestrol ban a high priority. It worried them "much less than the continued use of tobacco . . . and many other much more massive exposures to carcinogens for the population of the United States."[38]

These authorities' curious desire to outlaw the chemical although they had no real concern that its use in feeds was a potent threat stemmed from an acknowledgment that they did not know what caused cancer. Their experience told them that DES's carcinogenic properties were a concomitant of the drug's estrogenic effects, and they understood that the amounts of estrogenic substances normally found in the human and animal body dwarfed those in contaminated beef livers. They also recognized that physicians regularly prescribed doses of the drug thousands of times in excess of what a human might consume in beef. A healthy menstruating woman, not treated with DES, would have nearly twenty times as much estrogen in her body as she could receive eating a pound of liver contaminated with a 100-ppb DES residue. Cows slaughtered when in heat likewise showed estrogen levels well in excess of those of DES-tainted meat. This focused the question on "assessing the level of exposure," but a nagging realization that scientists had rarely studied DES as a unique substance, as distinct from a typical estrogen, plagued them. They opted to gloss over that omission by removing the drug from the food chain.[39]

The final physician called before the Fountain committee, Roy Hertz, former chief of the National Cancer Institute's endocrinology branch, offered the same essential argument, but in a more strident and explicit manner. He listed several factors that physicians needed to consider when they evaluated a drug's potential threat: the age of the exposed subject; the duration of exposure; individual genetic composition; and the dosage required to produce malignant change. For stilbestrol, researchers possessed few concrete answers to these questions. That ignorance portended a risk of some unknown dimension, and Hertz wanted that risk "evaluated in terms of the benefit derived." He then flatly asserted that the "economic benefits" of DES feeding are "not sufficient to out-weigh even the potential hazard of the constant daily exposure to these mate-rials." When quizzed about when the cost-benefit ratio might swing in favor of stilbestrol, Hertz responded that this would occur only in "a possible famine situation." He arrived at this conclusion precisely because he believed that DES's estrogenic properties produced its carcinogenic effects. Because of nat-ural estrogens circulating in women's bodies, he figured that one woman out of sixteen would develop breast cancer. Hertz knew that physicians could dras-tically slash this incidence simply "if we remove from her body the ovaries," the site of estrogenic production. He therefore thought it wrong to increase hu-man estrogenic exposure by ingesting a substance he deemed economically un-necessary. "Cows in heat" and "grains and vegetables of all kinds" contain es-trogens, he remarked, and humankind should not arbitrarily add to that natural "estrogenic dietary load."[40]

Fountain's physicians' less-than-ringing condemnation of DES feeding did not force the regulatory agencies to rebut new charges. Edwards noted that at most only 0.5 percent of American beef cattle livers contained measurable DES residues. He congratulated cattlemen for achieving "99.5 percent compliance with a voluntary program" and predicted success for the new mandatory pro-gram. Yeutter computed that an individual's chance of eating liver contami-nated with DES four days running was a minuscule 1 in 706 million, and main-tained that one-fifth of a pound of DES-polluted liver increased the estrogen normally found in healthy American males only by a five-hundredth. To receive daily a high therapeutic dose of the drug—a dose commonly given to women for up to seven months to prevent miscarriage—an individual needed to con-sume each day 137,000 pounds of liver containing a 2-ppb diethylstilbestrol concentration. Yeutter noted that it was "impossible to guarantee 100-percent safety on anything in this world," but confidently reported to Congress that the DES threat was "one in trillions and trillions."[41]

The only pessimistic note presented by the regulatory agencies was Van Houweling's almost incidental revelation that a recent radiocarbon isotopic measurement undertaken by the Agricultural Research Service in Fargo, North Dakoka, had indicated that it took as long as 132 hours—five and one-half days—for some individual cattle to eliminate DES from their bowels. This rather desultory hearing concluded quietly, but written accounts presented a different picture. *Feedstuffs* seized on Edwards's reassurances that the new withdrawal regulations would end the DES residue question, while Morton Mintz, in the *Washington Post*, highlighted Hertz's testimony under the head-line "Cancer Expert Attacks Use of Cattle Fattener."[42]

Livestock producers reacted warily to emotional appeals,[43] and this left them in "shocking low esteem with many consumers." They were blamed as a major cause of inflation, the very essence of unrestrained greed. These agri-businessmen and their agrochemical suppliers saw their methods increasingly questioned, a phenomenon some saw as "gratuitous McCarthyism." "I truly believe," commented Armistead M. Lee, director of economic research at the Pharmaceutical Manufacturers' Association, that a government official has "less to fear nowadays from a rumor that he is 'soft on Communism' than . . . that he is 'soft' on the pharmaceutical industry."[44]

A combative spirit marked the resumption of the Fountain committee hearings in mid-December. The FDA and Ohio's Clarence J. Brown, leader of the committee's Republican minority, contested every initiative by Fountain and the Democratic majority. Fountain relied on his special consultant, Gilbert S. Goldhammer, to frame penetrating questions and probe for inconsistencies, while the FDA's Edwards deferred to Peter Barton Hutt, recently appointed as the FDA's general counsel. Still in his thirties, the Yale- and Harvard-educated Hutt had formerly been a highly successful corporate lawyer.[45] The initial matter placed before the Fountain committee set the hearing's tone. Someone in the agency had leaked to Fountain a long internal FDA memorandum written just prior to the hearing. Penned by a mid-level scientist, M. Adrian Gross, and addressed to his immediate supervisor, it advocated the use of the Bryan-Mantel Virtual Safety approach to the regulation of carcinogens and argued that the agency needed to devise a DES residue detection test twenty thousand times more sensitive—able to detect 0.1 part per trillion—or ban the drug from cattle feed. Fountain considered Gross's statements as prima facie evidence that the FDA had been culpable in not outlawing DES feeds. Edwards contended that he had not yet had access to the Gross document, but indicated that his experts found Virtual Safety a specious concept. Furthermore, radioactive tracer studies completed about a month before the hearing seemed to demonstrate that cattle voided all DES—investigators could find no trace of the drug even in fecal matter—in about five days; and the tracer provided measurements in terms of atoms, not parts per anything. The seven-day withdrawal period, in light of a five-day elimination rate, made Virtual Safety considerations superfluous. Fountain asked Edwards whether he minded if the controversial memorandum was submitted for evaluation to the National Cancer Institute's Biometrics unit, the unit in which Mantel worked. Edwards countered by again noting that he was being asked to discuss a memorandum he had not yet seen, and asserted that he and his staff should have a formal opportunity to address its charges first. Brown then complained that only Fountain and his staffers had had access to the Gross memorandum and wondered whether other committee members were entitled to the same information. Fountain quickly changed the subject and accused Edwards of repeatedly stonewalling the committee's document requests, using delaying tactics, and participating in the hearing only grudgingly. Edwards disputed those assertions and pledged to continue to cooperate fully with Congress.[46]

Verbal sparring persisted in this quietly confrontational setting as each side offered its interpretation and held its ground. Fountain's relentless attempts to compare as "strikingly similar" the controversy over the 1959 DES chicken implant ban and the controversy over contemporary cattle feeding particularly galled the FDA chief. Fountain reminded Edwards that the caponette industry had declared its DES residues "so minuscule as not to pose a health hazard." Edwards argued that today no stilbestrol residue, no matter how infinitesimal, could legally exist in meat. Whether DES posed a threat or not remained outside the agency's immediate purview so long as none of the drug could be found in beef. Fountain next pointed to the fact that DES could be found in 0.5 percent of beef livers. Hutt defined that residue as "the result of unlawful use," and asserted that in the DES Proviso's "reasonably certain to be followed in practice" clause Congress had explicitly recognized that some unlawful use might occur. Brown then reminded the committee that the Natural Resources Defense Council and its allies were currently suing to end stilbestrol use in feed, and he decried the present hearing as jeopardizing the agency's court case. Fountain openly disagreed and cited the public's right to know, while Hutt indicated that Brown's argument might have merit.[47]

The committee next reviewed the Gass study and its regulatory policy implications. The agency admitted that its scientists did not know the minimum or threshold dose that would produce cancerous tumors in highly susceptible laboratory rodents. But Hutt contended that this ignorance was the reason that the agency refused to accept as legitimate any stilbestrol residue. FDA scientists, relatively silent at the hearing—their information lacked certainty and emotional appeal—noted that since DES allegedly produced cancer because of its estrogenic properties; and since mankind regularly ingested natural estrogens, and only a relatively few humans developed cancer, a threshold dose must exist and be quite high. Goldhammer finally seized on the longstanding practice of using a one-hundred-to-one safety factor as indicating that the agency should not permit DES residues in excess of 0.0624 ppb in beef. Hutt reminded his inquisitor that the agency did not allow any residue. Positive DES tests were a reason for "bringing law enforcement action against the individual, not [for] disapproving the drug." Fountain then asked whether the agency would "rule out" the idea that DES had special carcinogenic factors beyond its estrogenic properties. It had "no reason to rule it in," the FDA responded.[48]

The FDA-Republican alliance dueled with the suddenly defensive committee majority in several additional areas, asserting that each agency statement tapped some contemporary consumer concern and defending DES feeding as the means to meet the nation's explosive beef demand without having "to trust the American consumers to the tender mercy of foreign beef producers." Foreigners might presumably adulterate or doctor their product and ship to America a commodity much less wholesome than DES meat. The alliance contended that the bovine efficiency-increasing drug contributed to a cleaner environment by reducing American cattle excrement yearly by "a rather astonishing total of 17,650,000,000 pounds." The newly brash tandem vigorously defended FDA enforcement policy; assured the committee that GLC could distinguish between DES and other estrogens and would soon become the official analytical method, even as some agency scientists encountered difficulties in securing verifiable results; and dismissed the possibility of feed cross-contamination at feedlots.[49]

Adjournment did not end the new, openly adversarial relationship between the FDA and the committee majority. Two weeks later, Edwards provided Fountain his assessment of Gross's memo, which he fortified with statements by the directors of the agency's Bureau of Food, Bureau of Drugs, and Bureau of Veterinary Medicine; its Division of Toxicology and Nutrition Sciences; and its Office of Pharmaceutical Research. Edwards noted that "scientific freedom is essential" to the agency and its policy "encourage[d]" such freedom, but he termed Gross's effort "highly speculative and theoretical . . . not the type of information on which sound regulatory policy could be based." Fountain's mid-January 1972 follow-up requested other memoranda prepared by Gross. In these memoranda, Gross rejected assumptions of DES's similarity to naturally occurring estrogens as being "rather weak consolation to those mice (and perhaps humans) who would not ordinarily develop malignant mammary tumors if they were not exposed to DES." He called the FDA's stilbestrol-policing policies "truly pernicious" and asserted that they placed the agency "in a scientifically embarrassing and indefensible position." His "definite plea" for "some such rational kind of thinking" urged the FDA to determine "safety levels from experimental data" instead of continuing "to plod along with a discredited system of 'safety factors' about which no one . . . has even the vaguest idea of how safe exactly they are." Gross recognized that his questions took the agency into a new area, but realized that no "large-scale experiments" could provide the definitive data necessary for conclusive regulatory policy. Reliance on assumptions and extrapolations remained inevitable.[50]

Fountain sent Gross's initial memo to the National Cancer Institute's Biometrics Unit, where Mantel defended Gross's methodology and recommended that the FDA either ban stilbestrol or determine just how rapidly it disappears from animal tissue. He cautioned that the agency needed to estimate "disappearance rates" from "observations when DES levels were still measurable," a thinly veiled rebuke to those who attacked his Virtual Safety calculation as speculative, arbitrary, and dependent upon extrapolation. Ironically, Mantel's DES disappearance rate recommendation undercut Gross's position. Radiotracer data indicated that no DES remained in cattle or their waste products longer than about five days, nearly forty-eight hours less than the new seven-day withdrawal period. Despite Mantel's rather pessimistic comments, Mintz published the original Gross memo in the *Washington Post* and accentuated the "bitter quarrel among agency officials," which Proxmire interpreted as making "clear that DES is completely unacceptable as a livestock growth stimulant."[51]

Although Gross remained adamant that his superiors' analysis of his work was "replete with irrelevancies, scientific improprieties and errors, as well as with misinterpretations, distortions and misapprehensions," his memos evoked no further public interest. Shortly, however, two other veterinary staffers, A. J. Kowalk and R. L. Gillespie, mounted a public challenge to the FDA's regulatory policy on DES feed. Their statements, like Gross's, gained credibility because they worked for the FDA. According to their calculations, DES was far from a benign substance. Plant estrogens possessed less than a ten-thousandth of DES's power, and animal estrogens were only one-tenth as strong as the growth promoter. Nor did as much estrogen circulate in the animal body as was conventionally believed. They calculated that a single four-ounce portion of liver containing a 1-ppb stilbestrol concentration would nearly triple a postmenopausal woman's estrogenic load. This latter estimate appeared especially critical because recent studies of postmenopausal women had indicated that they showed a distinct biological response to diethylstilbestrol administered in doses equivalent to "approximately 20 ppb of the total daily diet." Gillespie and Kowalk also worried about the evidence on which the FDA based its regulation requiring a seven-day withdrawal period, finding that the agency had relied exclusively on a 1959 radiotracer test that employed but one animal fed a single stilbestrol dose. They urged the FDA to "obtain better data to more firmly establish . . . the adequacy of the 7-day withdrawal period" and to consider prohibiting DES use until it developed an analytical technique sensitive to 1 part per trillion. They wanted the FDA to "put the emphasis . . . on the side of safety," to acknowledge that it equated "less than 2 ppb with zero when in truth the body works physiologically at levels 1000X less," and to contemplate the consequences of "continuous lifetime exposure by young and old to low levels of compounds that in truth have effects in the molecular range."[52]

The Fountain committee secured the Gillespie-Kowalk memos, and the *Washington Post* published them.[53] The agency refused to revise its stilbestrol regulations in accordance with the two veterinarians' wishes, and it had considerable justification for its position. The FDA had based its seven-day withdrawal requirement not only on the 1959 study but also on the Agricultural Research Service's preliminary radiotracer results at its Fargo station. These ongoing, much more sophisticated experiments seemed to confirm that no DES remained in animals or their excrement after about five days. Moreover, Gillespie and Kowalk had not mentioned that their comparison between natural estrogens and DES depended on in vitro, not in vivo, calculations. Unlike natural estrogenic substances, DES lost considerable potency when ingested, for much of the substance was expelled undigested in fecal matter. Feeders needed to provide cattle ten milligrams of the chemical daily for almost two hundred days, a total of nearly two thousand milligrams, to achieve an effect roughly similar to the one implanters could achieve by merely placing one thirty-six milligram pellet in a steer's ear. Finally, Kowalk's and Gillespie's reference to "20 ppb of the total daily diet" as the threshold level for a biological response among postmenopausal women served to obscure. DES investigators and agency personnel conventionally discussed DES contamination in parts per billion in a particular item—beef. The amount of that item ingested determined the dosage of the drug. A total daily diet, all food consumed in a twenty-four-hour period, implied a volume perhaps ten or more times larger. The two veterinarians' failure to convert their information into the de facto standard format suggested an approach that was less than forthcoming.

No less than five congressmen, including Delaney, introduced bills in February 1972 seeking an immediate ban on stilbestrol in meat. Later that month, CMS announced that stilbestrol residues had been found in two animal livers. Producers had signed the mandatory certification form in both instances.[54] The agency quickly pinpointed feed cross-contamination as the cause of the illegal residues, and traced it to difficulties in cleansing machinery after using a liquid DES premix with a starch-urea-molasses ration. Feeders had dissolved the DES in the molasses, which apparently stuck to the mixing blades and contaminated withdrawal feeds. Within a week of recognizing cross-contamination as the culprit, the FDA announced its intention to ban liquid DES from the market. The liquid premix, almost always used with molasses, constituted an "inadvertent, but apparently unavoidable" source of DES beef residues.[55]

Federal law gave interested parties thirty days to demonstrate the need for a hearing. Thousands of Americans wrote to Congress or the FDA. Those not affiliated with the beef production industry overwhelmingly asked why the FDA hesitated to ban DES outright. They usually expressed a very real fear, as well as puzzlement over the agency's apparent lack of concern for the public weal; they *knew* the material constituted a grave danger. These individuals often sought to gain agency notice by professing to speak for a large organized body. For example, Mrs. Melvin A. Danielson identified herself as "Citizenship Chairman, Walworth County Homemakers' Council, Lake Geneva, Wisconsin." Ida Honorof, a self-styled "Consumer Advocate" at KPFK-FM, in Sherman Oaks, California, typed a newsletter entitled "A Report to the Consumer" in her home. She established an organization, the Committee to Get Drugs Out of Meat—known by the acronym DOOM—to enhance her credibility and to comment on FDA initiatives. Charging that twenty thousand Americans had signed petitions calling for an immediate stilbestrol ban, Honorof accused the FDA of "deliberately violat[ing] the Delaney Amendment." Asking the government, "Why do you continue to lie to the public?" and "Who are you protecting?" she termed the seven-day withdrawal period "a farce" and maintained that the government wanted to "aid and abet the use of . . . known carcinogens."[56]

The DES liquid premix feeding industry, constituting about one-quarter of DES feeders, cited potentially huge losses and declared the proposed ban discriminatory because other forms of DES remained on the market. The industry's advocates contended that careless feeders, not liquid premixes, were the problem, and urged the FDA to arrest violators rather than outlawing premixes. Corporations and trade organizations filed protests and lobbied legislators. Senator John Tower (R-Tex.), Senator Carl T. Curtis (R-Neb.), and Congressman Bob Price (R-Tex.) received numerous personal requests. Price even arranged for a meeting between premix users and Van Houweling, while Tower directly offered suggestions to Edwards.[57]

The FDA's decisive liquid premix proposal vanished when, after the thirty-day comment period, the agency made no move to hold a hearing or institute a ban. It certainly possessed potent incentives to act. The CMS had found liver residues at nearly double the rate seen under the superseded forty-eight-hour withdrawal regulation. The FDA, however, prosecuted few of these violators because the CMS had failed to preserve the contaminated liver tissue, necessary evidence for a sustainable court case. David Hawkins publicly lambasted the FDA for "procrastination," called its lack of initiative "an outrageous example of the failure to follow the dictates of science and common sense," and demanded an immediate total ban on stilbestrol. Another measure to outlaw the substance was introduced into Congress, and the Los Angeles City Council called for federal legislation ending the drug's use as a growth promoter. Difficulties also surfaced within the FDA. A crack research team repeatedly failed "to obtain consistent results" with the CMS's GLC residue detection method, and this failure cast additional doubt on the agency's capacity to punish wrongdoers.[58]

Agricultural scientists also pressured the agency. A special food additive committee of the American Society of Animal Science met with high-level FDA personnel to recommend that "no action be taken" against DES cattle feeds until the "true significance" of beef residues, "chemically minute and possibly biologically inactive" levels of the drug, were known. Committee members included William Hale, who had headed Pfizer's research on DES implants, and Tony Cunha, who had introduced Aureomycin into hog rations. The committee also urged the FDA to determine the significance or wisdom of declaring a substance carcinogenic just because it led to an increased incidence of tumor formation in the particular strain of mice used in Gass's 1964 experiment and many other similar studies. Almost 100 percent of these animals developed tumors when permitted to reach old age, and the committee hypothesized that even compounds "essential to human health" could be shown to produce cancer in these highly susceptible laboratory rodents.[59]

The committee's recommendations and those of consumers, consumer advocates, and others showed a continuing battle to direct the agency. But with respect to DES feeds, the individual interest–oriented contentiousness of the late 1960s took on a new immediacy, a new seriousness, after the Herbst study. Herbst's demonstration that human infants had an "exquisite sensitivity" to DES and bore its cancerous effects some two decades after exposure further personalized the matter, even among cattlemen. A hypothetical menace had become less hypothetical. Herbst's revelations reinforced concern among some that DES might be dangerous because of its unique chemical composition, as well as its estrogenic effects. Rediscovery of the Gass investigation further intensified feelings. To untrained individuals, the dosage required to cause cancer in some mice—6.25 ppb or less of the total daily diet—seemed remarkably close to the limits of FDA analytical sensitivity—2 ppb. These increased suspicions heightened anxiety about DES feeding and yielded converts to the anti-stilbestrol cause. It began to look to beef producers as if they might lose stilbestrol because an increasingly emotional and broad spectrum of the American middle class had decided it was about time.

6   Entitle and Victimize

As evidence of DES in beef livers mounted, cattle feeders faced the fact the the FDA might outlaw the drug. Livestock producers' response to that possibility far outweighed the threat to their welfare. By the early 1970s, several substances gave stilbestrol-like growth curves, and the American public could absorb a slight increase in the price of beef due to a ban. But cattlemen and many other agriculturists feared that a DES prohibition would be the first step in a ban on all chemical additive feeding. The government would remove DES, then antibiotics, and finally other additives from the nation's chemical-dependent, heavily mechanized agricultural system, finally attacking the reality of industrial agriculture itself. Regulation, when controlled by those ill-disposed to then-prevalent agricultural practice, seemed to be a weapon wielded against an entire industry: modern American agriculture.

An agriculture very different than that of the 1960s and early 1970s was the goal of those several dozen men and women who led the agitation against DES feeding. They demanded the use of federal regulatory power to fashion social change. In the eyes of these people, social engineering had replaced protection against particular substances as the appropriate function of government regulation. But although the threat of massive economic change following any dramatic agricultural reorganization frightened cattlemen, these agriculturists already felt victimized, even before such a program was in place. The lack of agreed-upon standards, which had formerly been provided by consensus about scientific results, stripped the appearance of objectivity from government regulation. The idea that regulation did not need to be based on "objective" science had personalized regulatory matters. It caused cattle feeders the deeply felt loss of something intimate: the pretense of autonomy. That injury led them to greet every initiative suspiciously and with considerable frustration, and to react with a vehemence far greater than the situation perhaps warranted. The FDA's repeated threats to ban DES unless residues completely disappeared played on these anxieties; livestock producers felt victimized by this "harrassment," irrationally singled out to bear the brunt of attacks by a disaffected few. These persistent rebukes were annoying reminders to cattlemen that they remained at the government's mercy. And the government hardly seemed to be a nonpartisan public arbitrator, free from outside pressures.

To be sure, beef cattle producers and other agriculturists had chafed at government regulation before. But in this era that cherished personal liberation, regulation was particularly despised. Cattlemen asserted that they had the right to pursue their chosen occupation and were entitled to protect their often substantial investments free from ill-considered or arbitrary interference. Notions of victimization and entitlement also served as potent motives for the cattlemen's faceless opponents, those middle-class individuals who were participants, but not necessarily leaders, of the movement to prohibit diethylstilbestrol feeding. Indeed, they very much resembled cattlemen in the way they used personal victimization and entitlement as rationales for their involvement. Those middle-class men and women who defined themselves as consumers and who articulated their concerns (and, presumably, many who did not express themselves publicly) saw contemporary cattle-growing practices as posing an unprecedented menace to their lives and liberties, a symptom of science and society run amok. To regain command, to restore the individual autonomy and choice to which they were entitled, these desperate men and women needed to seize control of their future. By attempting to prohibit artificial hormone use through governmental intervention, they sought to abandon their personal status as victims and to create a more desirable future.

Ironically, neither those seeking to ban stilbestrol nor those working to save it needed to achieve their stated goals to succeed. For many individuals, simply making the attempt and participating in the new public regulatory process was an antidote to dread impotence and anomie, and thus became the end, not the means. No matter what the regulatory outcome, no matter what the ultimate decision, those who had agitated on any side of the issue could revel in knowing that theirs was a job well done. They had made a difference—they had made themselves heard about a topic of great significance—and so preserved a measure of their individual initiative.

Feed and pharmaceutical manufacturers approached the DES matter with considerably more cynicism and less naiveté than did either cattlemen or consumers. Like their consumer advocate counterparts, they recognized the DES matter, aside from any health concerns, as part of a continuing battle for power. As with Nixon-era price controls and threats to increase overseas beef imports, the issues at stake were, who would dictate acceptability? what would constitute capitalism's parameters? and what would the market bear? DES simply was the site of the most recent skirmish, and regulation the newest locus. In this persistent dispute, distinctions between private and public lost meaning. Questions and debates were public. Attempts to conduct negotiations behind the scenes seemed like de facto admissions that parties had something to hide.

From the perspective of cattlemen and their agri-industrial supporters, the events of May 1972 confirmed the precariousness of their situation. Congress contemplated stripping the FDA of regulatory authority and placing it in a new consumer safety agency. Harrison Wellford filed a federal suit to prohibit the use of nitrate and nitrite coloring agents in cured meat products. The FDA announced its intention to remove from animal agriculture certain antibiotics frequently employed in human medicine, an apparent first step in a total antibiotic ban. By mid-May, the USDA's Consumer and Marketing Service had found stilbestrol contamination in eighteen carcasses (about 1.3% of those analyzed, and a nearly threefold increase over the previous year's figure), while the FDA made plans to develop an even more sensitive DES measure, a radioimmunoassay, potentially accurate to 1 part per trillion. The homespun, witty Jim Hightower released his long-awaited condemnation of contemporary agribusiness practice. Entitled *Hard Tomatoes, Hard Times,* the book investigated the incestuous relationship between the agribusiness elite and America's publicly supported scientific research establishment at the nation's agricultural colleges. Hightower damned this relationship, and the use of DES as a cattle feed additive, as "at once a service to industry and a disservice to consumers."[1]

Taken individually, none of the aforementioned occurrences proved of great portent. Yet they suggested to the cattle-feeding industry that its very essence was under attack.[2] Cattlemen and their advocates refused to concede the slightest point and rebutted the latest round of charges with uncharacteristic animosity. Hess and Clark claimed that the FDA's GLC test confused molasses with stilbestrol and produced numerous false positives. B. P. Cardon, president of Arizona Feeds, called for replacement of the Delaney clause's "chemical zero tolerance" with the concept of "biological zero tolerance," the maximum "amount of a substance that will have zero reaction in any biological system." Congressman Bill Scherle, a Republican from Council Bluffs, Iowa, noted that many foods naturally contain estrogens and urged Congress to modify the odious clause to allow small amounts of estrogens because "nature itself cannot comply with it."[3]

A persistent verbal harangue accompanied these feverish efforts. Viewing DES as "a precariously tilted domino that could bring down the whole stack," animal agriculture enthusiasts chose "aggressive challenge, rather than silence or meek denial." Refusing to "acquiesce without a fight," they called "editors or station managers to challenge the misinformation" put forth by opponents whose "noise often exceeds their knowledge." They saw these actions as necessary because "from DES to fertilizer—we're in a fight for survival." The forces lined up against cattle farmers included "self-styled ecologists," "organic farming cultists," and "misguided do-gooders," as well as "government bureaucrats who must regulate to survive and politicians who have eyes only for the next election." Cattlemen believed that "the attack on medicated feeds will not stop with hormones and antibiotics." Arsenicals and biologics seemed the next "target of protest" for foes whose arguments were "riddled with maybes." The future looked bleak. The "probable outcome" of the foes' aggressive crusade was the loss "of some of our valuable feed additives," yet cattlemen seemed to possess no viable alternative. Only by vociferously resisting at every turn could they hope to spare their chemicals.[4]

On June 16, the CMS announced that it had found DES residues in a startling 1.9 percent of sampled beef livers. These higher contamination rates left the politically sensitive FDA little choice but to act against the feed additive; regulatory policy and law now resembled a plebiscite. Charles Edwards called for a public hearing on the drug but took pains to announce that "DES clearly is a useful and effective product" and that the recent residue upsurge "does not indicate that the product cannot be used safely and effectively." The hearing would provide the FDA the "full information essential to balanced and reasonable judgment" about how to stem the residue tide, and that information would include the opinions of "competent scientists and concerned consumers" who possess "strong feelings" on the matter. Edwards intended to investigate questions such as the possibility of further residue-eliminating controls, the consequences of discarding all beef livers, and the desirability of permitting only those cattlemen capable of demonstrating regulatory conformance to use the hormone. He also wanted information on the environmental impact of a DES ban, the ability of other drugs to provide adequate alternatives, the residue-causing potential of DES implants, and the implications of more sensitive residue detection methods for continued stilbestrol feeding.[5]

Edwards couched his hearing call in the form required by law, by proposing to withdraw approval of the drug. As with DES liquid premixes, this proposal did not obligate the agency to do anything. Edwards's plea for public input virtually promised a hearing that would provide formalized discussion, not guaranteed conclusions. The FDA's foes characteristically claimed victimization and maintained that they were entitled to immediate action. David Hawkins, of the Natural Resources Defense Council, called the anticipated hearing "essentially a step nowhere," a formalized procedure to postpone a decision indefinitely. In the House, L. H. Fountain blasted Edwards for inventing "a mechanism for delaying the regulatory action the law requires." He professed himself "frankly amazed" that the FDA claimed to lack data when it already possessed all the "relevant factual evidence" necessary to enforce the law. Fountain also decried Edwards's "curious approach to law enforcement," especially since "the American people, acting through Congress, have declared total war on that most dreaded enemy—cancer," and he cited the call for a hearing as an explanation for the public's "wide distrust of government." James Delaney termed Edwards's action "unconscionable," "another in a series of incidents which endanger the public and undermine our food and drug laws"; he threatened that unless the FDA acted soon to protect the public, Congress would establish "an agency that will." Morton Mintz looked at the call for a hearing as evidence of a lack of seriousness in "the administration's talk about 'the conquest of cancer'" and urged the FDA to withdraw the drug first and then debate the matter. "Livestock may have to roam the range a bit longer (or [stay] penned in, to be precise)," he remarked, "but at least the consumer would be fully protected."[6]

The FDA's notice of opportunity for a hearing required the holders of stilbestrol New Animal Drug Applications to explain why they opposed the agency's proposal to terminate DES's growth-promoting use.[7] But the continued increase of the DES residue detection rate (which passed 2.25% by the end of June) and the realization that radioimmunoassay required at least several years of refinement before it could be used effectively in regulation pressured the agency to move forcefully.[8] Outpourings of public sentiment accompanied each week's DES residue figures and approached a crescendo when the CMS announced in early July that one-tenth of that week's analyzed livers had turned up DES-positive. Frank J. Rauscher, Jr., the newly appointed head of the National Cancer Institute, undercut the FDA's already tenuous public image. Contending that his "job [was] to protect people from cancer," Rauscher stated that "anything that increases the carcinogenic burden to man ought to be eliminated from the environment if at all possible." With stilbestrol, such elimination would be "possible," even "exceedingly simple," and he called on the FDA to outlaw DES pending a public hearing.[9]

Views such as Rauscher's, as well as the steady flood of DES-related stories, fed further the emotions of cancer-fearing Americans. Concerned over their "guinea pig status," terrified individuals contended that as the result of continued stilbestrol use, "vegetarians will inherit the earth." Jesse E. O. Berry personalized the matter when he asked point-blank if Edwards were "going to stop [stilbestrol feeding] now or wait until both of us get cancer." Marian Snuggs snidely accused Edwards of apathy toward stilbestrol regulation because "as a man you are not afraid of vaginal cancer." Lest he feel complacent, she reminded him of his considerable personal stake in the matter. "DES may be the cause of feminizing" human males, she noted, and "most men even old ones are dead frightened of IMPOTENCE." Some individuals argued that supermarket food "has reached the national disaster stage" and that "deliberate pollution" by additives had become "our no. 1 problem." Beef cattle, they continued, "are subjected, virtually from the hour of birth, to a bombardment of chemicals that produces radical changes in their basic physiology." Mrs. Joseph T. Hansford asserted that DES has "done more damage to America than LSD, etc. can ever do," while others claimed that "money is more important to some people than human lives." The continued legalization of DES constituted "mass medication of the U.S. population," which seemed tragic because "God knows there are enough chemicals in our world today without one used for the benefit of cattle growers." Mrs. C. Mosher expressed herself in a straightforward manner. She begged the federal government to see that "the meats we eat are not loaded with all kinds of hormons [sic] and all the other 'crap'—pardon the expression."[10]

These emotional, heartfelt petitions came from a frustrated population that considered itself victimized, and was desperate for reassurance and action.[11] The American Society of Animal Science's Committee on Additives typified the pro-DES response. The society contingent sought to shed its status as victim by demanding the right "to present evidence, file pleadings and cross examine witnesses" at the anticipated FDA hearing.[12] Those favoring continued stilbestrol usage received a boost when thirteen senators, generally from cattle-feeding states, implored Edwards to adopt some measure significantly short of an outright DES prohibition because the Delaney clause's "absolute prohibitions" have become "scientifically unsound in light of current knowledge." An Iowa cattle feeders' rally in support of Scherle's congressional initiative drew nearly six hundred cattlemen to petition the federal government to permit a "realistic" level of DES residue in beef carcasses. There Wise Burroughs termed "ridiculous" the idea that small quantities of DES constituted a health hazard, and urged the government to permit certain measurable quantities of the drug in food, as it did for pesticides and herbicides. Burroughs wanted animal scientists to establish tolerance levels and demanded that government grant them virtually complete regulatory autonomy; government bureaucrats would no longer victimize mature scientists.[13]

Agri-industry also responded to Edwards's call. Implant manufacturers objected to being lumped together with feeders. Feed and pharmaceutical companies accused the FDA of forsaking the drug rather than prosecuting those violating agency regulations; urged the discarding of beef livers to save DES; or dismissed the CMS's reporting of increased violations as conceivably inaccurate. They claimed that GLC analysis was notoriously unreliable and unverifiable, and cited as proof the FDA's persistent refusal to designate it the official method. Thus the true percentage of DES-contaminated livers remained unknown, and might actually be decreasing, even as the USDA's "faulty" methodology suggested otherwise. Representatives of one company declared that no DES had ever appeared in a beef carcass. They stated that what investigators measured in contaminated livers was DES-monoglucuronide—an ester of stilbestrol produced when the liver metabolized the drug, "a substance with chemical and pharmacological properties different than DES." Scientists knew little about this derivative and had no evidence of its possible carcinogenicity. Therefore, the FDA lacked grounds to invoke the Delaney Cancer Clause. The agency could outlaw the drug only by demonstrating the ester's carcinogenic potential or by proving that infinitesimally small quantities posed a human health threat.[14]

As the FDA mulled over the various briefs, petitions, and letters, the Senate Subcommittee on Health held a stilbestrol hearing to force a dramatic agency initiative. Chaired by Edward M. Kennedy (D-Mass.), the committee reinforced regulation's plebisicitary character by calling William Proxmire as its initial witness. He took special care to counter stilbestrol's economic claims, contending that the drug's legalization victimized anti-DES farmers, who were forced to adopt the drug to stay competitive with their less socially concerned rivals. Proxmire also explained that a prohibition of DES might actually save consumers money. America's grain surplus cost billions to store and forced up food prices. Thus, using the surplus to feed cattle—more feed would be necessary without DES—would eliminate these indirect agricultural subsidies.[15] Edwards followed Proxmire on the floor and almost immediately got into a disagreement with Kennedy.

The terms of the Kennedy-Edwards dispute neatly summed up the very real differences in perspective between consumer advocates and regulatory officials. Each felt victimized. Edwards referred to the "legal requirement" to declare a substance an imminent health hazard, and Kennedy immediately jumped on that conception. "Wait a minute. The Legal? What about health?" he demanded. "Are we getting bottled up on legalities," questioned Kennedy, "or are we talking about the health [of] the American people?" Edwards saw it as his responsibility to act in conformance with law. Kennedy wanted the FDA to do what was right, no matter what the law stipulated.[16]

An extraordinary, open-to-the-public meeting of the FDA's National Drug Advisory Committee a few days after the Senate hearing gave Proxmire another forum. Representatives of the Fountain committee and Nader's Center for the Study of Responsive Law joined Proxmire to reiterate their disgust with the agency's regulatory maneuvers regarding DES.[17] At almost the same instant, an article in *Science,* the official organ of the American Association for the Advancement of Science, denounced the FDA's handling of DES feeds.

The piece in the journal of the nation's largest scientific organization was entitled "DES: A Case Study of Regulatory Abdication" and was written by Nicholas Wade, who was not yet thirty years old. The United Kingdom–born, Cambridge-trained journalist had worked as an editorial writer for *Nature* for four years before emigrating to America to join *Science*. In his article, Wade described DES as "a chemical of bizarre and far-reaching properties, chief of which is that it is a spectacularly dangerous carcinogen"; he recounted the chemical caponization imbroglio of the late 1950s and cited the provocative Herbst study. He also pictured John N. S. White as a prescient hero, lauded A. J. Kowalk's and R. L. Gillespie's revelations, challenged the National Academy of Sciences as biased in favor of food and chemical companies, identified Frank Rauscher's statements as "courageous," and accused the FDA of pursuing "a silent evasion policy" for nearly seven years. Wade offered three damning explanations for the agency's apparently illogical defense of DES. He suggested that the Nixon administration's "visible" concern about "the rising price of meat in an election year," coupled with the belief that outlawing DES "would cause a small but perceptible [price] rise," had stopped the FDA from acting until after the November election. Wade also supposed that if the FDA failed to "hold the line with DES, which has a legal loophole tailored for it," the agency would find that "a lot of other chemicals might fall domino-like into the jaws of the Delaney clause." That description portrayed the FDA as a pawn of, and beholden to, the feed, food, and pharmaceutical special interests, but Wade also pictured the agency as arrogant, pompous, and pretentious, committed to DES's defense only because of selfish vanity and pride. In his view, the FDA had developed a bunker mentality: it considered itself victimized by the press and consumer advocates, and it was digging in its heels to fight back. The FDA's DES policy constituted "a self-sustaining activity," he noted, "from which the FDA can withdraw only at the price of admitting that the critics were right all along."[18]

The FDA made Wade's third explanation suspect when on August 2, 1972, it denied requests for a DES hearing and announced a ban of the drug as a cattle feed additive. The agency referred to preliminary results of a highly sensitive Agricultural Research Service radiotracer study it had received a few days earlier. ARS researchers still could measure about 0.2 ppb of the drug in tagged animal livers one week after ingestion. The rate at which DES left cattle livers apparently declined significantly after about three days. Investigators now extrapolated that the liver, rather than ridding the body of all DES in about five and one-half days, might retain minuscule amounts of the drug for as much as a full month.[19]

In announcing the ban, Edwards technically maintained that the holders of DES NADAs had failed to demonstrate that "genuine and substantial issues of fact" existed between themselves and the agency, and thus they did not merit a hearing. Their lack of response to the new ARS study, as well as their failure to "file a full factual analysis" of the evidence upon which their positions rested, served as proof. But while he articulated criteria to which no producer could conform—none knew of the ARS results—Edwards also claimed that DES remained "a regulatory, not a public health, problem." The agency was outlawing DES feeds only to fulfill its legal obligations under the Delaney clause; the drug's presence in beef livers constituted no imminent human health threat. That legalistic interpretation enabled the agency to avoid "an abrupt disruption" of the nation's beef production industry. It forestalled "unwarranted public concern" and prevented an "unjustified increase in meat prices" by implementing its ban in two stages. The FDA immediately prohibited the manufacture of DES for veterinary use but postponed terminating the drug's use as a growth promoter until January 1, 1973.[20]

The agency issued no new DES implant regulation as it awaited results from experiments then under way at the ARS's Metabolism and Radiation Research Laboratory. Letters from individuals and groups not directly involved in DES feeding or oversight generally supported the agency's action, although several expressed bewilderment over both the FDA's refusal to act against implants and its two-stage implementation of the DES feeding ban. One individual accused Edwards of imitating "Nixon and the Vietnam War," preferring "to wind [DES] down slowly rather than cut it off suddenly." To this victimized correspondent, the explanation seemed simple. "Cancer patients created in the interum [sic] don't matter as much as the farm vote." Another referred to the Herbst study and demanded an immediate cessation of stilbestrol usage because she objected to "being fed poisons . . . just to line some male chauvinist pigs' pocket."[21]

Legislators demonstrated no such relative harmony. The FDA's two-stage ban upset Kennedy, Proxmire, Ribicoff, Fountain, O'Hara, and Sullivan, who complained that FDA action should have been "total and immediate." Curtis, Scherle, and Price predictably lined up against the agency's "precipitous" move. The Kennedy subcommittee unanimously endorsed the Proxmire bill, which called for the immediate termination of DES feeding and the imposition of a complete implant ban no later than January 1, 1973. It allowed the FDA either to accelerate the implant prohibition if ARS experiments showed residues or to postpone it indefinitely if those tests found no extraneous drug. In a voice vote, the Senate agreed to the measure, but the bill stood no chance in the House. Scherle's proposal to modify the Delaney clause dominated there, although Fountain repeatedly tried to embarrass the agency and to declare its two-stage feeding ban illegal. Fountain secured a nonbinding legal opinion from the acting U.S. comptroller general stating that the FDA's phased withdrawal plan exceeded its legislative authority, but failed to mention that the Library of Congress's Congressional Research Service had issued a contrary legal brief more than a month earlier.[22]

Scherle's proposal secured attention as the rhetoric of victimization continued to unite antagonists. The measure quickly gained the favor of Earl Butz, then USDA secretary; as well as that of Edwards and other HEW officials, who identified the Delaney clause as "inflexible, although well-intended," the product of a simpler age in which regulators had lacked the technologies required to protect the American public. Now the situation had become far different. "Fantastically sensitive" techniques were uncovering infinitesimally small traces of carcinogens "virtually everywhere we look, bearing not the slightest relationship to the public health and safety," a fact testified to by "even the toughest of the truly qualified public health guardians." Thus the Delaney clause, carried to its logical conclusion and used in conjunction with exquisitely sensitive analytical methods, "would essentially ban all food ingredients and most foods." Everything employed to produce and give sustenance "will be in jeopardy, subjected to irrational attack and possible . . . utterly unjustified" prohibition. Consumer advocates hailed the clause as the "model consumer protection law" because it took the matter out of the hands of those allegedly beholden to agribusiness. Antagonists condemned it for its irrationality and its circumvention of legal precepts, claiming that it "defies reality" and dangerously "demands black-or-white judgments"—a particularly undemocratic approach. Decisions would be automatic, beyond human scrutiny, not the consequence of the electorate's "rule of reason." The clause treated the public as incapable of informed self-determination, needing to be told what to think and do. Kennedy's defense of the clause as essential on the grounds that the FDA has "consistently perverted the nature of the regulatory function and placed human health at repeated risk in order to enable industry to gain" provided matter for thought, as did Samuel Epstein's pithier statement. Then professor of environmental medicine at Western Reserve University, Epstein called for a congressional investigation of government food safety advisors, especially the National Academy of Sciences, because "in this country you can buy the data you want to support your case." Presumably, the public required self-appointed watchdogs like Epstein or fail-safe laws to protect public opinion from being misled.[23]

Cattlemen also staunchly backed the Delaney clause modification campaign but approached it in less grandiose, less cerebral terms.[24] They saw themselves as victims of a world gone mad. Individuals on society's political fringe, with markedly anti-American sentiments and tendencies, were using distortion and misinformation to seize control of the political process. Beef producers and their allies derisively called this "very vocal minority" "environmentalists and extreme consumer advocates," "Nader-type groups," "consumer agitators," and other, similar epithets, and claimed that this radical fringe hoped to undermine the structure of animal agriculture through piecemeal bans of crucial drugs and other chemical substances. Stilbestrol "was the first shoe to drop," according to this scenario, and cattlemen feared that antibiotics, selenium, nitrates, and nitrites were next. Concern over the DES feeding prohibition was of great moment not because beef producers lacked viable alternatives, including implanting, but because the prohibition seemed to foreshadow contemporary animal agriculture's possible fate. Livestock producers dug in to resist this first wave, while consumer advocates held little back as they sought to generate unstoppable momentum.[25]

Pharmaceutical and agrochemical giants such as Lilly and Merck expressed little concern over the DES feeding ban. Stilbestrol business comprised a tiny fraction—usually less than 1 percent—of their sales. These chemical enterprises produced other growth-promoting agents and would retain feeding customers for their other product lines. But smaller, more specialized chemical manufacturers felt the ban acutely. After asking the FDA to reconsider its action, three DES-dependent chemical companies challenged the agency in court. Dawe's Laboratory, Hess and Clark, and the Chemetron Corporation of Chicago asked a federal court to review the FDA decision, arguing that the agency had failed to show any imminent health hazard and therefore must conduct a hearing. In sum, the autocratic FDA had violated statutory procedures when it instituted a prohibition of DES feed usage.[26]

Soon after the companies initiated their suit, FDA attempts to standardize GLC analyses called the agency's stilbestrol decision into question. An agency-conceived field office trial demonstrated that only about half of its technicians could discern measurable differences between samples not dosed with DES and others dosed with 2 ppb.[27] USDA researchers also provided disturbing information. In early September, the ARS team reported no abnormal radioactivity in steers and heifers fed carbon 14–tagged DES ten days earlier. It confirmed these determinations at month's end and again in early November. Aggrieved small chemical companies did not wait for the second confirmation but petitioned the FDA in light of this new evidence to rescind its August 4 edict and establish a ten-day or even a fourteen-day withdrawal period. The FDA refused their request on rather surprising procedural grounds. It maintained that the NADAs had been revoked and that plaintiffs would have to submit and defend new NADAs. That placed the burden of proof on the chemical companies, who would need to devise "a more sensitive regulatory surveillance method for DES than is now available." The agency also denied the companies' request for a public hearing on the new determinations on the grounds that such a hearing was not "useful at this time."[28]

By setting stipulations that no group could meet, the FDA made it impossible for the chemical concerns to reinstitute DES feeding. The FDA's hesitancy to restore DES probably stemmed from public and congressional pressure,[29] and the agency boldly employed that outside momentum for new initiatives. During the fall of 1972, ARS researchers detected radiation in excess of background radiation in the livers and kidneys of animals slaughtered 30 and 60 days after radioisotope-tagged DES implantation. These government studies disagreed completely with those simultaneously pursued by scientists at Hess and Clark. Using both GLC and the mouse uterine method, the company's chemists found no residue in animals slaughtered more than two days after implantation. The FDA ignored the company's results and in mid-December instituted a 120-day DES implant withdrawal period, which prohibited cattlemen from sending animals to slaughter for at least 120 days after implantation. The agency further stipulated that if the ARS tests then under way demonstrated stilbestrol residues after 120 days, the agency would prohibit implant usage immediately.[30]

Even as the FDA announced its implant withdrawal protocol, the small chemical companies questioned the legality of using the ARS radioisotopic assays to establish a DES ban. The DES Proviso's requirement that "no residue . . . be found . . . (by methods of examination prescribed or approved by the Secretary by regulations . . .)" led Chemetron and the others to maintain that the secretary-approved mouse uterine test constituted the only legally binding DES assay. The HEW secretary could prescribe a different and presumably more sensitive test, of course, but not cavalierly. Approval required proof that the new test produced accurate, verifiable results superior to those yielded by the presently approved method. Both the FDA and the AOAC had long sought such a test, and both had failed. GLC and its predecessors had proved insufficient, and radioimmunoassay was far from a reality. There existed no test that would withstand scientific scrutiny and legal challenge.[31]

The court's continued deliberations occurred against a backdrop of exacerbated tensions. The FDA's stilbestrol feeding ban had further polarized partisans. For cattlemen, it reified the arrival of the threat they had long feared. The anti-DES forces apparently sensed weakness, vulnerability, and opportunity and rushed to consolidate their advantage. Consumers capitalized on their feelings of victimization to increase pressure on the FDA. Many fretted about the "increasing chemicalization" of meat and championed the DES ban even if it raised beef prices, because "people would rather be broke (or vegetarians) than DEAD." Only a few viewed the DES decision as "another of many spawned by the disaster lobbies, and hysterics mongers"—ne'er-do-wells who were "systematically bankrupting our country and our citizenry of the standards of living which our country is capable of providing and has provided." Edwards urged Delaney clause reform. Kennedy persistently protested the FDA's two-stage phase-out of stilbestrol feeding, while Proxmire railed against the agency's stand on implantation. Fountain confidently predicted the banning of other animal feed additives as carcinogens, while some demanded "an end to chemical farming." *Sowing the Wind,* Wellford's indictment of the USDA, the FDA, and the meat production industry, appeared on best seller lists, and the *Washington Post* had it reviewed by James Turner, a former colleague of Wellford's at the Center for the Study of Responsive Law. In short, even as apprehension reached new heights, the same handful of men and organizations that had been publicly involved in stilbestrol regulatory battles almost from their beginning continued to dominate public forums.[32]

These several spokesmen for radically different positions had many characteristics in common. Always accessible, they articulated specific, clearcut, eminently quotable opinions. They quickly had become media darlings, even if some lacked even a semblance of charisma. Their fame, as well as newspaper sales, depended on relentless antagonism and left no room for compromise. But the media's reliance on the same very few individuals limited ideas and tended to rigidify stances as the same impassioned advocates repeatedly trotted out the same inflamed rhetoric. It suggested that these zealots represented the full range of public opinion and, more important, that they accurately reflected public sentiment.

A notion of the talented few was implicit in concepts of expertise and professionalism. But these individuals seemed chosen not for their knowledge or merit but for their dependability. They had put forward controversial views, which served their own purposes or those of third parties. Two public forums held ostensibly to consider the Delaney clause demonstrated the power of dependability. A new medical journal, *Preventive Medicine,* sponsored a symposium on the Delaney clause in late 1972. Although the journal was open to the professions and the public, in this symposium it published almost exclusively position papers by individuals already active in the clause dispute. James Delaney, Samuel Epstein, Thomas Jukes, Umberto Saffiotti, and Nathan Mantel repeated their well-known arguments. A similar recognition of partisans led Edwards to refuse to attend a two-day workshop sponsored by the New York Academy of Sciences or send any high-level agency officials when he learned that the other delegates opposed modification of the Delaney clause. Edwards's rebuff did not prohibit Mintz from reporting that an informal poll of ten scientists at this meeting could not produce "the name of one other [scientist] anywhere, who had no commercial ties and who wanted to modify the clause."[33]

Ironically, at almost the same time the New York group met, agricultural scientists and their supporters held a conclave in Moline, Illinois, to establish the Council for Agricultural Science and Technology (CAST). There scientists expressed alarm that "adverse propaganda about the quality of food" had created an "increasingly difficult" "professional environment." State legislatures and Congress enacted law, and the public formulated opinion, "on the basis of inadequate and sometimes inaccurate information about agriculture and agricultural science"; and CAST sought to redress the problem by adopting a more aggressive posture. The new organization "would mobilize agricultural scientists to assemble and interpret factual information" and disseminate it "to those who have decision-making responsibility" as well as to the "news media and the public." Through publications, movies, television, and radio pieces, CAST would become "the voice of agricultural science on the national scene." Agriculturally related scientific societies and individuals would comprise the new organization's membership and completely control its agenda and activities, but financial support would come from "agricultural industries, trade associations and foundations."[34]

CAST did not seek to introduce new issues into the political arena. Rather, it sought to liberate its victimized members by mobilizing them to present their case. Charles Black, an Iowa State University agronomist, served as CAST's first president, and he structured the organization to compete with media and consumer advocates simply by mimicking their strategy of publishing short "white papers." The cattle industry certainly welcomed CAST support. A series of disasters seemed to affect the industry. High beef prices had encouraged the Nixon administration to permit additional imports, and the antichemical trend had raised domestic production costs. Consumers rebelled because of beef's escalating retail price and boycotted the product. Nixon added to the boycott's strength by proclaiming it "patriotic to eat fish." Regulations calling for an environmental impact statement for each new animal drug promised to delay the introduction of new substances for as long as six months, to increase research costs, and to raise commercial prices. The FDA's new requirement that manufacturers prove subclinical antibiotic feed doses effective and not a human health hazard threatened to inflate the consumer prices and production costs further. Two Nader's Raider reports also stung the meat production industry. The first declared nitrites, commonly found in hot dogs, hams, and bacon, to be a carcinogen, while the second proclaimed that the National Academy of

Sciences was in league with food producers. The latter report proposed the creation of a consumer consultant force, a move to which the FDA reluctantly agreed.[35]

The FDA experienced even greater pressure than the meat industry. Consumer advocates and other citizens paradoxically deemed the agency inadequate and unresponsive both for not testing drugs sufficiently before legalization and for requiring unnecessarily extensive premarket testing, which denied individuals immediate access to new drugs. The Nobel Prize–winning economist Milton Friedman even accused the agency of murder for not licensing at once certain proposed new medications. These conflicting positions produced demands both that consumers or their self-appointed advocates serve on FDA panels and that Congress dismantle the agency. The creation of a consumer protection agency seemed especially desirable. FDA decisions rarely went unchallenged. Not only did partisans question virtually every action, but when the FDA ruled against them, they went to court. These persistent legal suits regularly made judges final arbiters. Rule by judiciary was twofold. Not only did judges' decisions set FDA policies, but the mere possibility of such rulings became a potent force. In this milieu, the best agency regulatory decisions were those that were not challenged in court or those that were sustained by jurists. This placed a certain cast on the FDA's work and transformed the scientific regulatory agency's legal counsel, Peter Hutt, into its most important officer. His abilities to interpret regulatory law and to justify the agency's actions to members of the judiciary dictated whether its edicts survived legal scrutiny. But crafting regulatory decisions primarily to conform to legalistic opinion almost guaranteed that the agency's initiatives would frequently run afoul of a frustrated, anxious, and demanding public or of Congress, which responded to its emotion-driven constituents. This de facto policy seemed likely to fuel another round of vituperative attacks and threats of legislative reprisals. If, however, the FDA attempted to follow the popular will and placate its critics, it would probably violate some legal precedent, and a court would rebuke the agency and countermand its action. The agency would look foolish and ineffectual and have its regulatory authority more closely circumscribed.[36]

These unpleasant prospects faced the FDA in mid-April as it confronted the latest DES implant studies. Edwards had left the agency, and Sherwin Gardner was acting commissioner. The results of two experiments lay on his desk. The first showed that free, nonconjugated DES made up a substantial proportion of the residue found in the livers of implanted steers. This discovery destroyed the idea that the possibly noncarcinogenic DES-monoglucuronide comprised the bulk of residues, which might exempt DES use from the Delaney clause. The second was more ominous. Radiotracer-adept ARS investigators had found stilbestrol concentrations as high as 0.70 ppb in cattle livers 120 days after implantation. That Hutt remained active in his post as agency counsel no doubt provided crucial regulatory continuity, and Gardner interpreted these results as if he were Edwards. Gardner announced an immediate ban on DES implantation, but maintained that implants were outlawed not "because of possible danger, but because . . . requirements of the law" could not be "fully satisfied." Gardner saw "no reason to disrupt the present meat supply," and the agency planned to permit already implanted animals to be sold at market.[37]

Consumer activists and sympathetic legislators quickly moved on to other causes. The cattle industry responded strongly to the FDA implantation ban but acted with resignation, gravitating to natural estrogen additives and implants. These natural substances promoted animal growth almost as well as DES but were nearly ten times as expensive. Researchers suspected them of being carcinogenic in doses large enough to produce anatomical and physiological changes. But unlike stilbestrol, which was artificial, these drugs seemed to be beyond the law because of their naturalness. Because they were commonly found in plants and animals, regulatory officials puzzled about how to differentiate how much of these substances occurred naturally in meat and what quantity producers added. But although the substitution of natural estrogens for artificial ones solved the cattlemen's immediate problem, the threat that led to the removal of DES remained unchecked. The cattle industry claimed that it needed to retrench, reorganize, and refocus simply to keep the regulatory plague from wiping out chemical-based animal agriculture entirely.[38]

The new campaign used DES as a symbol of what could be, and concentrated much ire and attention on the dreaded Delaney clause. Stilbestrol was memorialized as "a fallen idol, . . . a challenge that must not be forgotten." The industry needed to remember that "once the [disaster lobby's] bandwagon started rolling, the producers of DES and those who used it had little to say about [the] outcome." Cattle trade publications identified the Delaney clause as the industry's foremost menace and emendation of the clause as its only effective option. AFMA's Technical Advisory Committee urged the FDA to reinterpret the clause as requiring biological zero, and the American Veterinary Medical Association's Council on Biological and Therapeutic Agents concurred. Representatives of the nation's three major animal science organizations—the American Society of Animal Science (ASAS), the Dairy Science Association, and the Poultry Science Association—met to coordinate their efforts against the Delaney clause. The Agricultural Research Institute scheduled and sponsored a public workshop to vent antagonisms toward the clause. The ASAS's subcommittee on hormones took the unprecedented action of seeking to have its position paper on clause revision published in *Science*. Purdue's distinguished

William Beeson, who generally shunned agricultural politics, spoke out forcefully for clause modification, as did Cunha and Jukes.[39]

Support for revision of the Delaney clause came from numerous sources. The giant pharmaceutical manufacturers Pfizer, Upjohn, and Lilly lent their considerable weight. R. M. Henderickson, the president of Pfizer's agricultural division, referred to the DES prohibition as an example of the "burden of politics and political pressure, which has its own high degree of self-interest." It marked "a compromise with, or a downright surrender to, completely non-scientific factors." Continued blind conformance to the Delaney clause "could lead us straight to disaster." The AHI agreed, as did several congressmen, including Guy Vander Jagt (R-Mich.), who termed the clause "more like fanaticism than intelligent public policy." The *New York Times* unexpectedly called for a reevaluation of the Delaney clause. The FDA's handling of DES angered the *Times*, which asked, how "real is the risk" of stilbestrol, measured in "parts per trillion?" How does it "compare with the risk of breathing normal polluted air in Manhattan . . . or with the risk of having a chest x-ray or smoking a single cigarette?" The problem remained that the clause was "an all or nothing affair," and that was a situation Congress must revise.[40]

The FDA persistently prosecuted owners of DES-contaminated cattle and labored to develop an assay suitable for regulation. But it also capitalized on the sentiments expressed in the *Times*'s editorial, and proposed a new standardized assay protocol to "bring reason and scientific stability" to regulatory questions. The agency's ability to measure a quantity regularly exceeded its ability to interpret that quantity, and it proposed to grade carcinogens by their tumor-generating propensity and eliminate the Delaney clause. Weak carcinogens, like stilbestrol, would not be victimized when the state of the art—the limits of the most modern methodology's sensitivity—changed, bringing about a spate of "false positives," the product of assays beyond the FDA's capacity to analyze. The FDA would establish for each substance a single, scientifically determined value, expressed in parts per billion or maybe even parts per million. This value would incorporate "the maximum level of exposure," not the average amount, and would err on the side of caution, using some criteria similar to those employed in the Bryan-Mantel Virtual Safety model. The agency also identified a series of practical criteria, such as cost, speed, and convenience, that each individual assay technique must meet to gain approval.[41]

The agency noted that it was not discarding or undermining the Delaney clause or setting tolerances for carcinogenic substances. It merely sought to establish scientific and systematic procedures to determine methods sufficiently sensitive to be used in applying that much-maligned clause.[42] That disclaimer fooled no one, and the FDA sifted through responses to its proposal. Although it claimed that its effort would stand up to scientific and legal scrutiny, the agency finally declined to put its plan into action at that time.

Part of that decision may have stemmed from the proliferating legal challenges to the agency.[43] Hess and Clark appealed the FDA's DES implant order and asked for a stay. Vineland Laboratories, a small New Jersey biologics firm, joined the suit "to end damage" that the agency's preemptive act caused to the company, beef producers, consumers, and the environment. The District of Columbia Court of Appeals quickly agreed with the plaintiffs' general complaint that the agency had violated its statutory procedures and caused the companies to suffer "irreparable harm." The court permitted a stay, but because of the "compelling and difficult context" of the case it withheld that stay temporarily, granting the implant manufacturers twenty days to present materials for a hearing, and giving the agency twenty additional days to start proceedings. Only if the agency withdrew the offensive order or if the plaintiffs neglected to seek a hearing would the stay order be vacated.[44]

The court's order supposed that issues of fact separated the contending parties. The FDA circumvented the court's will and denied a hearing on the grounds that the plaintiffs had failed "to identify any scientific evidence to demonstrate the existence of any genuine and substantial issue of fact sufficient to warrant a hearing." That concluded the matter of the stay, but the companies' challenge to the withdrawal order remained.[45] The Fountain subcommittee launched two new broadsides. It first compared feeding statistics from the early 1950s to 1971 and claimed that the amount of formulated feed required to bring a steer to slaughter weight had increased over that period. Fountain then questioned the efficiency of chemically dependent animal agriculture. Few feeders had used hormones, antibiotics, and other drugs in earlier years, yet cattlemen got animals to slaughter with less manufacturer-enhanced feed per steer.

The committee's analysis ignored two critical facts. Feeders raised cattle to a much greater slaughter weight in 1971 than they had done a decade and a half earlier, and they scrupulously restricted intake of nonformulated feedstuffs, which they had not done earlier. The subcommittee's second complaint was more damning. Fountain's staff had combed the several congressional hearings on the FDA since February 1971 in search of inconsistencies to use in formulating an indictment of the agency's handling of the stilbestrol matter. The result was a report entitled *Regulation of Diethylstilbestrol (DES) and Other Drugs Used in Food Producing Animals*. This December 1973 accusation identified alleged FDA wrongdoing and demanded immediate procedural and policy changes. It alerted the newly appointed FDA commissioner, Alexander M. Schmidt, formerly dean of the University of Illinois College of Medicine, that the committee considered itself entitled to a prominent regulatory role. The report termed the FDA's stilbestrol regulation "inadequate because of errors in judgment and deficiencies in administration" and lambasted agency legal opinions, especially about congressional intent, prolonged DES use, withdrawal periods, and the like, arguing that the FDA "excessively delayed" regulatory decisions on animal feed.[46]

The subcommittee's reform proposals cut right to the matter's heart. They demanded strict enforcement of the Delaney clause, more stringent premarket testing, quicker drug withdrawals, immediate evaluation of new evidence, and closer cooperation between the FDA and the USDA. Fountain explicitly pointed to the agency's October implant edict as a guide to and a model for appropriate FDA regulatory action. The report was a blatantly partisan document. Five of the six subcommittee Republicans voted against the report (the eight subcommittee Democrats all voted for it) and filed their own partisan summation and recommendations. The Republican brief absolved the FDA of virtually all faults, endorsed the agency's deliberate, cautious approach, and applauded its diligence.[47]

The Fountain subcommittee report quickly became moot[48] as the U.S. Court of Appeals for the District of Columbia Circuit announced its DES decisions soon after Congress's Christmas recess. In unusually blunt language, the court vacated both FDA orders. It charged the agency with acting in a manner inconsistent with "basic fairness," employing a "palpably impermissible procedure," citing the Delaney clause as "only a scare tactic," and ignoring procedures "on the basis of a public health hazard that has not been declared." The court also accused the FDA of deciding important matters "in a patchwork of legal theory that is sewn in a confusion inconsistent with responsible review" and of asking the court to discard legal precedent to defer to the commissioner's "expertise and judgment as the agent charged with responsibility for the public health."[49]

The FDA's refusal to hold hearings without adequate grounds to declare a summary judgment drew the court's ire. So too did the agency's failure to provide plaintiffs with radioisotopic tracer results, as well as the agency's unwillingness to permit or accept comment about these questionable studies. The court also chided the FDA for using as a basis for regulation or denial of a hearing a method of analysis not approved by the HEW secretary. It overruled the agency's argument that it became the plaintiff's obligation to prove liver residues safe, and gave the FDA the initial responsibility to demonstrate a possible health threat from DES residues.[50]

The court reestablished the status quo ante and held that the bans violated administrative and legal procedures. Ironically, the court singled out as particularly odious the two acts that were approved by the Fountain subcommittee's Democratic majority, as well as by Kennedy, Proxmire, and consumer advocates. Although ruling only about stilbestrol, the court established precedent, which set sweeping new limits on FDA authority. The court's decisions inaugurated a new era in FDA history and placed chemical-based agriculture on a slightly more secure legal footing. The agency had to conform to the new court-ordained stipulations whenever it engaged in regulatory action. In the specific case of diethylstilbestrol, several courses stood open to DES's opponents and the FDA. Congress could enact new legislation to prohibit the drug's use as a growth promoter, or the FDA could begin to marshal the massive, complex evidence necessary to hold definitive hearings on withdrawing approval of stilbestrol NADAs. Both would take time. Until one or the other happened, stilbestrol would remain a part of cattlemen's repertoire.

7 The New Synthesis

The court's decision did not revive DES. During the year-long prohibition, beef producers had substituted other growth promoters. None worked quite as well or was as cheap, but feeders hesitated to use a once-banned substance and risk further disruption of their enterprises.[1] Persistent FDA hostility to the drug discouraged large pharmaceutical companies from refitting already converted plants. Only Dawe's Laboratory immediately resumed stilbestrol production, but that relatively small biologics firm could supply only a fraction of the beef industry.[2]

Virtually all parties to the imbroglio over DES expected the FDA to start the process of withdrawing approval of DES NADAs as soon as possible. Although the procedure might take substantial time, the issue of stilbestrol's continued legalization seemed resolved. The FDA would ban the drug. Now each party in the controversy needed to explain why the situation had ended as it had and to declare some measure of victory, even while developing strategies to prevent similar occurrences. Posturing remained an important part of these victory declarations. Consumers focused on the court ruling as confirmation of a business-controlled bureaucracy and the anticipated ban as evidence of the individual's ability to triumph over that bureaucracy. Consumer advocates saw the DES controversy as justification for their own and their programs' existence. Livestock producers pointed to the repeated assaults on the drug as an indication of governmental highhandedness and the ruthlessness of consumer advocates. The producers' trade organizations and agricultural scientific bodies saw the controversy as the first instance in which they had defended their rights, and as the beginning of a new, more assertive—and presumably more successful—style.

Ironically, the spate of explanations and self-justifications masked something much more significant. As these men and women prepared for a world without stilbestrol to unite them, they (unconsciously, I think) helped to pioneer a new regulatory synthesis, one that maintained the decision-making prerogatives of each of them. To be sure, partisans continued to embrace a philosophy of regulation as opposition and held resolutely to discourses on victimization and entitlement. And they remained steadfast in their particular opinions about stilbestrol. But for the absolutism of the Delaney clause they substituted the relativism of the cost-benefit approach, and they defined cost and benefit in a myriad of nonspecific ways—economic, environmental, social, or health-related. Individuals could measure both cost and benefit in their own personal terms and from their own individual perspectives; cost and benefit and the relative merits of each were in the eye of the beholder.

The emergence and establishment of cost-benefit analysis as the regulatory sine qua non permitted parties to disagree, often vehemently, but offered them a measure of immortality. The nebulousness of the cost-benefit approach also preserved each party's integrity and veneer of importance and utility. Each party persisted as a vital force, as an integral facet of the new regulatory nexus. Your foe could never be vanquished from this participatory regulatory schema, but neither could you. Ironically, scientists had long advocated a version of this regulatory approach. But they had claimed that their training and method, their expertise, entitled them to weigh cost against benefit without outside intervention; they had subscribed to this peculiar concept to retain what had historically been their unquestioned regulatory dominance. The institutionalization of cost-benefit analysis in the 1970s enabled them to continue to hold a formal regulatory place. Indeed, their "expertise" could rarely offer the kind of absolute answers and definitive evidence demanded by an intensely interested public. Before stilbestrol was finally declared illegal, court decisions, congressional mandates, and executive orders had firmly entrenched cost-benefit analysis within the regulatory machinery.

The court's demand that the FDA conform to strict legal guidelines granted Peter Hutt additional agenda-setting authority even as one critic maintained that the FDA was "run by legal sharpshooters whose goal is power." Hutt meticulously built the antistilbestrol case by proposing to decertify the mouse bioassay as the official analytical method. The agency then would charge that stilbestrol, lacking an official method, failed to conform to the DES Proviso's requirement for "methods of examination prescribed or approved by the Secretary." Since there would be no approved method, any carcinogenic residue at any detectable level violated the Delaney clause.[3]

During the mandatory comment period for Hutt's proposal, the FDA and the USDA reinstituted earlier rules and techniques to regulate the use of DES.[4] A broader reexamination of the Delaney clause followed. As part of the animal scientists' earlier cost-benefit crusade, the House had funded a Scherle-sponsored, FDA-conducted study of the clause. The document explored the clause's legal history and its partisans' various positions, provided an elaborate discussion of the difficulties of translating animal results to human situations, described new, highly sensitive analytical techniques, and recounted theories of carcinogenesis. It also considered several mathematical risk assessment models, as well as benefit appraisal formulas and ethical guidelines based in the social sciences.[5] Scherle sent Wise Burroughs and Charles Black, CAST president, the FDA report in May. The two Iowa State scientists complained to Scherle that the report tended "to confuse the situation" by offering support to "any point of view that an individual might choose to champion." Burroughs noted that it only considered the possible risks of agricultural chemicals, and suggested that the USDA prepare a companion report on their benefits. Black arranged a CAST panel of twelve agricultural college, federal, and industrial scientists and a lawyer to attack the idea of zero tolerance. Apparently without Black's encouragement, the acerbic, combative Thomas Jukes simultaneously attacked the Delaney clause. In several provocatively titled articles, he stressed consumer benefits from chemical agriculture and downplayed or dismissed possible social costs. George Gass also joined the discussion. His 1964 study provided, after the Herbst revelations, the most pervasive evidence against stilbestrol's use as a growth promoter and granted him a certain cachet among advocates of the clause. He publicly reconsidered his earlier study, disavowing interpretations that suggested that a 6.25-ppb concentration of DES in the daily feed caused cancer in highly sensitized mice, and estimating that a person who each day consumed an average amount of liver tainted with 0.5 ppb had at least a 5,280-fold margin of safety.[6]

Gass redefined his earlier work's significance because an experiment of several years' duration had led him to consider DES-generated tumor formation far more complicated than he had previously suspected. In this experiment he used mice similar to those used in the 1964 experiment, but now he recognized a tumor genesis cofactor, a virus commonly known as mammary tumor virus. A group of these highly susceptible animals had been fed a daily ration containing a 250-ppb concentration of stilbestrol. Those fed the diet but lacking the virus, and those that had the virus but were not fed stilbestrol, had shown virtually no increased incidence of tumor formation. Those with the DES diet *and* the virus had proven thirty-six times more likely than controls to develop tumors. These refined results called into question whether stilbestrol was in fact a carcinogen, and provided a possible explanation for the nonlinearity of Gass's original results. Neither the virus nor the drug alone in this mouse experiment caused cancer. Only the two in combination produced tumors in numbers significantly higher than were seen in the control group. What was DES's role in this process? Did this experiment indicate that diethylstilbestrol was not carcinogenic?

Did other apparently innocuous substances possess similar effects? Was the Delaney clause an intellectually flawed concept?[7]

The prestigious *Journal of the National Cancer Institute* published Gass's research. Gass also participated in the ASAS Hormone Subcommittee's mid-1974 symposium on hormone use in livestock production. As a result of that symposium, the subcommittee (of which Gass and Hale were members) wrote a CAST-supported review essay on the safety of estrogenic hormone residues in edible animal products, which *Bioscience* published in January 1975. In the essay, subcommittee members developed the legally dubious position that because hormones at low levels "are non-carcinogenic" (conventional wisdom suggested that the carcinogenic properties of estrogens stemmed from their estrogenic effects, and Gass had shown that carcinogenesis might require several factors) then "these low levels are not subject to the Delaney amendment." The subcommittee maintained that the FDA and the courts must adopt this new formulation because "legislation is a living, adaptable system which evolves as our knowledge of the scientific principles underlying its application develops"; and they proposed two ways to establish the upper boundaries of those low levels. For animals (such as heifers capable of estrus) that at some point in their lives normally exhibited considerable estrogenic activity, the subcommittee wanted to fix those heightened levels as natural and permit residues up to those amounts. Steers clearly posed a problem, and the subcommittee based its recommendation for those castrated animals on a 1968 study, ironically also by Gass, that showed an estrogen's carcinogenic potential to be partially dependent on its continuous long-term application. Gass's 1968 research on a specific strain of cancer-susceptible mice (C_3H mice) demonstrated that when these animals received intermittent DES doses over long periods, they did not develop tumors at a rate significantly higher than that seen in controls. And the experimental doses were substantial, involving concentrations far in excess of the 2 ppb found in contaminated beef livers or the 6.25 ppb that had originally drawn attention. The subcommittee agreed to use these highly sensitive mice as appropriate level-setting subjects, but only if the mice were fed in a manner consonant with the way in which humans ate liver. More precisely, it wanted them to eat DES-enhanced liver only at one meal a day, and to be given that modified substance only as often as the average person ate that organ meat. The subcommittee was so sure that the allowable DES level determined by such testing would be much higher than the level found normally in the livers of stilbestrol-treated cattle that it urged the FDA to employ the Mantel-Bryan Virtual Safety method at a risk of one in one hundred million.[8]

The subcommittee's clever "real world" regulatory solution showed its partisans' certainty that DES was harmless in the quantities used to promote cattle growth. It reaffirmed their right and responsibility to determine regulatory questions for an unscientific public. The possession of scientific training had tempered them, enabling them to view situations clearly and logically. Despite the scenario's datedness, the FDA deferred to the subcommittee as it would to any proven scientists and scientific organizations. The agency's respect stemmed from the fact that a significant percentage of FDA personnel were scientists, had worked in the private sector, and would do so again when their agency service ended. This bias influenced even the agency's legal department and helped determine the order in which it tackled issues. The decision to subsume the question of stilbestrol within a more general examination of carcinogenic regulation and to delay final action on that drug no doubt reflected the pervasive scientific opinion that stilbestrol residues posed a minor threat. By emphasizing the larger issue of developing a protocol and clear rules for regulation of carcinogens, the agency sought to prevent further disruptions like the stilbestrol conflict and to reposition the agency as the rightful regulatory leader. A single clearly articulated policy based on cost-benefit calculations as well as carcinogenic potency would rationalize regulatory action.

Any general rule on carcinogens would circumvent the spirit of the Delaney clause, and the "methods of examination prescribed or approved by the Secretary" provision of the DES Proviso offered the means. But constructing general parameters in which to place each particular carcinogen proved no easy matter.[9] During mid-1974 and 1975, agency personnel continued efforts to develop an effective stilbestrol-specific regulatory method, and the USDA monitored livers for residues. Radioimmunoassay research, methodologies to distinguish among estrogenic substances, and examinations of DES conjugates and metabolites persisted, often with AOAC assistance.[10] None of these efforts lessened congressional concern about DES. Delaney framed a House measure to outlaw stilbestrol, which provided for fines of up to fifty thousand dollars and imprisonment for up to five years for anyone introducing the drug into a food product or an animal used for food. Edward Kennedy introduced into the Senate a similar bill, cosponsored by Richard Schweiker (R-Pa.), the ranking minority member on the Senate Committee on Labor and Public Welfare.[11]

The Senate bill received HEW secretary Caspar Weinberger's endorsement. Carl Curtis, Henry Bellmon (D-Okla.), James Eastland (D-Miss.), Walter Huddleston (D-Ky.), and Strom Thurmond (R-S.C.) cosponsored an amendment to postpone the ban's implementation until the FDA completed a series of complicated scientific investigations. Continued legalization was, according to Roman Hruska (R-Neb.), "the only sound way to approach this issue until more is known as to the results of eating beef from cattle that have been fed DES." The full Senate took up the matter on September 8 and 9, 1975. Kennedy eloquently expressed his reasons for framing the measure when he remarked that in concerns of health, "the benefit of the doubt must go to the American people." Curtis asserted that government must work on "confidence," not "rumor or suspicion or fear," and that confidence required a logical, fact-based case. In the instance of DES, he maintained, no evidence existed of the harmfulness of any conceivable residue of the drug. Passions became inflamed during the discussion. Schweiker claimed that DES proponents were "condemning a number of young ladies to death"—a not very oblique reference to the Herbst study—simply because beef "will cost a few cents more." The situation was far too grave to wait for scientific proof of the drug's harmlessness. Twenty years ago scientists had offered assurances of DES's safety. "Now, all of a sudden, we find out, 20 years later, that the female offspring of the women who used it are now dead." Curtis complained that Schweiker's zeal was misplaced: rather than attacking DES, he should attack cigarettes. Curtis's amendment gained the cosponsorship of Gale McGee (D-Wyo.), Dick Clark (D-Iowa), Paul Fannin (R-Ariz.), John Tower (R-Tex.), Bennett Johnston, Jr. (D-La.), Paul Laxalt (R-Nev.), Barry Goldwater (R-Ariz.), Clifford P. Hansen (R-Wyo.), and Hruska. "Why not ban all food, then, on the basis that something might show up?" Curtis asked opponents. Huddleston supported the Curtis amendment for its health benefits. On average, the carcasses of DES cattle yielded thirty-three and one-half pounds of lean meat more per one thousand pounds of animal than untreated bovines did. DES-fed steers also had less carcass fat and produced meat lower in saturated fat, a factor implicated in colon cancer and heart disease. Gary Hart (D-Colo.) added a friendly amendment to the Kennedy bill which would prohibit DES use in cattle until the National Cancer Institute could guarantee that the drug was safe in the quantities found in tainted livers. Bellmon's hostile amendment sought to add the word *harmful* before the term *residue* in the DES Proviso. This change would permit the FDA to remove an additive from market only after it proved that residues caused harm.[12]

The Curtis and Bellmon forces failed in the Senate. By more than two to one, that body approved the Hart-amended Kennedy bill. Urban and eastern liberals generally supported the legislation, while opposition came from the Plains, the Midwest, and the South. Party affiliation figured only peripherally.[13] Farm and feeder periodicals decried the Senate's action as "a victory for 'Stilbestrophobia,'" and Jukes expressed particular outrage. In a column in *Nature*, he estimated that a stilbestrol ban would cost an extra seven million tons of corn yearly. He compared that figure to Kennedy's professed sympathy for those starving in Bangladesh. Jukes reconciled these two apparently dissonant facts by noting that Bangladeshis "don't vote in Massachusetts!"[14]

A House health subcommittee chaired by Paul Rogers (D-Fla.) scheduled

hearings for mid-December. During the interim, stilbestrol-related research initiatives did not cease. Investigators attempted to pinpoint stilbestrol's mode of
action as an animal growth promoter, to consider how the animal body removed
the drug, and to assess the substance's carcinogenic potential.[15] The ASAS's
Hormone Subcommittee recommended that CAST form a task force to consider
similar questions. Black agreed, and O. D. Butler, chairman of the ASAS's
Regulatory Agency Committee—the Hormone Subcommittee's parent body—
chaired. FDA researchers conducted several large-scale experiments at the National Center for Toxicological Research. There scientists sought to determine
the long-term effect of minute DES doses on C_3H mice, rats, and hamsters;
develop a functional radioimmunoassay; and evaluate the suitability of chim-
panzees as laboratory animals for DES toxicity studies. The center also held a
mid-November symposium to discuss recent DES animal studies. The fledgling
Journal of Toxicology and Environmental Health devoted a special issue to conference presentations, but the *Washington Post* reported only one investigation.
North Carolina researchers had dosed pregnant mice with large quantities of
stilbestrol and found that 60 percent of the male offspring were sterile, presumably because of fibrotic lesions or nodular masses. This study indicated that
there might be DES sons as well as DES daughters; although DES was not tied
to increased cancer levels in human males, sons, too, might bear the consequences of their mothers' medical regimen. Sidney Wolfe, a member of
Nader's Public Citizens' Health Research Group, cited this research to demand
that the FDA ban DES in cattle as an imminent health hazard.[16]

The FDA refused Wolfe's demand, but its head, Alexander Schmidt, revealed in testimony before the Rogers subcommittee that the agency was now
prepared to revoke its selection of the mouse bioassay as the official DES assay
method and to schedule a hearing to withdraw its approval of stilbestrol
NADAs. Schmidt coupled that announcement with opposition to the Senate's
attempt to ban DES. He feared that the passage of laws dealing only with
stilbestrol would encourage Congress to enact similar bills for other drugs, and
thus would Balkanize regulation.[17] Schmidt's decision to begin action against
stilbestrol corresponded to a precipitous increase in the rate of detection of
DES-contaminated livers. The USDA was detecting residues at a rate not encountered during the last three years, and the FDA seemed no closer to developing an adequate regulatory tool for stilbestrol. These factors, and Schmidt's
desire to avoid fragmented regulation, led to his statement on January 9, 1976,
beginning the revocation of agency approval of stilbestrol NADAs.[18]

The *Federal Register* published the official hearing notice three days later. In
it, Schmidt maintained that the present official method of analysis was inadequate, denied that law required the agency to identify a substitute method, re-

fused to designate livers as inedible products because no evidence demonstrated "that DES residues in livers may not serve as an indicator of residues present, albeit at lower levels, in other edible tissue," and rejected economic arguments as irrelevant. He then justified the FDA's proposed withdrawal of approval by claiming that the drug was a carcinogen, that residues had been detected in edible tissue, and that no practical, satisfactory method of analysis existed. In short, DES had not been shown to be safe. Information revealed since the approval of the NADAs some twenty years earlier had raised serious questions about the drug, including its apparent violation of the Delaney clause. As per executive office edict, the agency analyzed the proposition's inflationary impact and determined that a ban would cost consumers $503 million dollars yearly, which was not a "major inflation impact." The agency listed among the potential benefits of a ban the "elimination of any risk of any cancer associated with consumption," but admitted that those benefits were "impossible to quantify . . . since the risk of any cancer from ingesting DES through the food supply is unknown."[19]

These were halting first steps toward the use of cost-benefit analysis as the regulatory standard. That no one knew the relationship between DES feed and cancer, or even knew whether a relationship existed, was the regulatory dilemma, the sort of paradox in which cost-benefit analysis could seem a suitable resolution. The FDA's action mollified Rogers, however, who kept the Kennedy bill in subcommittee to allow the FDA latitude. Farm and feed newspapers showed scant interest in the proposed ban. DES was "a goner," and they worried only about the "effect on the status of other hormones and drugs." Several smaller pharmaceutical companies—Dawe's Laboratory, Hess and Clark, Vineland, and American Home Products, the last three of which decided to produce the drug in limited amounts during the long delay—submitted hearing requests. The American Society of Animal Science, the American National Cattlemen's Association, and the AHI also asked for a hearing. The Natural Resources Defense Council and the Environmental Defense Fund demanded a formal voice in the FDA stilbestrol revocation process, as did the Pacific Legal Foundation, founded in 1973 as a counterpoise to such groups. The foundation labored to bring Jukes, CAST, and the American National Cattlemen's Association before the FDA. In letters to Mintz and CBS's Lesley Stahl, in a speech before the Maryland Nutrition Conference, and in an article in *Preventive Medicine*, Jukes's message and reasoning remained the same. No evidence existed that DES meat "causes *any* increased risk of cancer . . . and some indication [existed] that it may be useful in diminishing the risk" by producing leaner beef. A ban on agricultural uses of DES would raise consumer prices an estimated $503 million annually.[20]

Potential litigants prepared data for the hearing, marshaled evidence, and secured expert scientific testimony. During this period, the nature of the regulatory process repeatedly made news, though stilbestrol did not. Kennedy resurrected charges that high FDA officials, especially Van Houweling, exhibited a profound proindustry bias, and the senator brought disgruntled agency employees before his highly visible subcommittee. Consumer advocates compiled exposé after exposé about nitrofurans, antibiotics in feeds, red dye no. 2, and numerous other chemical substances. Schmidt's July announcement of his resignation, effective later that year, added to the scrutiny.[21] Ironically, the FDA moved against stilbestrol soon after Schmidt left office. Acting Commissioner Gardner declared the DES NADAs held by companies that had not requested a hearing voided, and set January 5, 1977, as a prehearing to establish procedures and rules of evidence for a full hearing on the other DES NADAs.

In his late November message, Gardner presented several "factual issues" in a manner that echoed the cost-benefit approach. Manufacturers needed to provide answers to the questions of whether DES was a carcinogen, whether it had a known no-effect level, and whether it had any other biological effects that might affect human health. They also had to respond to queries on whether DES feeding and implanting resulted in residues in edible tissue, whether all residues "had been identified, evaluated and shown to be safe," and whether the mouse bioassay was "adequate and practicable for regulatory purposes and capable of detecting and identifying" illegal residues. Cattlemen experienced a "sense of resigned disappointment" when they learned of Gardner's announcement but expected no disruption in operations. Most had been "weaned of DES usage through lack of availability."[22]

Administrative law judge Daniel J. Davidson conducted the January prehearing. There the NADA holders and their allies—the Pacific Legal Foundation, the American National Cattlemen's Association, and the National Livestock Feeders' Association—tried to modify the issues the hearing would cover. They urged a more complete cost-benefit accounting, asking the judge to examine DES's "economic, environmental, and public health benefits" and to compare those benefits to actual, proven hazards. They demanded consideration of the dosage issue in all determinations; proof that the status quo constituted a public health menace; and exploration of whether new restrictions short of a ban offered the public adequate protection. According to the four NADA holders, who presented their case jointly, if the FDA was to declare DES unsafe under the General Safety Clause of the 1938 Food, Drug, and Cosmetics Act, it must show that the drug's risks—its costs, social and otherwise—outweighed its benefits. That constituted the only possible grounds to outlaw the drug, they argued, because invoking the Delaney clause violated the 1974 ruling by the U.S. Court of Appeals for the District of Columbia. The agency had never found stilbestrol in edible parts of slaughtered animals using officially approved methods of analysis when livestock producers employed the drug in accordance with specified conditions of use.[23]

As outlandish as the last statement seemed, it was substantially correct. The mouse bioassay's deficiencies as a regulatory tool did not excuse the agency from complying with the DES Proviso. The NADA holders' united nonparty supporters then agreed to defer scientific questions to the ASAS because of its "recognized expertise . . . in matters involving DES and animal health." The ASAS president appointed two members, including Jukes, to represent animal science in the hearing and directed them to "rely upon" a still-incomplete document, tentatively entitled "Hormonally Active Substances in Foods: A Safety Evaluation," as a full expression of animal agriculture's views. A draft of that CAST report had been circulated. Jukes sat on the task force, as did the Pacific Legal Foundation's Glenn E. Davis.[24]

This interlocking web of DES-supporting nonparty participants gained the Agricultural Research Institute's support later that year. CAST had become animal agriculture's voice, and the organization's much-anticipated report accentuated issues of cost versus benefit. The task force claimed that it found "no evidence of a cancer hazard" from hormonal growth promoters, especially stilbestrol, and stressed that drug's economic benefits. Its assertion that "an estrogen is an estrogen is an estrogen" identified dosage as the key determinant. The group readily acknowledged that all estrogens produced tumors in certain highly susceptible laboratory rodents, but asserted that this result was only seen when the drugs were given in doses massive enough to yield estrogenic effects. Task force members unveiled their report at a luncheon press conference at which they served beef, a green salad, whole wheat bread, and green peas. They then pointed out that the salad greens, the peas, and the bread "each contained significantly more estrogenic hormone than the beef did." The force finally asked the assembled journalists to help the public place DES in "proper" perspective. It claimed that healthy women of child-bearing age normally produced daily about 15,000 times as much estrogen as had been found in tainted livers, and that some birth control pills contained more than 657,000 times as much estrogen as did contaminated organ meat.[25]

The CAST report's socioeconomic focus, its emphasis on comparisons to natural processes, and its dependence on statistical models, no matter how crude, placed it squarely within the emerging regulatory milieu.[26] Others sought an apt regulatory methodology. Efforts ranged from pursuing a workable GLC to the *Journal of Toxicology and Environmental Health*'s symposium on regulatory problems associated with radioimmunoassay and tracer studies.

Considerable doubt had surfaced about the meaning of radioactivity levels in tagged meat which were higher than background radiation. The symposium's participants generally concluded that these heightened levels did not necessarily indicate the presence of unauthorized residue.[27]

Doubts about radioimmunoassay further undercut the idea of scientific absolutism, but the FDA's threatened saccharin ban unleashed a persistent public clamor and invited congressional intervention. So too did the FDA's handling of low-level antibiotic animal feeding. In striking contrast to the DES imbroglio, a special investigative committee accused high agency officials of stifling debate and harassing employees as the FDA made preparations to prohibit antibiotic usage. The agency's regulation of carcinogens also caused it trouble. Consumer advocates criticized the continued legalization of nitrate and nitrite food preservatives,[28] while two Denver medical researchers undertook an ethically questionable but highly revealing "experiment" to expose what they considered FDA regulatory folly. They claimed that the agency's dependence on the Delaney clause "protected the rat, reduced the credibility of cancer scientists and kept their string of inane pronouncements on cancer dangers intact." These investigators inserted sterilized dimes into the peritoneal cavities of thirty-five rats to dramatize their objections. Following and parodying formalized procedures, they found that more than 50 percent of the rats developed malignancies, which led them to demand facetiously that the FDA declare all coins carcinogenic and remove them from circulation. The scientists threatened to test paper money, credit cards, eyeglass fragments, bottle tops, and beer-can tabs next to make the point that "foreign body tumorgenesis" was among cancer researchers a well-known phenomenon. The introduction of large quantities of any substance into an animal would likely produce tumors. Since few if any of these situations occurred in nature, being the product only of well-intentioned but misguided laboratory investigations, they "fortunately are more disastrous to rodents than to man."[29]

This report appeared in the *Journal of the American Medical Association* under the title "Money Causes Cancer: Ban It." It did little to ease very real fears. Two DES-related incidents, neither of which specifically touched on residues, seemed to confirm the stilbestrol menace. In the first, three women, including Patsy Mink, then assistant secretary of state, filed a class action suit against a Chicago hospital and Eli Lilly and Company in late April. They claimed that during a twenty-month period in 1951–52 they and 1,078 other pregnant women had been put at risk when without their knowledge they were given DES to assess the drug's complication-preventing potential. The second event stemmed from the first. Through the Freedom of Information Act, the anti-DES activist Sidney Wolfe had secured the 1951–52 research protocols. His early December analysis—at variance with the hospital's and Lilly's—suggested that these DES mothers developed breast cancer earlier and at a markedly higher rate than their control counterparts.[30]

The administration of President Jimmy Carter, especially his appointments to high-level government posts and his modest reshaping of the federal bureaucracy, eased some of consumers' fears. But those same acts disappointed livestock producers and other agribusiness people. Bob Bergland, the new secretary of agriculture, provided a sharp contrast to Butz, and the selection of Carol Tucker Foreman as assistant secretary of agriculture for consumer affairs seemed especially galling. As the Consumer Federation of America's executive director, Foreman had "been considered by many in the red meat and poultry industries to be a serious threat to their businesses." The choice of Donald Kennedy to head the FDA struck some as being of the same character. A neurophysiologist at Stanford University, he was dynamic and had impeccable basic science research credentials but little direct experience with drugs or drug policy. Kennedy had been involved in National Academy of Sciences-sponsored studies on the effects of pesticides on health and the environment, as well as in other "green" activities.[31]

Even before Kennedy took office in early April, the agency published its long-anticipated sensitivity of method (SOM) final order. Not subject to debate or hearing, this "refinement" of the Delaney clause defined levels of agency-required analytical accuracy for carcinogenic substances; this cost-benefit assessment provided "and operational definition of the no-residue requirement of the . . . 'DES Proviso.'" Drug companies bore the financial burden of proving that a drug met the order's stipulations. Using a well-drawn-out six-step data collection an evaluation procedure, a manufacturer under FDA guidance first determined a carcinogen's relative strength and then used a revised Mantel-Bryan formulation. The agency defined the Virtual Safety level of a carcinogen as the dose that would produce one chance in one million of developing cancer, assuming daily ingestion of a similarly dosed foodstuff for a lifetime and no interference from competitive or additive carcinogens or cancer promoters or retardants. Sponsoring drug companies needed to develop practical assay methods appropriate for detecting drug residues at the Mantel-Bryan-determined level, and to convince the FDA that these methods met agency requirements.[32]

The AHI, with support from the four stilbestrol litigants and the Pacific Legal Foundation, challenged this extraordinary regulation in early May. It sued the FDA for substantially changing proposed SOM procedures without providing opportunity for a hearing. It also denied that the Mantel-Bryan model determined real-world safety—the FDA offered no evidence that it did—and demanded that the agency publish a revamped SOM proposal, one based on "realistic" cost-benefit calculations.[33]

The start of the DES hearing coincided with the lawsuit. The FDA called twenty-six witnesses in mid- and late May, while manufacturers presented their case at the end of October. Judge Davidson requested summary briefs in early 1978 and then took the matter under advisement. This process brought the DES issue closer to resolution, but much remained before an FDA ban could be implemented. Davidson lacked authority to outlaw the drug. His opinion was advisory. Only the commissioner could ban the substance. After Davidson's judgment, the commissioner would have to sift through the evidence, listen to the various parties' rebuttals, and frame a final decision, which could be challenged in court.[34]

The court of appeals ruled in the SOM case during Davidson's deliberations. It found for the AHI, and urged the FDA to begin the SOM process again.[35] Almost simultaneously, reports appeared questioning the DES-cancer linkage in humans, suggesting that the drug's possible social costs were not nearly as great as its most vocal critics had contended. These documents were written by longtime opponents of DES, which enhanced their credibility. A National Cancer Institute study examined one-thousand-five-hundred randomly selected daughters of women who had received large doses of the drug during pregnancy, and found not a single cancer. A special NIH-HEW task force on DES, created to investigate Wolfe's charges that DES mothers suffered from breast cancer at a statistically significant higher rate than women who had never taken the drug, reported that such a relationship was "unproved." A National Cancer Institute–American Cancer Society conference concluded that the public's fear "about the cancer-causing potential of food additives" was misplaced. The nation's high-fat diet, predilection for broiled, fried, or charcoal-broiled foods, exposure to a toxin-producing mold often found in grains, and excessive alcohol and tobacco consumption accounted for the vast majority of nonenvironmental cancers. Additives contributed little, if at all.[36]

A widely distributed preliminary feed additive report issued by the Congressional Office of Technology Assessment furthered cost-benefit analysis. The allegedly nonpartisan group found feed additives to be a boon to consumers because they enabled farmers to produce food more cheaply and at virtually no additional risk "as far as science can tell us today." It specifically asked Congress to compel the FDA to consider a drug's economic benefits when making regulatory decisions. In the case of DES, the office saw no health advantage to banning the substance, as that action "would not reduce the cancer risk significantly." It favored more vigorous enforcement of the withdrawal period, claiming that "risks from DES residues can be virtually eliminated by withdrawal of this chemical from animals 10–12 days prior to slaughter."[37]

These newest studies and reports, which portrayed DES as less menacing than previously suspected, came from facets of the consumer coalition. Which of the studies compiled over the previous several decades were correct, how they related to one another, and what the concrete terms of adjudication ought to be were questions that went unasked. No one knew the answers to these puzzles, but more significantly, few seemed to confront the issue of their ignorance. The lack of concern for that issue suggests that it just was not seen as very important. Support for or opposition to DES was and long had been an article of faith, both among scientists and among their nonprofessional brethren. Scientists could provide only data that demanded interpretation; they could not provide self-evident, conclusive proof of the correctness of any given view. The manner of interpretation—cost-benefit analysis—remained at the heart of the matter.

The emphasis on the manner of interpretation set up two distinct regulatory situations and provided two different kinds of solace. Debates among consumer advocates, beef producers, agricultural scientists, legislators, and regulators revolved around winning rather than arriving at absolute assessments of truth, accuracy, or safety. These groups focused on gaining, maintaining, or restoring authority and respect—control—for their cohort. The individual's role in American society provided a potent argument around which to rally support. Each side understood that public backing was crucial to its cause and used a motif of individual assertiveness and empowerment to gain this backing: that is banning or absolving the drug would provide a public demonstration that each individual could make a difference.

Recognition that public support mattered led to increased efforts to encourage segments of the public to participate. These efforts met with success as men and women, generally from the middle class, engaged in regulatory agitation in unprecedented numbers. The benefits they received from their activities were far different than the benefits received by those seeking outright control of regulatory mechanisms. Agitation—participation—itself was the reward. It furnished individual's a sense of belonging and of potency, a sense that they were guarding their own and their families' future at a time when the traditional arbiters—government, churches, and schools—seemed unwilling or unable to perform that function. Whether the cause for which one agitated succeeded or failed was not the point. Every participant shed his or her status as victim and received identity as part of a group, an identity that served as a hedge against anomie. Only those individuals not taking sides, not trying, not participating, were "losers."

This intellectual context had been the background of DES debates for years. Now, however, the concept of cost-benefit analysis legitimized previously entrenched interests and opened the regulatory process to newly self-defined partisans. Equally important, all retained their status in the regulatory process no matter what the decision. Judge Davidson's fifty-two-page opinion, unveiled on September 21, 1978, reflected these new conventions. His decision that the agency could not invoke the Delaney clause against DES because of the "methods of examination prescribed or approved by the Secretary" exception—the DES Proviso—was vitiated by his assertion that agents that in some form or dosage produced cancer could violate the General Safety Clause. He also ruled against the validity of risk-benefit analysis (a version of cost-benefit analysis), arguing that Congress had intended only consideration of therapeutic benefits, but he backtracked by claiming that the evidence "does not establish that the societal and economic benefits of DES outweigh the risks to public health." Davidson called "questionable" the manufacturers' contention that stilbestrol feeding produced beef that was more proteinaceous and less fatty and therefore lowered human exposure to colon cancer, heart disease, and diabetes. While admitting that beef from DES-fed cattle contained more protein, the judge maintained that according to one study the drug decreased fat content by only 1 percent, an insignificant reduction. He also viewed any world grain shortage as irrelevant because the United States was experiencing no such deficiency, and he termed the inflationary impact of a ban negligible since alternative growth promoters existed. He disallowed as not "conclusively demonstrated" the argument that only estrogenic effects cause cancer, claimed that there was "no foundation for the assumption that demonstrable estrogenic activity must be displayed before ill effects will occur," and demanded that manufacturers contrast the drug's hypothetical benefits with a full-range calculation of possible DES-related health hazards. The substance's "true" potential risk was far greater than the sum of its vaginal cancer and male mutagen risks, and the regulatory process needed to determine and evaluate that larger risk.[38]

From that pronouncement Davidson moved to a consideration of stilbestrol under the General Safety Clause. He cited isolated studies indicating the sons as well as the daughters of DES-treated women suffered a variety of anatomical and sexual ills. These results in humans were consistent with established transplacental animal studies. Other, more recent animal studies had suggested a relationship between transplacental stilbestrol administration and cardiovascular defects, cleft palate, and breast cancer in offspring.

Davidson then asked rhetorically, of "what relevance" are the aforementioned studies—which were based on huge DES doses given over comparably long periods—to the issue of whether "the small amount of DES ingested by the American public [is] a threat to its health?" He admitted that "at first blush, there appears little correlation between these two areas of inquiry." But Davidson contended that "the appearance of the rare tumors [the vaginal adenocarcinomas noted by Herbst and others] cannot be completely accounted for by the therapeutically administered DES." The judge thought it "conceivable that postnatal as well as prenatal dietary exposure to small amounts of DES is contributing to any or all of these increases."

Davidson also examined the manufacturers' contention that DES's similarity to other estrogens made special regulation unnecessary. He supported instead "the possibility of a double mode of cancer production by DES" because of "the wide array of tumors associated with DES exposure." In any case, he recognized that "possible non-estrogenic mechanisms for the adverse effects of DES have not, as yet, been ruled out." He noted that DES did not occur naturally and was structurally different from other estrogens, which yielded different metabolic products than stilbestrol did. Most of DES's known metabolites—not all had been evaluated—were not estrogenic, and as a consequence were not measurable by the mouse bioassay. He therefore speculated that "DES may be producing carcinogenic and other adverse effects . . . through the action of its non-estrogenic metabolites."

Davidson argued that regulatory law placed "the onus . . . upon the manufacturers to establish that DES, its conjugates, and its metabolites occur only at safe levels in the meat. If the manufacturers cannot submit a reliable method by which to detect the drug at those levels, the drug has not been shown to be safe and must be withdrawn from use." His reference to method of analysis proved to be no accident. In an argument more sophisticated than the single-molecule thesis (which holds that one molecule of a substance may prove harmful), Davidson deemed it "difficult to establish the safety of use of DES by showing that it reaches the consuming public only in minute amounts." The difficulty rested "in establishing what the no-effect level for [each of] these various observed pathologies—the pathologies mentioned above—would be." The situation was exacerbated "because many of such pathologies have just recently been discovered, and many of the studies are still ongoing," factors that would make "an attribution of safety to any specific level . . . necessarily speculative." He sought to protect future Americans by relying on "common sense [to] dictate when there is a sufficiently large safety margin to accommodate later positive results which could occur at lower exposure levels."

The 1964 study by George Gass at Southern Illinois University colored Davidson's thinking. He complained that "the present detection methods are extremely limited in their sensitivity when compared to levels which show effects in animals." Equally important, the mouse assay only tested estrogenic activity, "a quality enjoyed by only one of DES's many metabolites." Other noxious or hazardous properties could reside in metabolites that were non-estrogenic and therefore not measured. For all these reasons, the judge recommended that approval of the stilbestrol NADAs be withdrawn. The FDA, Davidson stated, "must only raise reasonable questions as to the safety of the drug to justify a revocation."

Davidson's creative justifications of a stilbestrol ban made it impossible for the DES manufacturers to refute his arguments absolutely. Nonetheless, Davidson had acknowledged cost-benefit considerations and therefore legitimated them as de facto agency policy. Commissioner Kennedy's then-current activities against nitrites and nontherapeutic antibiotic usage certainly suggested that Kennedy would use cost-benefit analysis to support a stilbestrol prohibition.[39] Before he could officially take up that question, the several parties had thirty days to file exceptions to Davidsons's ruling and an additional twenty days to reply to those exceptions. The manufacturers bluntly claimed that no scientific evidence existed to back up any of Davidson's significant claims. They accused the judge of misinterpreting data and used a comparison calculated by Jukes to estimate that a typical DES mother's daily dosage and the average amount of DES eaten in tainted liver differed by "more than 160,000,000-fold," an irrelevant basis on which to establish deterministic comparisons. Such rationales led Jukes to characterize Davidson's decision as "a massive exercise in futility, a travesty of science, and an affront to the intelligence." Nonparty participants also objected to the judge's occasional reference to expert scientific opinion, arguing that the individuals Davidson sanctioned did not represent sound science and that CAST and ASAS members, not Wolfe and other Naderites, were the "true" experts. The bald insistence that bovine nutritionists and other animal scientists had qualifications, experience, and knowledge uniquely suited to decide questions of human oncology, toxicology, and tumor genesis was no less ludicrous than Davidson's explanations in support of a stilbestrol prohibition. The judge's decision to include carcinogenicity under the General Safety Clause was the single substantive issue raised by DES's supporters. They contended that Congress had intended to restrict regulation of carcinogens to the Delaney clause. To also consider DES under the General Safety Clause placed the drug in a sort of unconstitutional double jeopardy. The FDA's sole major disagreement with Davidson stemmed from his decision to dismiss the Delaney clause as not applicable to this case.[40]

At the FDA, Kennedy's team of lawyers and office staff began to review Davidson's opinion and the responses to it to recommend appropriate action.[41] The Delaney clause itself continued to draw public comment as Thomas S. Foley (D-Wash.) submitted a House bill to revoke the clause, capitalizing on the outrage over the recent cyclamate ban and the pending proposal to ban saccharin. James Martin (R-N.C.) became the bill's most articulate and credible spokesman. Holding a doctorate in chemistry from Princeton University, Martin possessed credentials likely to sway congressional colleagues. His arguments differed little from those espoused by like-minded individuals over the previous two decades, or even from the FDA's SOM proposition. What marked Martin's arguments as distinct from the SOM proposition was his refusal to establish a hypothetically objective scale, such as the Mantel-Bryan method, to quantify risk. He thought such an approach "a political response to unknown fears," foisted on the public by "those trained in the fields of law and public administration," simply to fulfill "their penchant for organization orderliness." Martin preferred to rely exclusively on the messier process of weighing and measuring evidence and other factors to derive an individual assessment for each drug.[42]

In his relativistic cost-benefit formulation Martin rejected in a rather straightforward way an absolutist science in which consensus reigned and agreement was achieved on nearly every question. If such a science had ever existed, its time was well past. What had made the stilbestrol debates so arresting was that their very existence suggested the fiction of a monolithic science. That individuals made their own decisions and felt free to side with one party or the other, that individuals thought themselves capable of assuming and advocating positions, indicated that they had lost reverence for "the expert" and for expertise. These positions, which individuals now often identified as political or philosophical, occurred outside the realm of expertise. They seemed to be the consequence of individual or group experience—the social and economic forces acting on and buffeting a person or persons—rather than products of special knowledge. The idea of expertise had become anachronistic, or at best irrelevant.

The Kennedy-guided FDA tapped this sentiment when it marketed its revamped SOM procedure as allowing for "acceptable risk." The new procedure would vitiate the Delaney clause by permitting the "weighing [of] health risks against health benefits." The Institute of Medicine (IOM) went even further. The medical complement to the National Academy of Sciences, the IOM found the Delaney clause and other similar clauses "confusing, cumbersome, and not always related to risk." It demanded qualitative, not quantitative, risk assessment, noting that "when we deal with issues where the science can only take you so far, whether you like it or not, you have to make judgments." To that end, it proposed to establish three categories of risk: low, moderate, and high, and wanted these "health risks . . . balanced against the economic benefits." "Warning labels, logos and educational campaigns" would replace most

FDA bans as the IOM placed "greater reliance on consumer awareness and intelligence."[43]

Such strong propositions had gained a certain cachet during the uproar over the proposed saccharin ban, as the individual proclivity to wish to be thin clashed with a potentially minute disease risk. That simple, immediate desire personalized the Delaney clause in a way that stilbestrol never had done and helped undermine the clause's public credibility. Only a few years earlier, Edward G. Feldmann, longtime editor of the *Journal of Pharmaceutical Sciences,* would not have chastened consumer advocates for "crying wolf" and unleashing "reckless charges." Discussing the immediate past, Feldmann concluded that "it is almost as if the political excesses of the French Revolution, or of the Senator McCarthy era in the United States, had been revived in the guise of misdirected science and runaway regulation." *Industrial Research Development,* an industry-based journal, conducted an unsystematic readership poll and determined that 88 percent favored repeal or amendment of the Delaney clause. Only 3 percent urged rigorous enforcement.[44]

Ironically, rejection of the absolutism of the Delaney clause and institutionalization of the cost-benefit approach occurred even as those involved in making decisions on public policy were being attacked. A *Bioscience* exposé attacked CAST, portraying the organization as a tool of agribusiness and agri-industry, suggesting that it was posing as a scientific body, and charging that it was an agent of advocacy that pretended to cherish objectivity. The essayist objected to CAST's dependence on agricultural corporations for financial support, claiming that it stacked its task forces to get specific proindustry results and that it refused to brook dissent even on the most controversial issues.[45]

CAST had done little to hide its agribusiness connections. Congress and the FDA certainly had long understood the organization's slant. Indeed, during the DES hearings in 1977, an FDA attorney, Robert Spiller, had termed the CAST hormone report "a conclusion in search of citations," marred by "circular cites, phantom references and contradictory statements." This "purportedly scientific document . . . in fact is a collection of conclusions by a number of cattle-oriented people, followed by a list of disconnected references and appendices."[46] CAST withstood these assaults, but its prestige, and that of agricultural scientists, suffered. The National Cancer Institute (NCI), long regarded as the bastion of scientific competence within HEW and a voice of caution, also was attacked. The General Accounting Office criticized the agency for its inept handling of its expensive multigenerational carcinogenesis testing program and charged that the institute's researchers did not exercise quality control over mouse bioassays, a full 51 percent of which proved so deficient that the agency jettisoned them as unreliable. Scientific auditors doubted the entire program's credibility when investigators found NCI bioassay facilities in a state of disrepair. Gaps beneath doors, for instance, "permitted test animals to roam from room to room," a most serious violation of experimental protocol. Congressman Henry Waxman (D-Calif.) added to the indictment by claiming that the agency had deliberately withheld more than two hundred test results during the past three years.[47]

The NCI, agricultural scientists, consumer advocates, and the mythologies of expertise and science had all lost some luster in the period since Davidson issued his DES opinion. These occurrences, the modified climate accompanying them, and the de facto institutionalization of the cost-benefit approach made it convenient to ban stilbestrol. On June 29, 1979, when Kennedy issued his decision banning stilbestrol as a cattle growth promoter, newspapers and other periodicals published little commentary. The commissioner felt no need to file his reasons in the *Federal Register* until nearly three months later. His detailed explanation cited many of the same arguments that Davidson had used, but provided additional documentary evidence and speculations. Equally significant, the ban's terms permitted manufacturers a grace period to rid themselves of stock and gave cattlemen a substantial period to get animals to market. Livestock producers and their chemical suppliers requested extensions of both. The agency permitted the implantation of cattle until November 1, additional evidence that it did not consider DES residues in liver an imminent health threat.[48]

Hess and Clark and Vineland Laboratories predictably petitioned the FDA for a stay (they were turned down almost immediately) and petitioned the U.S. District Court for the District of Columbia to overturn the stilbestrol ruling. American Home Products and Dawe's Laboratory decided to drop the matter because the "FDA's decision seems to be a careful and thorough piece of work": the agency had accepted the principle of cost-benefit analysis within the decision-making process.[49] In any case, the FDA's banning of stilbestrol no longer seemed as much of a menace to livestock producers and agrochemical manufacturers as it had just a few years earlier. Reliance on cost-benefit analysis, reflected by the failed prohibition of saccharin as well as by the constant "discovery" of new cancer hazards in American staples, such as fried foods or barbecued beef, presented federal regulation in a fundamentally different way. To a broadening segment of the American middle class, saccharin regulation and the regulation of foods and drugs generally seemed to be of intense private personal interest or to be a matter in which each individual deserved a clear choice. Individual assessment, not legal or scientific absolutism, had become the regulatory sine qua non.[50]

Epilogue

The USDA began perfunctory DES residue testing in early November 1979 but soon stopped that practice when it failed to uncover positives. High-level USDA and FDA personnel believed that an overwhelming majority of American cattlemen were faithfully complying with the ban and that attempts to deceive would easily be exposed. Skilled observers would recognize the characteristic ear implants even on routine feedlot tours. Violators were subject to fines of up to ten thousand dollars and jail terms of up to three years for each illicitly implanted animal.[1]

The assumptions on which the agency's voluntary compliance program were grounded were shaken in late March 1980, when a dissatisfied feedlot employee informed the FDA that an agricultural subsidiary of Allied Chemical had ignored the November deadline and showed no sign of relenting. Subsequent FDA surveillance revealed that this was not an isolated instance. In early April, FDA officials estimated that some seventy thousand cattle had been illegally implanted. Two weeks later, the FDA had found more than four hundred thousand cattle in 156 feedlots in eighteen states implanted with the drug. It also identified twenty-five firms that supplied DES pellets for implantation. It implicated no DES manufacturer, although several had apparently sold large quantities of the drug to suppliers just prior to the legal deadline. In the wake of these discoveries, the USDA reinstituted DES residue monitoring. By late June, the agency had detected 318 cattlemen who had illegally used the hormone.[2]

The scale of the problem caught everyone off guard. The American National Cattlemen's Association (ANCA), the nation's largest beef cattle producers' association, disclaimed any knowledge of widespread cheating and demanded amnesty for individuals willing to turn in stockpiled DES and remove cattle implants. The FDA refused, and invited Hess and Clark and Vineland Laboratories to recall stockpiled DES voluntarily and reimburse those who came forward. The two pharmaceutical firms accepted no responsibility for the fiasco and refused to assume financial liability. The ANCA finally acted as a central stilbestrol repository, collecting the contraband material and furnishing participants certificates to prove their feedlots were DES-free. The FDA also had to determine how to deal with what it contended was "the biggest veterinary drug scandal in U.S. history" and its most costly regulatory action ever. The agency lacked funds to prosecute every violator. It considered confiscating all DES-enhanced meat but feared a consumer backlash. As a high-ranking USDA official noted, the contraband "meat is no worse than any that was on the market prior to the ban."[3] The FDA decided to prosecute the largest feedlots as a lesson to their smaller counterparts and required each offender, large or small, to pay a USDA-accredited veterinarian to remove ear implants surgically, and to withhold animals from slaughter for forty-one days after the minor operation. The withholding period was extended to sixty-one days if the cattlemen wanted to market kidneys and livers. Violators had to pay for extra feed and holding costs and had their capital tied up for as much as two more months, a distinct disadvantage in the highly competitive feeding business. Producers failing to recondition beef had their meat confiscated and condemned.[4]

This complicated regulatory approach begged the question of why so many "ordinarily law abiding citizens . . . [would] just thumb their nose at the law" by continuing to use stilbestrol. Small feeders, such as Les Crow, of Williamsburg, Iowa, and Ronald Martin, of Checotah, Oklahoma, ascribed their illegal action to ignorance of the DES ban. But at least one livestock producer "went so far as to develop a system to hide the use of DES in their [sic] computer records," hardly an act of naiveté. Other cattlemen were more forthcoming. The FDA and media "haven't convinced me DES isn't safe," admitted Jay Crofoot, operator of a large Texas feedlot. Beef producers "felt [it] was a bad law, so [they] did not observe it." Why break old habits and substitute an inferior, more costly product for a superior, cheaper one just because a bunch of "radicals" insist? asked another producer. John McClung, the Washington editor of *Feedstuffs*, reported that many feedlot managers attributed continued DES use to pressure from sales representatives, who encouraged cattlemen to implant the drug illegally just to stay even with their equally corrupt neighbors. Wayne Anderson, the editor of *Feedstuffs*, and Burton Eller, lobbyist for the ANCA, claimed that livestock producers have always been gamblers. This time they had stockpiled massive amounts of the drug, confident that the agency would again extend the deadline. When that failed to happen, the producers then gambled on not getting caught, a move that "makes about as much sense as drawing to an inside straight." Lester Crawford, Van Houweling's replacement at the FDA's Bureau of Veterinary Medicine, explained the fiasco as a manifestation of the "myth of the pioneering spirit," a rugged individualism that empowered cattlemen to do anything they pleased with their land and livestock. Yet Charles Ball, executive vice-president of the Texas Cattle Feeders' Association, wondered why anyone raised a ruckus. He found it hard "to say that these people were involved in willful, criminal activity." All they had done was place "a little investment in DES, and [want] to utilize their investment."[5]

Charles Olentine, editor of *Feed Management*, refused to romanticize or trivialize a blatant disregard for law or a craven selfishness. He decried violators' "public be damned" attitude, which showed "a contempt for our regulatory enforcement agencies and more importantly a total disregard for public welfare." "The animal industry had its day in court and lost." Yet "numerous self-righteous distributors and feedlot operators felt that their profits were more important than abiding by the system." Whenever "profits take a higher priority than the law or the public welfare," concluded Olentine, "it is time to re-evaluate the system." His views were hardly typical. Small and large beef producers expressing dismay over the cattlemen's lawlessness did so not because the cattlemen's action was wrong per se, but because it undercut the entire industry's regulatory position. The "real tragedy" of the situation, wrote Chuck Leavitt, industry analyst for Shearson Loeb Rhoades, was that it "goes a long way to destroying efforts toward voluntary restrictions and compliance, long sought by industry and government alike." Crawford lamented that "sadly, some innocence has died and some resolve has hardened. The FDA will emerge more vigilant and somewhat less conciliatory." The agency is "going to come down hard on them," promised Theodore Rotto, investigations director of the Dallas FDA office.[6]

The shock and hand wringing that accompanied the discovery of massive illegal DES use should not have been surprising. Cattlemen had merely applied their own form of cost-benefit analysis. They took a chance that they could profit without detection, aware that if they were caught the FDA was unlikely to penalize them severely. Nor should it be surprising that there existed such striking parallels between the new regulatory synthesis and the cattlemen's actions. The new synthesis-creating movement stemmed from precisely those notions that manifested themselves in the cattlemen's lawlessness. The breakdown of a sense of absolute authority based on faith in science and scientists, and the abandonment of the progressive partnership in the late 1950s and early 1960s, had yielded an individual-based, regulation-as-opposition philosophy in which the former partners—government, industries, universities, producers, and consumers—accused each other of vile self-interest and brooked no consensus, seeing only antagonism and antagonists. Institutions such as the FDA, which had been celebrated a decade earlier, now seemed narcissistic, self-interested, unable or unwilling to provide protection or even assistance to those outside.

By about 1970, however, even the former partners' tacit accord to work through government to control the regulatory apparatus gave way to direct emotional appeals to the public, and accompanying discourses of victimization and entitlement. A lack of knowledge, in conjunction with a willingness to assert definitive knowledge, and a propensity to blame others or forces beyond anyone's control for personal failings, also were aspects of this milieu. Agreements or even myths of agreement about what constituted definitive evidence, long the province of scientists, seemed no longer to exist. What was left was terror-driven and frustration-driven individual opinion, a matter of personal psychology. In that world, the cost-benefit approach made tremendous sense, but not because it offered adequate protection or even a measure of protection. The cost-benefit approach provided a multitude of emotional rewards. It opened the regulatory process to those most likely to feel the need to participate (i.e., middle-class suburbanites), enabling anyone who wished to do so to become involved and to declare a sense of importance, even to claim victory. The cost-benefit approach also allowed established interests such as business, industry, and scientists—even government—to continue to trumpet their own significance and gain public approbation even when they were no longer relevant. Indeed, what counted was immediate individual gratification, born of individual prerogative. Coalitions—consumer federations, trade associations, industry-science consortiums—were countenanced and tolerated only so long as they seemed likely to achieve some form of that desired end. To the "victor," and usually also to the "loser" (since anomie-banishing participation was a product), went the spoils.

Industry, business, government, and college-based scientists—groups that had long been dominant in regulation—pioneered implementation of the new regulatory synthesis. As they ceased to cite their science as a justification for their regulatory role and began to cite their personal experience as scientists, they abandoned the pretense of a homogenized scientific community and acknowledged that regulation was an art, a matter of interpretation, rather than a science. Their recognition of relativism and indeterminacy succeeded in retaining for these scientists a regulatory role, but they lost their specialness; they became just one of the new regulatory elements. This fractionalization of interests and perspectives, reflected in the new regulation, certainly gave new meaning to established terms such as *public interest*. In an individuated society, a public and a public's interest ceased to exist. Outsiders looking at the system might conclude that the public interest could be measured in maximization of immediate self-gratification, but that hardly was the view of individual actors as they pursued their activities. A lack of persistence, a quick shifting of interests

or attention, accompanied and characterized this vainglorious search for immediate individual gratification. Patience, a virtue few could afford, went by the
board.

The cost-benefit approach framed both the new regulatory synthesis and the
cattlemen's unlawful response. It persists as the order of the day, making any
form of sustained, reasoned government activity—including regulation—
virtually impossible. Journalists emphasize extremes, regularly citing putative
public interest groups declaring a public health catastrophe, and/or industry
representatives describing the same condition as harmless.[7] The controversies
about Alar and silicone breast implants demonstrate the pattern's continuance
in the immediate past. The implant dispute revolves around rights, not simply
those of the regulated and the consumer, but those of the regulated and various
groups of consumers. Women's right to health and safety is weighed against
their right to control their bodies and their self-image. Personal revelations of
disfigurement and terrible autoimmune disease emerge, but little evidence is
presented beyond the anecdotal; there is "no body of scientific knowledge to
prove or disprove the charges." Science is—and has been—for practical purposes silent, not germane.[8] In the case of Alar, the situation is even clearer.
Labeling Alar, a growth regulator used in the apple industry, a "life threatening
chemical," the media, led by the redoubtable "Sixty Minutes," reported that the
public faced "intolerable risks" from ingesting this substance. The most menacing of these wildly hypothetical Virtual Safety–like statistical estimates placed
the danger as slightly in excess of one case of cancer per million Americans.
That claim galvanized Congress to hold hearings and offer its members' assessments of the matter. The Environmental Protection Agency, the Food and Drug
Administration, and the Department of Agriculture each testified that Alar-
contaminated apples did not constitute a "an imminent hazard," but that opinion was overshadowed by Meryl Streep. Not generally recognized as knowl-
edgeable in health science, the Academy Award–winning actress testified to
Congress regarding her concern and dominated, even monopolized, television
and press coverage. Streep had worked with the Natural Resources Defense
Council on Alar, an indication that the NRDC truly understood how to use the
cost-benefit approach. *Washington Post* columnist Hobart Rowen called on consumers not to "take these outrages any longer," and a highly effective consumer
boycott of apples and apple products marked the culmination of the matter. A
half-dozen large supermarket chains and several major apple processing companies and apple growers quickly engaged in what best can be considered
damage control. The chains claimed to purchase apple products only from suppliers that did not use Alar, and the companies simultaneously announced that
their products were Alar-free—a fact in considerable dispute. For their part,

the apple growers renounced Alar, maintained that they had long been phasing it out, and pledged to discontinue its use immediately. With these dramatic acts, the "crisis" quickly abated and was soon just a memory.[9]

In cost-benefit analysis, each instance or situation is unique, not a matter of precedent. Faith, belief, confidence, even authority, the basis for persisting societal agreements, seem no longer to exist. Quaint ideas of the past, these ideas have become relics locked in past times. In late twentieth-century regulatory questions, only individuals have the right to decide. And in every case, each individual as purchaser and modern citizen, each interest group as advocate, and each governmental unit as regulator always knows for certain the correct action that must be taken.[10]

Notes

1. Orville Schell, *Modern Meat* (New York: Random House, 1984), 181–320; Terry G. Summons, "Animal Feed Additives, 1940–1966," *Agricultural History*, 42 (1968): 305–13; K. E. McMartin, K. A. Kennedy, P. Greenspan, S. N. Alam, P. Greiner, and J. Yam, "Diethylstilbestrol: A Review of Its Toxicity and Use as a Growth Promotant in Food-Producing Animals," *Journal of Environmental Pathology and Toxicology*, 1 (1978): 279–313; and Joseph V. Rodricks, "FDA's Ban of the Use of DES in Meat Production: A Case Study," *Agriculture and Human Values*, 3 (Winter–Spring 1986): 10–25. For other studies of DES, see, for example, Kenneth L. Noller and Charles R. Fish, "Diethylstilbestrol Usage: Its Interesting Past, Important Present, and Questionable Future," *Medical Clinics of North America*, 58 (July 1974): 793–810; Roberta J. Apfel and Susan M. Fisher, *To Do No Harm: DES and the Dilemmas of Modern Medicine* (New Haven: Yale UP, 1984); and Richard Gillam and Barton J. Bernstein, "Doing Harm: The DES Tragedy and Modern American Medicine," *Public Historian*, 9 (Winter 1987): 57–82. The term *cattlemen* is used only because no gender-neutral term could be found, and because saying "cattlemen and cattlewomen" would have hampered the flow of the narrative.

2. On the rise of the expert, see, for example, Louis Galambos, "The American Economy and the Reorganization of the Sources of Knowledge," in Alexandra Oleson and John Vos, eds., *The Organization of Knowledge in Modern America, 1860–1920* (Baltimore: Johns Hopkins UP, 1979), 269–82; Hugh Hawkins, "University Identity: The Teaching and Research Functions," in ibid., 285–312; Daniel J. Kevles, *The Physicists: The History of a Scientific Community in Modern America* (New York: Knopf, 1977); and Edward H. Beardsley, *The Rise of the American Chemical Profession, 1850–1900* (Gainesville: U of Florida P, 1964). I also have worked extensively on this question. See especially Alan I Marcus, "Professional Revolution and Reform in the Progressive Era: Cincinnati Physicians and the City Elections of 1897 and 1900," *Journal of Urban History*, 5, no. 2 (Feb. 1979): 183–207; and idem, "Back to the Present: Historians' Treatment of a City as a Social System during the Reign of the Idea of Community," in Howard Gillette, ed., *American Urbanism: A Historiographic Review* (Westport: Greenwood, 1987), 7–26; as well as Alan I Marcus and Howard P. Segal, *Technology in America: A Brief History* (San Diego: Harcourt Brace Jovanovich, 1989), 165–253. On agricultural scientists, see Alan I Marcus, *Agricultural Science and the Quest for Legitimacy: Farmers, Agricultural Colleges, and Experiment Stations, 1870–1890* (Ames: Iowa State UP, 1985); idem, "The Wisdom of the Body Politic: The Changing Nature of Publicly Sponsored American Agricultural Research since the 1830s," *Agricultural History*, 62, no. 2 (Spring 1988): 4–26; idem, "Constituents and Constituencies: An Overview of the History of Public Agricultural Research Institutions in America," in Donald F. Hadwiger and William B. Browne, eds., *Public Policy and Agricultural Technology: Adversity despite Achievement* (London: Macmillan; and New York: St. Martin's, 1987), 15–30; and idem, "Setting the Standard: Fertilizers, State Chemists, and Early National Commercial Regulation, 1880–1887," *Agricultural History*, 61, no. 1 (Winter 1987): 47–73. Also see Margaret W. Rossiter, "The Organization of the Agricultural Sciences," in Oleson and Voss, eds., *Organization of Knowledge*, 211–48.

3. The passage and formulation of the 1906 act are explained expertly in James Harvey Young, *Pure Food: Securing the Federal Food and Drugs Act of 1906* (Princeton: Princeton UP, 1989); Oscar E. Anderson, Jr., *The Health of a Nation: Harvey W. Wiley and the Fight for Pure Food* (Chicago: U of Chicago P, 1958); and Peter Temin, *Taking Your Medicine: Drug Regulation in the United States* (Cambridge: Harvard UP, 1980), 27–37.

4. This intellectual reorganization has come to be known among historians as associationalism. See, for example, Ellis W. Hawley, *The Great War and the Search for Modern Order: A History of the American People and Their Institutions, 1917–1933* (New York: St. Martin's, 1979); David E. Hamilton, *From New Day to New Deal: American Farm Policy from Hoover to Roosevelt, 1928–1933* (Chapel Hill: U of North Carolina P, 1991); and Ellis W. Hawley, "Three Facets of Hooverian Associationalism: Lumber, Aviation, and Movies, 1921–1930," in Thomas K. McCraw, ed., *Regulation in Perspective: Historical Essays* (Cambridge: Harvard UP, 1981).

5. See, for example, Wallace E. Huffman, "The Supply of New Agricultural Scientists by U.S. Land Grant Universities, 1920–1979," in Lawrence Busch and William Lacy, eds., *The Agricultural Scientific Enterprise* (Boulder, Colo.: Westview, 1986), 108–28; Marcus, "Wisdom of the Body Politic"; and Margaret W. Rossiter, "Graduate Work in the Agricultural Sciences," *Agricultural History*, 60, no. 2 (1986): 37–57. For the proliferation of agricultural chemicals after World War II, see, for example, Suzanne White, "Chemicals in Food: The Delaney Committee, 1950–52" (Ph.D. diss., Emory U, in progress). A useful book is Larry Wherry, *The Golden Anniversary of Scientific Feeding* (Milwaukee: Business, 1947).

6. See both Charles O. Jackson, *Food and Drug Legislation in the New Deal* (Princeton: Princeton UP, 1970), especially 192–200; and Temin, *Taking Your Medicine*, 38–50. For the pressure on government to force fringe food, drug, and cosmetic industries to "scientize," see, for example, Arthur Kallet and F. J. Schlink, *100,000,000 Guinea Pigs: Dangers in Everyday Foods, Drugs, and Cosmetics* (New York: Vanguard, 1933); and Ruth Deforest Lamb, *American Chamber of Horrors: The Truth about Foods and Drugs* (New York: Farrar and Rinehart, 1936). For the emergence and centrality of advertisers as a discrete group specializing in market formation and manipulation, see James Harvey Young, "Food and Drug Enforcers in the 1920s: Restraining and Educating Business," *Business and Economic History*, 21 (1992): 119–28. For an indication of the widespread belief in the ability of experts working cooperatively to solve virtually all of society's problems, see Marcus and Segal, *Technology in America*, 255–311.

7. Important monographs considering aspects of this new regulation include Thomas R. Dunlap, *DDT: Scientists, Citizens, and Public Policy* (Princeton: Princeton UP, 1981); Sam P. Hays, *Beauty, Health, and Permanence: Environmental Politics, 1955–85* (Cambridge: Cambridge UP, 1987); James T. Patterson, *The Dread Disease: Cancer and Modern American Culture* (Cambridge: Harvard UP, 1987); and John H. Perkins, *Insects, Experts, and the Insecticide Crisis: The Quest for New Pest Management Strategies* (New York: Plenum, 1982).

8. For the chemicalization of America, see Hugh D. Crone, *Chemicals and Society: A Guide to the New Chemical Age* (Cambridge: Cambridge UP, 1986).

CHAPTER I. PARTNERS IN PROGRESS

1. Jay L. Lush, Damon Catron, and Norman Jacobson were among the leaders in their fields. See, for example, Jay L. Lush, *Animal Breeding Plans* (Ames, Iowa: Collegiate, 1937); "Dr. Catron Wins AFMA Award," *Flour and Feed*, 56 (Feb. 1955): 24; "Two Iowans Receive AFMA Award," ibid., 56 (July 1955): 7; and Theodore P. Thery, "Iowa State College Ag Setup Accents Three-Phase Plan," ibid., 57 (Feb. 1956): 16–17.

2. The college's neglect of beef cattle research is graphically demonstrated in *Report on Agricultural Research for the Year Ending June 30, 1949, for the Iowa Agricultural Experiment Station*, 146–49; and *Report of Iowa Agricultural Experiment Station for the Two Years July 1, 1949 to June 30, 1951*, 5–12, 70. Iowa would remain the nation's cattle-feeding leader until the early 1970s.

3. Harold Lee's *Roswell Garst: A Biography* (Ames: Iowa State UP, 1984) is the definitive work on Garst; pp. 143–58 deal with issues raised in this section. Also see Floyd Andre to Roswell Garst,

Sept. 19 and Oct. 20, 1949; and Garst to Andre, Oct. 14, 1949, Roswell Garst Papers, box 8, folder 7, Special Collections Department, Parks Library, Iowa State University, Ames, Iowa; Wise Burroughs to Garst, Mar. 15, 1951, Garst Papers, box 9, folder 2; "Interview, Wise Burroughs by Harold Lee," n.d., Garst Papers, box 90, cassette 32; and "Animal Nutrition Authority to Join Iowa State Staff," *Ames Daily Tribune*, Jan. 16, 1951, 1. In Ohio, Burroughs worked in a research group headed by Paul Gerlaugh. For the work of Gerlaugh's group with corncobs, molasses, and urea, see, for example, Wise Burroughs, Paul Gerlaugh, A. F. Schalk, E. A. Silver, and L. E. Kunkle, "The Nutritive Value of Corn Cobs in Beef Cattle Rations," *Journal of Animal Science* (hereafter cited as *JAS*), 4 (1945): 373–86. For the DuPont Company's involvement with urea research, see E. I. DuPont De Nemours and Co., *Digest of Research on Urea and Ruminant Nutrition* (Wilmington: E. I. DuPont De Nemours, 1958), especially 26–28. At least one urea-molasses-corncob experiment was contemplated at Iowa State College prior to Burroughs's appointment. See "Ames Meet Held to Discuss New Experiments," *Corn Belt Lamb Feeder*, 8 (Oct. 7, 1948): 1, 4. For Garst's remembrances, see Roswell Garst, "The Use of Corn Cobs, Corn Stalks, and Grain Sorghum Stubble for Cattle Feed," *Garst and Thomas Bulletin*, no. 5 (n.d.): 2–5.

4. On these questions, see, for instance, Wise Burroughs and Paul Gerlaugh, "The Influence of Soybean Oil Meal upon Roughage Digestion in Cattle," *JAS*, 8 (1949): 3–8; Wise Burroughs, Paul Gerlaugh, B. H. Edgington, and R. M. Bethke, "The Influence of Corn Starch upon Roughage Digestion in Cattle," ibid., 8 (1949): 271–78; Wise Burroughs, Paul Gerlaugh, and R. M. Bethke, "The Influence of Alfalfa Hay and Fractions of Alfalfa Hay upon the Digestion of Ground Corncobs," ibid., 9 (1950): 207–13; Wise Burroughs, H. G. Headley, R. M. Bethke, and Paul Gerlaugh, "Cellulose Digestion in Good and Poor Quality Roughages Using an Artificial Rumen," ibid., 9 (1950): 513–22; G. E. Stoddard, N. N. Allen, W. H. Hale, A. L. Pope, D. K. Sorensen, and W. R. Winchester, "A Permanent Rumen Fistula Cannula for Cows and Sheep," ibid., 10 (1951): 417–23; Wise Burroughs, Carlos Arias, Peter Ed Paul, Paul Gerlaugh, and R. M. Bethke, "In Vitro Observations upon the Nature of Protein Influences upon Urea Utilization by Rumen Microorganisms," ibid., 10 (1951): 672–82; and Carlos Arias, Wise Burroughs, Paul Gerlaugh, and R. M. Bethke, "The Influence of Different Amounts and Sources of Energy upon In Vitro Urea Utilization by Rumen Microorganisms," ibid., 10 (1951): 683–92.

5. See, for example, Charlie F. Peterson, "Supplying APF in Poultry Rations," *Feedstuffs*, 22 (Mar. 11, 1950): 22–23; "Aureomycin Greatly Accelerates Animal Growth, Scientists Report," ibid., 22 (Apr. 15, 1950): 1, 65; T. H. Jukes, E. L. R. Stokstad, R. R. Taylor, T. J. Cunha, H. M. Edwards, and G. B. Meadows, "Growth-Promoting Effect of Aureomycin on Pigs," *Archives of Biochemistry and Biophysics*, 26 (1950): 324–25; L. L. Bucy, U. S. Garrigus, R. M. Forbes, and W. H. Hale, "Effect of Arsenical Supplements on Growth-Fattening Lambs," *JAS*, 12 (1953): 909–10; and E. A. Kline, M. E. Ensminger, T. J. Cunha, W. W. Heinemann, and W. E. Ham, "Effect of Adding Drugs to the Ration of Fattening Cattle," ibid., 8 (1949): 411–24.

6. E. C. Dodds, L. Goldberg, W. Lawson, and R. Robinson, "Oestrogenic Activity of Certain Synthetic Compounds," *Nature*, 141 (1938): 247; and Edward A. Doisy, *Sex Hormones* (Lawrence: U of Kansas, University Extension Division, 1936). Prior to stilbestrol's synthesis, endocrinologists had extracted estrogenic substances from animal ovaries, an expensive, time-consuming process. Diethylstilbestrol, produced at a cost of pennies per unit and manufactured in relatively pure form, enabled investigators to launch new endocrinological initiatives and soon came to be the compound against which all estrogenically active substances were measured. Also of use on these points is Albert Q. Maisel, *The Hormone Quest* (New York: Random House, 1965), 28–37; and Victor Cornelius Medvei, *A History of Endocrinology* (Boston, Mass.: MTP, 1982).

7. F. W. Lorenz, "Fattening Cockerels by Stilbestrol Administration," *Poultry Science*, 22 (1943): 190–91; idem, "The Fattening Action of Orally Administered Synthetic Estrogens as Compared with Diethylstilbestrol Pellet Implants," ibid., 24 (1945): 91–92; and idem, "The Influence of Diethylstilbestrol on Fat Deposition and Meat Quality in Chickens," ibid., 24 (1945): 128–34.

8. Rollin H. Thayer, R. Georg Jaap, and Robert Penquite, "Fattening Chickens by Feeding Estrogens," *Poultry Science*, 24 (1945): 483–95; and John L. Adams, W. H. McGibbon, and L. E.

Casida, "The Effect of Orally Administered Synthetic Estrogens on Single Comb White Leghorn Pullets," ibid., 29 (1950): 666–71.

9. F. N. Andrews and B. B. Bohren, "Influence of Thiouracil and Stilbestrol on Growth, Fattening, and Feed Efficiency in Broilers," *Poultry Science*, 26 (1947): 447–52, is the key transition. The earlier beef work includes F. N. Andrews and J. F. Bullard, "The Effect of Partial Thyroidectomy on the Fattening of Steers," *Proceedings of the American Society for Animal Production*, 1940, 112–16; J. F. Bullard and F. N. Andrews, "Experimental Alteration of Thyroid Function in Cattle," *Journal of the American Veterinary Medical Association* (hereafter cited as *JAVMA*), 102 (1943): 376; and W. M. Beeson, F. N. Andrews, and P. T. Brown, "The Effects of Thiouracil on the Growth and Fattening of Yearling Steers," *JAS*, 6 (1947): 16.

10. Destructive or debilitating effects included raised tailheads, restlessness and nervousness, pronounced mammary development, swayback conformation, engorged genitalia, recurrent mounting, severe prolapse of the anus, and similar prolapses of the vagina and uterus in immature ewes and heifers. See, for example, F. N. Andrews, W. M. Beeson, and Claude Harper, "The Effect of Stilbestrol and Testosterone on the Growth and Fattening of Lambs," *JAS*, 8 (1949): 578–82; W. E. Dinusson, F. N. Andrews, and W. M. Beeson, "The Effects of Stilbestrol, Testosterone, Thyroid Alteration, and Spaying on the Growth and Fattening of Beef Heifers," ibid., 9 (1950): 321–30; F. N. Andrews, W. M. Beeson, and W. E. Dinusson, "The Effect of Certain Hormone Treatments on the Growth and Fattening of Heifers," Purdue University Mimeo no. A.H. 37, Nov. 12, 1948; F. N. Andrews, W. M. Beeson, and F. D. Johnson, "Effect of Hormones on the Growth and Fattening of Yearling Steers," Purdue University Mimeo no. A.H. 46, Apr. 21, 1950; "The Use of Hormones in Meat Animals," Purdue University Mimeo no. A.H. 56, Dec. 18, 1950; and F. N. Andrews, W. M. Beeson, and F. D. Johnson, "Effect of Hormones on the Growth and Fattening of Yearling Steers," Purdue University Mimeo no. A.H. 60, Apr. 27, 1951, Burroughs Papers, box 3; and "Progress of Agricultural Research in Indiana," *Sixty-fourth Annual Report of the Director, Purdue University Agricultural Experiment Station, Lafayette, Indiana, for the Year Ending June 30, 1951*, 57–58. Also see Wise Burroughs to Timothy L. Tilton, Oct. 14, 1960, and Jan. 11, 1961; Tilton to Burroughs, Nov. 16, 1960; "Deposition of Frederick N. Andrews, June 5, 1959," 5–6, 8–9, 15–25, 33–40, 46–52, 92–95, in *Iowa State College Research Foundation v. John J. Vanier, O. Burr Ross, B. K. Smoot, the Western Star Mill Company, and Kansas Soya Products Company, Inc.*, Civil Action no. W-1220, U.S. District Court for the District of Kansas (hereafter cited as *ISCRF v. Vanier*); and "Deposition of Frederick N. Andrews, May 28, 1960," 9–12, 15–18, 26–27, and 31–33, in *Iowa State College Research Foundation v. Dawe's Laboratories, Inc.*, Civil Action no. 57c 2075, U.S. District Court for the Northern District of Illinois (hereafter cited as *ISCRF v. Dawe's*), Burroughs Papers, box 2. The Wise Burroughs Papers, housed in the Special Collections Department of the Parks Library, are uncatalogued and fill three Page boxes. Most of the collection is about several lawsuits that emerged from the stilbestrol patent. The testimony at these trials, and especially the pretrial depositions, are a gold mine of historical information. So too are the long letters Burroughs wrote Timothy L. Tilton, counsel for the Iowa State College Research Foundation. These materials constitute the bulk of the Burroughs collection.

11. "Deposition of Wise Burroughs, April 29, 1960," 6–10, in *ISCRF v. Dawe's*, Burroughs Papers, box 2; and "Testimony of Wise Burroughs, December 22, 1960," 127–28, 137–39, 246–48, in *ISCRF v. Dawe's*, Burroughs Papers, box 3. Burroughs was not the only investigator to attempt to confirm or extend the Purdue hormone work. See, for example, R. M. Jordan, "The Effect of Stilbestrol on Fattening Lambs," *JAS*, 9 (1950): 383–86; C. C. O'Mary, A. L. Pope, G. D. Wilson, R. W. Bray, and L. E. Casida, "The Effects of Diethylstilbestrol, Testosterone, and Progesterone on Growth and Fattening and Certain Carcass Characteristics of Western Lambs," ibid., 11 (1952): 656–73; M. T. Clegg and H. H. Cole, "The Action of Stilbestrol on the Growth Response in Ruminants," ibid., 13 (1954): 108–30; "Deposition of William H. Hale, March 12, 1959," 205–6, in *ISCRF v. Dawe's*, Burroughs Papers, box 2; J. H. Galloway, L. J. Bratzler, L. H. Blakeslee, and J. Meites, "Effects of Stilbestrol-Progesterone Implants on the Growth and Carcass Quality of Lambs," *Quarterly Bulletin of the Michigan Agricultural Experiment Station*, 35 (Aug.

1953): 68–74; "University of Minnesota Reports on Recent Lamb Feeding Tests," *Corn Belt Lamb Feeder*, 10 (Feb. 8, 1951): 8–11; and "Hormone Testosterone Boosts Cattle Gains," *Farm Journal*, 77 (Dec. 1953): 51. For the Australian work, see, for example, H. W. Bennetts, E. J. Underwood, and F. L. Shier, "A Specific Breeding Problem of Sheep on Subterranean Clover Pastures in Western Australia," *Australian Veterinary Journal*, 22 (1946): 10–11; D. H. Curnow, T. J. Robinson, and E. J. Underwood, "Oestrogenic Action of Extracts of Subterranean Clover (T. Subterraneum L. VAR. DWALGANUP): The Preparation and Assay of Extracts," *Australian Journal of Experimental Biology*, 27 (1949): 297–305; and A. B. Beck and A. W. Braden, "Studies on the Oestrogenic Substance in Subterranean Clover (Trifolium Subterraneum L. VAR. DWALGANUP)," *Australian Journal of Experimental Biology and Medical Science*, 29 (1951): 273–79. For some of the American studies, see F. C. Dohan, E. M. Richardson, R. C. Stribley, and P. Gyorgy, "The Estrogenic Effects of Extracts of Spring Rye Grass and Clover," *JAVMA*, 118 (1951): 323–24; and R. B. Bradbury and D. E. White, "The Chemistry of Subterranean Clover," *Journal of the Chemical Society*, pt. 4 (1951): 3447–49. Also see Wise Burroughs to Timothy L. Tilton, Apr. 6 and 24, 1959, Burroughs Papers, box 2.

12. "Iowa Lamb Feeders Meet at Ames," *Corn Belt Lamb Feeder*, 11 (Feb. 28, 1952): 1, 4; "Deposition of Wise Burroughs, December 8, 1958," 92–105, in *ISCRF* v. *Dawe's*, Burroughs Papers, box 2; "Deposition of Burroughs, April 29, 1960," 6–17; Edmund W. Cheng, Charles D. Story, C. C. Culbertson, and Wise Burroughs, "Detection of Estrogenic Substances in Alfalfa and Clover Hays Fed to Fattening Lambs," *JAS*, 11 (1952): 758; W. H. Hale, C. D. Story, C. C. Culbertson, and Wise Burroughs, "The Value of Low Levels of Stilbestrol in the Rations of Fattening Lambs," ibid., 12 (1953): 918; and Wise Burroughs to Timothy L. Tilton, Feb. 6, 1959, Burroughs Papers, box 2. Burroughs knew of plant estrogen studies as early as March 1951. See Ralph McCall to Burroughs, Mar. 1, 1951, Burroughs Papers, box 2.

13. C. C. Culbertson, C. W. MacDonald, Wise Burroughs, P. S. Shearer, and W. E. Hammond, "Different Amounts of Corn, Stilbestrol, and Rumen Inoculations for Fattening Lambs," Iowa State College Animal Husbandry Leaflet no. 181, Feb. 1952, Burroughs Papers, box 1; Wise Burroughs to Timothy L. Tilton, May 8, 13, and 26, 1959, Burroughs Papers, box 2; "Deposition of Hale, Mar. 12, 1959," 210, 212; John S. Evans, Roger F. Varney, and F. C. Koch, "The Mouse Uterine Weight Method for the Assay of Estrogens," *Endocrinology*, 28 (1941): 747–52; Betty L. Rubia, Adeline S. Dorfman, Lila Black, and Ralph I. Dorfman, "Bioassay of Estrogens Using the Mouse Uterine Response," ibid., 49 (1951): 429–39; "Deposition of Charles D. Story, March 12, 1959," 21–23, in *ISCRF* v. *Dawe's*, Burroughs Papers, box 2; "Deposition of Burroughs, December 8, 1958," 106–8; and Charles Dean Story, "Estrogenic Substances in Certain Livestock Feeds and Their Influence upon the Nutrition of Growing and Fattening Lambs" (Ph.D. diss. Iowa State Coll., 1954), 12–14.

14. Burroughs is quoted in Edmund Cheng, Charles D. Story, Loyal C. Payne, Lester Yoder, and Wise Burroughs, "Detection of Estrogenic Substances in Alfalfa and Clover Hays Fed to Fattening Lambs," *JAS*, 12 (1953): 513. Also see Edmund Cheng, Charles D. Story, Lester Yoder, W. H. Hale, and Wise Burroughs, "Estrogenic Activity of Isoflavone Derivatives Extracted and Prepared from Soybean Oil Meal," *Science*, 118 (1953): 164–65; Wise Burroughs to Timothy L. Tilton, Mar. 31 and Apr. 16, 1959, Burroughs Papers, box 2; "Deposition of Lester Yoder, December 8, 1958," 131–38, 142, in *ISCRF* v. *Dawe's*, Burroughs Papers, box 3; "Deposition of Story, March 12, 1959," 53–56; E. D. Walter, "Genistin (an Isoflavone Glucoside) and Its Aqulcone, Genistein, from Soybeans," *Journal of the American Chemical Society*, 63 (1941): 3273–76; and Edmund Cheng, Lester Yoder, Charles D. Story, and Wise Burroughs, "Estrogenic Activity of Some Isoflavone Derivatives," *Science*, 120 (Oct. 8, 1954): 575–76.

15. "Testimony of George Browning, December 21, 1960," 44–46, in *ISCRF* v. *Dawe's*, Burroughs Papers, box 3; and "Deposition of Burroughs, December 8, 1958," 197–200.

16. The quotations are from "Statement of Policy with Reference to Patents, Copyrights, and Trade-marks Effective at Iowa State College, Nov. 1, 1950," Burroughs Papers, box 3. Also see "Deposition of Quincy C. Ayres, December 8, 1958," 34–36, 39–40, in *ISCRF* v. *Dawe's*, Bur-

roughs Papers, box 3. For the Wisconsin Alumni Research Foundation, see Edward H. Beardsley, *Harry L. Russell and Agricultural Science in Wisconsin* (Madison: U of Wisconsin P, 1969), 155–71.

17. "Deposition of Ayres, December 8, 1958," 39–40; Quincy C. Ayres to Horace Dawson, Apr. 2, 1953, Burroughs Papers, box 2; and Ayres to Dawson, Apr. 28, 1953, Iowa State University Research Foundation Papers (hereafter cited as the ISURF Papers), Burroughs General Correspondence from Apr. 28, 1953, folder, Special Collections Department, Parks Library. Burroughs's patent application, no. 359,449, was entitled "Method of Treating Growing Beef Cattle and Sheep and Feed Materials for Use Therein."

18. The quotations are from "Packer Buyers Grade Cattle Fed Stilbestrol," *Ames Daily Tribune*, Feb. 19, 1954, 1. Also see "Low Quality Roughage Can Produce Choice Beef," *Iowa Farm Science*, 8 (Jan. 1954): 15–16; Wise Burroughs to Roswell Garst, Dec. 7, 1953, Garst Papers, box 23, folder 5; and Wise Burroughs, C. C. Culbertson, Joseph Kastelic, W. E. Hammond, and Edmund Cheng, "Hormone Feeding (Diethylstilbestrol) to Fattening Cattle II," Iowa State College Animal Husbandry Leaflet no. 189, Feb. 1954; and C. C. Culbertson, Wise Burroughs, Joseph Kastelic, and Roger Yoerger, "Using a 65-35 Corn and Cob Mixture for Fattening Cattle," Iowa State College Animal Husbandry Leaflet no. 190, Feb. 1954, Burroughs Papers, box 3.

19. Wise Burroughs to Ray N. Ammon, Jan. 28, 1954, ISURF Papers, Burroughs General Correspondence from Apr. 28, 1953, folder.

20. Wise Burroughs to Albert Guggedahl, Jan. 25, 1954; and Quincy Ayres to J. C. Decesare, Feb. 4, 1954, ISURF Papers, Burroughs General Correspondence from Apr. 28, 1953, folder; "Cattle Nutrition to Be Prime Topic of Midwest Program," *Feedstuffs*, 26 (Feb. 6, 1954): 1; J. S. Russell, "Will Talk on Hormones in Cattle Feed," *Des Moines Register*, Feb. 18, 1954, 5; and Wise Burroughs, "Special Cattle Feeders' Day," ISURF Papers, Burroughs General Correspondence from Apr. 28, 1953, folder; This last item was a memorandum sent to the Iowa Feed Institute and the Western Grain and Feed Dealers' Association on January 29, 1954, describing the DES work.

21. "Testimony of Browning, December 21, 1960," 47–49; J. C. Petersen, "Pete's Bleats," *Corn Belt Lamb Feeder*, 13 (Mar. 8, 1954): 1–4; Jim Evans, "Is Stilbestrol the Answer?" *Iowa Agriculturist*, 52 (Mar. 1954): 10–12; "Iowa Cattle Feeders' Day," *Flour and Feed*, 54 (Mar. 1954): 24; 54 (April 1954): 21; and "Feeders' Day Is Biggest Ever," *Iowa State Daily*, Feb. 19, 1954, 1.

22. The quotations are from Iowa State College Information Service, "Stilbestrol Experiments Promise Faster, Cheaper Beef Production"; idem, "Hormone Feeding Speeds Up Gains in Beef Steers; Reduces Costs"; idem, "Steers Gain Two Pounds Daily on Stalk Silage with 'Hormone'"; idem, "New Feeding Research Means More Beef at Lower Cost for Consumer in Long Run"; and idem, "Packer Buyers Place Top Grades and Top Price on Stilbestrol-Fed Steers," ISURF Papers, Burroughs General Correspondence from Apr. 28, 1953, folder.

23. See, for example, "Can Hormones Increase Cattle Gains?" *Shorthorn World*, 39 (Apr. 1, 1954): 11; "Hormone Steps Up Steer Gains, Midwest Feed Manufacturers Told," *Feedstuffs*, 26 (Feb. 27, 1954): 6; "Putting Feed Research to Work," ibid., 26 (Mar. 6, 1954): 62, 64, 66; Wise Burroughs, "New Developments in Beef Cattle Nutrition," *Proceedings of the Distillers' Feed Conference*, 1955: 64–67; "Hormones—New Beef-Gain Booster," *Successful Farming*, 52 (Apr. 1954): 15; "Faster, Cheaper Beef Gain," *Nebraska Farmer*, 96 (Mar. 6, 1954): 22; "'Dynamite' Drug Boosts Gain," *Wallaces Farmer and Iowa Homestead*, 79 (Mar. 6, 1954): 32; J. S. Russell, "Hopes High on Hormone in Beef Feed," *Des Moines Register*, Feb. 19, 1954, 11; Roger Blobaum, "Synthetic Hormone May Revolutionize Stock Feeding," *Ames Daily Tribune*, Feb. 18, 1954, 1, 4; and Wise Burroughs, "Beef Supplements Containing Diethylstilbestrol Hormone," talk delivered to Midwest Feed Manufacturers' Association, Feb. 19, 1954, Burroughs Papers, box 3. Plans were made for a special radio broadcast about stilbestrol one month prior to the February 1954 announcement. See R. M. McWilliams to Wise Burroughs and Friends, Jan. 19, 1954, ISURF Papers, Burroughs General Correspondence from Apr. 28, 1953, folder. For a sample of the letters sent to Burroughs, see Guy Cooper, Jr., to Burroughs, Feb. 23, 1954; and H. H. Draper to Burroughs, Feb. 25, 1954, ISURF Papers, Burroughs General Correspondence from Feb. 1, 1954, folder. For

announcements in scientific or technical journals, see "Sex Hormone in Feed Fattens Steers Faster," *Science News Letter*, July 17, 1954, 39; Wise Burroughs, C. C. Culbertson, Joseph Kastelic, Edmund Cheng, and W. H. Hale, "The Effects of Trace Amounts of Diethylstilbestrol in Rations of Fattening Steers," *Science*, 120 (July 9, 1954): 66–67; and Wise Burroughs, C. C. Culbertson, Edmund Cheng, W. H. Hale, and Paul Homeyer, "The Influence of Oral Administration of Diethylstilbestrol to Beef Cattle," *JAS*, 14 (1955): 1015–24.

24. See, for example, P. S. Jordan, "What a Pill . . . Stilbestrol," *Feed Age*, 2 (Nov. 1952): 22–25; M. E. Coates, "The Mode of Action of Antibiotics," ibid., 2 (Nov. 1952): 36–37, 58; Wise Burroughs, "Using Urea Effectively in Cattle Supplements," ibid., 3 (Aug. 1953): 20–23; Sidney W. Fox, "What Are Antibiotics?" *Iowa Farm Science*, 7 (Oct. 1952): 3–4; "Urea," *Farm Quarterly*, 8 (Spring 1953): 57, 94, 96, 98, 126; Daryl F. Visser, "Feeding Buys Which Produce High Quality Beef," *Iowa Agriculturist*, 50 (Feb. 1952): 10–11, 16; "Wonder Drug Stops Enteritis," *Successful Farming*, 48 (June 1950): 125; A. V. Nalbandov, "Hormones," *Successful Farming*, 50 (Nov. 1952): 31, 58–59; "Low-Cost Roughage Cuts Beef Feed Cost," *Nebraska Farmer*, 94 (Aug. 15, 1953): 38; "Ton of Cobs Feeds Steer Four Months," *Wallaces Farmer and Iowa Homestead*, 74 (May 21, 1949): 24–25; "He's Making the Corncob Famous," *Des Moines Register*, Mar. 2, 1952, 7-F, 8-F; Earl C. Richardson, "Research Shows Hormones Help Lambs Grow Faster," *Farm Journal*, 77 (Jan. 1953): 40; J. F. Sykes, "Hormones for Farm Animals," *Flour and Feed*, 54 (Aug. 1953): 30; and Charlie F. Petersen, "Supplying APF in Poultry Rations," *Feedstuffs*, 22 (Mar. 11, 1950): 22.

25. For histories of these means of transmitting information to the agricultural public, see Roy V. Scott, *The Reluctant Farmer: The Rise of Agricultural Extension to 1914* (Urbana: U of Illinois P, 1971).

26. The *Proceedings of the Cornell Nutrition Conference for Feed Manufacturers* and the *Proceedings of the Maryland Nutrition Conference for Feed Manufacturers* contain much useful information on these points. Also see, for example, "Nutritional Symposiums Feature Midwest Feed Men's Convention," *Feedstuffs*, 22 (Feb. 18, 1950): 1, 4; J. S. Hughes, "Hormones in Animal Production," ibid., 23 (Feb. 24, 1951): 30, 32; Tom Brinegar, "Concerted Effort of All Poultry and Livestock People Urged at Purdue Conference," *Feed Age*, 2 (May 1952): 33; "Outlook . . . Effects of Drouth and Dramatic Nutrition Development Reviewed at Midwest Meeting," ibid., 4 (Apr. 1954): 39–40; "Glennon Discusses Industry Status and Problems at Michigan Meeting," ibid., 4 (Dec. 1954): 48; J. T. Reid, "Protein Replacements for Ruminants," *Flour and Feed*, 55 (June 1954): 14–16; W. M. Beeson, "Convert High Cellulose Feed for Top Beef Production," ibid., 55 (Dec. 1954): 17–18; Earle W. Klosterman, V. R. Cahill, and L. E. Kunkle, "The Influence of Sex Hormones upon Feed Lot Performance and Carcass Quality of Fattening Cattle," *Proceedings of the Cornell Nutrition Conference for Feed Manufacturers*, 1955, 60–66; and Orville G. Bentley and Earle W. Klosterman, "Recent Developments in Studies on Growth Factors for Rumen Microorganisms," *Proceedings of the Distillers' Feed Conference*, 1956, 11–13.

27. The American Feed Manufacturers' Nutrition Council, formed in 1945, is of particular relevance for this study. See its *Proceedings*, May 1954, 6; ibid., Nov. 1954, 4–6; ibid., May 1956, 7–8; ibid., Nov. 1956, 7; and ibid., May 1958, 6–7. Sometimes college scientists were invited to address special symposia. See, for example, Wise Burroughs, "Principles and Theory of Rumen Function and Rumen Nutrition," paper delivered as part of a Symposium on Rumen Function and Ruminant Nutrition, and published in *Proceedings of the Nutrition Council of the American Feed Manufacturers' Association*, Nov. 1954, 14–18. For the Agricultural Research Institute and its ambitions, see "Agricultural Research Institute," *Proceedings of the Agricultural Research Institute*, 1 (1952): 1. The quotations in the text are from that article. Also see C. L. Cheim, "Industry's Stake in the Agricultural Research Institute," ibid., 3 (1954): 11–15; Willard M. Field, "Role of Federal and State Institutions in the Agricultural Research Institute," ibid., 3 (1954): 16–20; E. C. Stakman, "The National Science Foundation in Relation to the National Research Council and the Agricultural Research Institute," ibid., 3 (1954): 59–61; and E. C. Elting, "Panel Discussion on Industry-Government Relations in Agricultural Research," ibid., 3 (1954): 66–85.

28. Harvey E. Yantla, "Another Milestone," *Feedstuffs*, 26 (Feb. 27, 1954): 2.

29. See, for example, Charles F. Geschickter, "Mammary Carcinoma in the Rat with Metastasis Induced by Estrogen," *Science*, 89 (Jan. 13, 1939): 35–37; and Michael B. Shimkin and Hugh Grady, "Mammary Carcinomas in Mice following Oral Administration of Stilbestrol," *Proceedings of the Society for Experimental Biology and Medicine*, 45 (Oct. 1940): 246–48.

30. For the changes in federal law and FDA policy in the years immediately prior to the discovery of stilbestrol's usefulness as a cattle feed additive, see *Federal Register*, 13 (Dec. 4, 1948): 7403–4. Also see Erwin E. Nelson, "New Drug Requirements of the Federal Food, Drug, and Cosmetic Act," *Food Drug Cosmetic Law Quarterly* (hereafter cited as *FDCLQ*), 4 (1949): 227–32; Arnold J. Lehman, Edwin P. Laug, Geoffrey Woodard, John H. Draize, O. Garth Fitzhugh, and Arthur A. Nelson, "Procedures for the Appraisal of the Toxicity of Chemicals in Foods," *FDCLQ*, 4 (1949): 412–34; Bryant W. Brennan, "Proof Required to Establish the Safety of a Chemical Additive for Use in Food," *Food Drug Cosmetic Law Journal* (hereafter cited as *FDCLJ*), 8 (1953): 561–69; "Report of the Food and Drug Administration," in *Annual Report of the Federal Security Agency*, 1950, 19; and "Report of the Food and Drug Administration," in *Annual Report of the Department of Health, Education, and Welfare*, 1954, 216. The *FDCLQ* became the *FDCLJ* in 1950. For the proliferation of chemical agents, see Hugh D. Crone, *Chemicals and Society: A Guide to the New Chemical Age* (Cambridge: Cambridge UP, 1986); and Thomas R. Dunlap, *DDT: Scientists, Citizens, and Public Policy* (Princeton: Princeton UP, 1981). The FDA's approval of a NADA constituted permission to use the drug in the manner specified in the NADA, but this permission was revokable: the agency could withdraw its approval of a NADA.

31. For the Delaney committee, see, for example, "Use of Chemicals with Food Products," *FDCLJ*, 5 (1950): 276, 373, 452–53, 798–99; "Chemicals in Food Products," ibid., 6 (1951): 4; Charles Wesley Dunn, "Statement before the House Select Committee to Investigate the Use of Chemicals in Food Products," ibid., 6 (1951): 72–78; Vincent A. Kleinfeld, "Congress Investigates Chemicals in Foods," ibid., 6 (1951): 120–28; and "Report of House Select Committee to Investigate the Use of Chemicals in Foods," ibid., 6 (1951): 149–58. Stilbestrol-related testimony and discussions before the committee can be found in the transcript of the Delaney hearing, published as *Chemicals in Food Products* (Washington: GPO, 1951), 47–50, 59–60, 85–86, 740–44, 811–29. Few historians have examined the hearings; however, see Phyllis Anderson Meyer, "The Last Per Se: The Delaney Cancer Clause in United States Food Regulation" (Ph.D. diss., U of Wisconsin, 1983); and Suzanne White, "Chemicals in Food: The Delaney Committee, 1950–52" (Ph.D. diss., Emory U, in progress), are exceptions.

32. For the history of the initiatives outlined in this section, see Charles Wesley Dunn, "The Food, Drug, and Cosmetic Law in the United States," *FDCLQ*, 3 (1948): 308–31; and idem, "The Food Law Institute," ibid., 4 (1949): 471–77. For some of the positions espoused regarding chemicals in food, see, for instance, W. B. White, "The Addition of Chemicals to Foods," ibid., 2 (1947): 475–89; Paul B. Dunbar, "Chemicals Used in Manufactured Foods," ibid., 4 (1949): 296–303; J. L. St. John, "Chemical Additives," ibid., 6 (1951): 269–81; Vincent A. Kleinfeld, "Is There a 'Chemical in Food' Problem," ibid., 6 (1951): 824–39; C. W. Crawford, "The Food and Drug Administrations's Viewpoint on Chemicals in Food," ibid., 7 (1952): 85–95; Bernard L. Oser, "The Interdependence of Science and Law," *Journal of Agricultural and Food Chemistry*, 2 (1954): 118–21; and Fred Bartenstein, "The Role of Law in Food Safety," ibid., 2 (1954): 122–24.

33. The quotations are from Emanuel Kaplan, "A Review of Chemicals in Food," *Quarterly Bulletin of the Association of Food and Drug Officials*, 17 (July 1953): 109. Also see "Urge More Industry Study of Food Additive Problem," *Journal of Agricultural and Food Chemistry*, 1 (1953): 546–48; and National Research Council, Food Protection Committee, *Use of Chemical Additives in Foods* (Washington, D.C.: National Research Council, Food and Nutrition Board; and National Academy of Sciences, 1951); and idem, *Safe Use of Chemical Additives in Foods* (Washington, D.C.: National Research Council, Food and Nutrition Board; and National Academy of Sciences, 1952).

34. For congressional action, see, for example, John J. Powers, Jr., "Some Aspects of Certification of Antibiotics under the Federal Food, Drug, and Cosmetic Act," *FDCLQ*, 4 (1949): 337–52;

"In Congress," *FDCLJ*, 8 (1953): 67–68, 126–28; R. C. Newton, "The Chemical-Additive Amendment to the 1938 Federal Food, Drug, and Cosmetic Act," ibid., 9 (1954): 587–93; Fredus N. Peters, "The O'Hara Bill (H.R. 9166)—Its Effect on Food Research," ibid., 9 (1954): 594–602; "Chemicals in Foods," *Chemical and Engineering News*, 31 (1953): 639; "Feed Industry Involved in Pending Legislation on Chemical Additives," *Feedstuffs*, 25 (Feb. 28, 1953): 4; and "New Philosophy in Administration of Food and Drug Laws Involved in Miller Pesticide Bill," *Journal of Agricultural and Food Chemistry*, 1 (1953): 601. For Harvey's statements, see John L. Harvey to Frederick N. Andrews, Oct. 23, 1953; and Harvey to Wise Burroughs, Jan. 21, 1954, U.S. Food and Drug Administration Papers (hereafter cited as FDA Papers), Accession no. 88-82-57, box 21, vol. 280, Washington National Record Center, Suitland, Md., For suggestions about the FDA, see, for instance, H. H. Draper to Burroughs, Mar. 10, 1954, ISURF Papers, Burroughs General Correspondence from Feb. 1, 1954, folder. For some further indications of public concern, see, for instance, "Report of the Food and Drug Administration," in *Annual Report of the Federal Security Agency*, 1951, 2–3, 5, 16–17; and "Report of the Food and Drug Administration," in *Annual Report of the Federal Security Agency*, 1952, 2, 7, 21. State and local officials also were wary about chemicals in foods. See, for instance, Charles W. Crawford, "Some Current Problems Facing Food and Drug Officials," *Quarterly Bulletin of the Association of Food and Drug Officials*, 1 (July 1953): 89–97.

35. Wise Burroughs, C. C. Culbertson, R. M. McWilliams, Joseph Kastelic, and H. W. Reber, "Hormone Feeding (Diethylstilbestrol) to Fattening Heifers," Iowa State College Animal Husbandry Leaflet no. 192, Burroughs Papers, box 1; Wise Burroughs, C. C. Culbertson, Glen Hall, Kenneth Barnes, Joseph Kastelic, and W. E. Hammond, "Cornstalk Silage Fed with Different Cattle Supplements Followed by a High-Corn Finishing Ration for Yearling Steers," Iowa State College Animal Husbandry Leaflet no. 193; and C. C. Culbertson, Wise Burroughs, Joseph Kastelic, W. E. Hammond, and Edmund Cheng, "Different Feeding Levels of Diethylstilbestrol for Fattening Steers," Iowa State College Animal Husbandry Leaflet no. 194, Burroughs Papers, box 3; Wise Burroughs, C. C. Culbertson, Joseph Kastelic, Bruce Taylor, and William H. Hale, "Ground Ear Corn Full-Fed Using Iowa Corncob Supplement and No Hay versus Iowa Supplement Four with Limited Hay and Barley or Malt Additions," Iowa State College Animal Husbandry Leaflet no. 195; Wise Burroughs, W. H. Hale, Bruce Taylor, R. M. McWilliams, J. M. Scholl, Kenneth Barnes, and Robert Zimmerman, "Antibiotic vs. Hormone Additions to Cattle Supplements Fed in Conjunction with High-Roughage Wintering Rations," Iowa State College Farm Service Report no. 107, Burroughs Papers, box 1; Iowa State College Information Service, "Cattle Gains Cost Seventeen Cents a Pound on Cornstalk Silage with Stilbestrol"; and idem, "Experiments Indicate Best Cattle-Feeding Level for Stilbestrol Is Ten Milligrams," ISURF Papers, Burroughs General Correspondence from Feb. 1, 1954, folder; "Winter Gains for Seventeen Cents a Pound," *Nebraska Farmer*, 96 (Sept. 4, 1954): 16; "Iowa State Reports Further Evidence of Benefits from Feeding Hormone to Cattle," *Feedstuffs*, 26 (July 24, 1954): 59; and Joseph Kastelic to Duane Acker, Apr. 4, 1954; Wise Burroughs to A. L. McBride, May 10, 1954; and McBride to Burroughs, May 11, 1954, ISURF Papers, Burroughs General Correspondence from Feb. 1, 1954, folder.

36. The quotation is from John C. Beasley to Wise Burroughs, Mar. 30, 1954, ISURF Papers, Burroughs General Correspondence from Feb. 1, 1954, folder. Also see, for example, H. H. Draper to Burroughs, Feb. 25, 1954; and L. R. Rubin to Iowa State College Research Foundation, Feb. 26, 1954, ISURF Papers, Burroughs General Correspondence from Apr. 28, 1953, folder; R. W. Bennison to Burroughs, Mar. 1, 1954; Curtis R. Gray to Quincy Ayres, Mar. 1, 1954; Frank Mangelsdorf to Iowa State College Research Foundation, Mar. 3, 1954; E. E. McInroy to Iowa State College Research Foundation, Mar. 8, 1954; D. D. Martin to Burroughs, Mar. 9, 1954; F. J. Bouska to Burroughs, Mar. 13, 1954; R. E. Ahlin to Burroughs, Mar. 16, 1954; Ralph Kail to Burroughs, Apr. 1, 1954; and J. L. Gabby to Burroughs, May 4, 1954, ISURF Papers, Burroughs General Correspondence from Feb. 1, 1954, folder; and D. F. Green to Ayres, Aug. 3, 1954; and P. C. Hereld to Burroughs, Aug. 18, 1954, ISURF Papers, Burroughs General Correspondence from Aug. 3, 1954, folder. Also see H. H. Cortelyou to Burroughs, Apr. 2, 1954, ISURF Papers, Burroughs General Correspondence from Feb. 1, 1954, folder; R. M. Robinson to Burroughs,

Aug. 18, 1954, ISURF Papers, Burroughs General Correspondence from Aug. 3, 1954, folder; J. C. Decesare to Ayres, Feb. 4, 1954, ISURF Papers, Burroughs General Correspondence from Apr. 28, 1953, folder; Decesare to Ayres, Mar. 4 and 18, 1954; J. J. Feldmann to Ayres, Apr. 21 and May 26 and 28, 1954; and Beasley to Burroughs, Mar. 12, 1954, ISURF Papers, Burroughs General Correspondence from Feb. 1, 1954, folder.

37. Quincy C. Ayres to J. J. Feldman, May 24, 1954; and Wise Burroughs to Timothy Tilton, May 5, 1954, ISURF Papers, Burroughs General Correspondence from Feb. 1, 1954, folder; Tilton to Ayres, Apr. 30, 1954, ISURF Papers, Burroughs General Correspondence from Dec. 28, 1954, folder; and Wise Burroughs, "Diethylstilbestrol as a Cattle Feed Additive: A Technical Report Submitted for Use of Food and Drug Administration, Washington, D.C., June 25, 1954," 106–25, Burroughs Papers, box 2. For Lilly's thinking on this issue, see E. J. Kahn, Jr., *All in a Century: The First One Hundred Years of Eli Lilly and Company* (N.p., 1975), 172–74.

38. "Report concerning Diethylstilbestrol Patent, June 7, 1954," ISURF Papers, Burroughs General Correspondence from Feb. 1, 1954, folder; Timothy T. Tilton to Quincy C. Ayres, June 14, 1954, ISURF Papers, Burroughs General Correspondence from Dec. 28, 1954, folder; "Minutes— Conference concerning Procedures with Reference to Stilbestrol Patent, July 20, 1954"; and Iowa State College Information Service, "College Licenses Stilbestrol Premix Manufacture, July 29, 1954," ISURF Papers, Burroughs General Correspondence from Feb. 1, 1954, folder; and "License Stilbestrol Premix Manufacture," *Wallaces Farmer and Iowa Homestead*, 79 (Aug. 21, 1954): 98.

39. Burroughs, "Diethylstilbestrol as a Cattle Feed Additive," especially 98–105. Also see Gray H. Twombly and Erwin F. Schoenwaldt, "Tissue Localization and Excretion Routes of Radioactive Diethylstilbestrol," *Cancer Journal*, 4 (1951): 296–302; Oscar Jones and F. F. Deatherage, "The Diethylstilbestrol Content of the Meat from Chickens Treated with This Estrogenic Substance," *Food Research*, 18 (1953): 30–34; and Martin Stob, F. N. Andrews, and M. X. Zarrow, "The Detection of Residual Hormone in the Meat of Animals Treated with Synthetic Estrogens," Purdue University Agricultural Experiment Station Journal Paper no. 740, Burroughs Papers, box 2.

40. Harvey expressed his sentiments in John L. Harvey to Wise Burroughs, Aug. 25, 1954, ISURF Papers, Burroughs General Correspondence from Aug. 3, 1954, folder. Also see Ernest J. Umberger to Harvey, July 27, 1954, FDA Papers, Accession no. 88-82-57, box 21, vol. 280; "Early Use of Stilbestrol in Feed Sought," *Feedstuffs*, 26 (July 31, 1954): 46; and "Cattle Feeders Day Held at Ames," *Corn Belt Lamb Feeder*, 13 (July 15, 1954): 6.

41. "Memorandum of Interview re: Diethylstilbestrol, September 16, 1954," FDA Papers, Accession no. 88-82-57, box 21, vol. 280.

42. The FDA position is spelled out in Jack M. Curtis to J. Hauser, Oct. 28, 1954, FDA Papers, Accession no. 88-82-57, box 21, vol. 280. Also see Wise Burroughs, "Diethylstilbestrol as a Cattle Feed Additive: Supplementary Information for Use of Food and Drug Administration, October 9, 1954"; and Paul G. Homeyer, "Statistical Procedures of Estimation and Validity in Assay of Stilbestrol in Tissues from Cattle Fed Stilbestrol: Supplementary Information for Use of Food and Drug Administration, October 9, 1954," Burroughs Papers, box 2; Iowa State College Information Service, "Stilbestrol Cattle Premix Gets 'Green Light,' November 5, 1954," ISURF Papers, Burroughs General Correspondence from Aug. 3, 1954, folder; John Cipperly, "FDA Authorizes Sale of Stilbestrol Pre-Mix for Use in Cattle Feeds," *Feedstuffs*, 26 (Nov. 6, 1954): 1, 8; John A. Rohlf, "Stilbestrol OK'd for Fattening Beef," *Farm Journal*, 77 (Dec. 1954): 34–35; "FDA OKs Hormone Mixes for Cattle, Poultry Feeds," *Feed Age*, 4 (Dec. 1954): 32–33; "Cattle Gains Are Faster, Cheaper," *Wallaces Farmer and Iowa Homestead*, 79 (Nov. 20, 1954): 29; A. V. Nalbandov, "Stilbestrol," *Successful Farming*, 53 (Dec. 1954): 23. Andrews kept abreast of FDA actions. See Ralph F. Kneeland, Jr., to F. N. Andrews, Nov. 3 and 18, 1954; and Andrews to Kneeland, Oct. 18 and Nov. 10, 1954, FDA Papers, Accession no. 88-82-57, box 21, vol. 280.

43. For Lilly's efforts in tabular form, see "Report to the Feed Industry," *Feed Age*, 6 (Jan. 1956): 74. Also see "Plans Proceed for Sale, Use of Stilbestrol Feed Pre-Mix," *Feedstuffs*, 26 (Nov. 13, 1954): 6; "Stilbestrol Feeds Hit Market," ibid., 26 (Dec. 18, 1954): 1; "Stilbosol," *Feed Age*, 4

(Dec. 1954): 49; "Lilly Markets Hormone Premix," *Flour and Feed*, 55 (Dec. 1954): 23; "Newsletter," *Chemical Week*, 75 (Dec. 4, 1954): 15; "Lilly in Feed Field," *Flour and Feed*, 56 (Jan. 1955): 19; "Lilly Expands Agricultural Products Unit," *Feedstuffs*, 27 (Jan. 1, 1955): 34–35; "Eli Lilly to Expand Ag Products Division," *Journal of Agricultural and Food Chemistry*, 3 (1955): 20; "Cattle on Feed Near Record," *Wallaces Farmer and Iowa Homestead*, 80 (Feb. 5, 1955): 16; and "Feed Steers Gain-Boosters," ibid., 80 (Feb. 19, 1955): 12.

44. "New Beef Gain Booster—Stilbosol," advertisement, *Flour and Feed*, 55 (Dec. 1954): 13–16; "Increase Beef Gains 15% to 33%!" advertisement, *Nebraska Farmer*, 96 (Dec. 18, 1954): 3; and "Stilbosol Digest," advertisement, ibid., 97 (Apr. 2, 1955): 48.

45. Journalists expressed their excitement over DES in John A. Rohlf, "Two Million Head on Stilbestrol!" *Farm Journal*, 78 (Mar. 1955): 38, 165; and Tom Leadley, "Science in the Feedlot," *Nebraska Farmer*, 96 (Dec. 18, 1954): 6. Also see, for instance, John A. Rohlf, "Stilbestrol OK'd for Fattening Beef," *Farm Journal*, 77 (Dec. 1954): 34–35; "F and DA OKs Hormone Mixes for Cattle, Poultry Feeds," *Feed Age*, 4 (Dec. 1954): 32–33; "Stilbestrol Fed Cattle Means More Beef for Less," *Nebraska Farmer*, 96 (Dec. 18, 1954): 24; Glenn Cunningham, "What about Stilbestrol?" *Des Moines Register*, Jan. 2, 1955, 5-H; and Dick Hanson, "Answers to Your Questions on Feeding Stilbestrol," *Successful Farming*, 53 (Feb. 1955): 59, 142–43.

46. See, for example, W. S. Wilkinson, C. C. O'Mary, G. D. Wilson, R. W. Bray, A. L. Pope, and L. E. Casida, "The Effect of Diethylstilbestrol upon Growth, Fattening, and Certain Carcass Characteristics of Full-Fed and Limited-Fed Western Lambs," *JAS*, 14 (1955): 866–77; Earle W. Klosterman, V. R. Cahill, L. E. Kunkle, and A. L. Moxon, "The Subcutaneous Implantation of Stilbestrol in Fattening Bulls and Steers," ibid., 14 (1955): 1050–58; E. S. Erwin, I. A. Dyer, and M. E. Ensminger, "Digestibility of Steer Fattening Rations as Affected by Quality of Roughage, Fat, Chlortetracycline, and Stilbestrol," ibid., 14 (1955): 1201–2; G. E. Mitchell, W. W. Albert, D. L. Staheli, and A. L. Neumann, "Performance of Steers with or without Stilbestrol, When Fed to a Constant Weight and When Fed Equal Amounts of Total Concentrates," ibid., 14 (1955): 1218; Marion Simone, Floyd Carroll, Elly Hinreiner, and M. T. Clegg, "Effect of Corn, Barley, Stilbestrol, and Degree of Finish upon Quality of Beef," *Food Research*, 20 (1955): 512–29; Milton A. Madsen, "Stilbestrol Implants Produce Harmful Aftereffects," *Utah Farm and Home Science*, 16 (Dec. 1955): 75, 83–84; "Reactions to Feeding Stilbestrol," *JAVMA*, 127 (1955): 253; A. B. Schultze, "Stilbestrol in Dairy Cattle," *Nebraska Farmer*, 97 (May 21, 1955): 66; "On the Research Front," *Shorthorn World*, 40 (Aug. 1, 1955): 10; "Dairy Bull Calves on Stilbestrol Gain Less than Controls in Vermont Study," *Feedstuffs*, 27 (Jan. 22, 1955): 6; "Stilbestrol Cattle Feeding Experiments at Agricultural Colleges Summarized," ibid., 27 (Oct. 15, 1955): 58; "Illinois Researchers Report on Results of Stilbestrol, Other Cattle Experiments," ibid., 27 (Dec. 10, 1955): 7; "College Roundup," *Flour and Feed*, 56 (Nov. 1955): 6–7; Homer Fine, "Spring Feeder's Day," *Nebraska Farmer*, 97 (May 21, 1955): 10; "More Information on Stilbestrol," ibid., 97 (Oct. 15, 1955): 76; and P. T. Marion, C. E. Fisher, and J. H. Jones, "Diethylstilbestrol for Fattening Yearling Steers," *Cattleman*, 41 (Jan. 1955): 133–34.

47. "Latest Feeding Results," *Farm Journal*, 79 (June 1955): 14; "Quaker Tests Stilbestrol," *Flour and Feed*, 56 (June 1955): 5; Jerome Foster, "Antibiotic Ups Stilbestrol Gain," ibid., 56 (July 1955): 18; "USDA Research Confirms Value of Stilbestrol Feeding," *Feedstuffs*, 27 (Oct. 1, 1955): 6; George Gates and Roger Berglund, "Feed Industry Experience with Stilbestrol 'Very Satisfactory,'" ibid., 27 (Oct. 1, 1955): 12–13, 66, 69, 72; "Stilbestrol Gets Favorable Decision," *Journal of Agricultural and Food Chemistry*, 3 (1955): 809; "USDA Approves Stilbestrol Feeding," *JAVMA*, 127 (1955): 456; "Feeding Stilbestrol," *Agricultural Research*, 4 (Nov. 1955): 16; and "Stilbestrol for Beef-Cattle Feeding," *Feed Age*, 5 (Dec. 1955): 76.

CHAPTER 2. CRACKS IN THE FACADE

1. T. W. Perry, W. M. Beeson, F. N. Andrews, and Martin Stob, "The Effect of Oral Administration of Hormones on Growth Rate and Deposition in the Carcass of Fattening Steers," *JAS*, 14 (1955): 329–35; T. W. Perry, W. M. Beeson, F. N. Andrews, and Martin Stob, "The Effect of Oral

Administration of Hormones on Growth Rate and Deposition in the Carcass of Fattening Steers," *Feedstuffs*, 27 (Jan. 1, 1955): 12–13, 16; "The Latest on Hormone Feeding," *Shorthorn World*, 40 (May 1, 1955): 9; and "Feeders' Day," *Nebraska Farmer*, 97 (Apr. 2, 1955): 77.

2. The expressions of manufacturers' fears are from C. B. Knodt to Wise Burroughs, Mar. 18, 1955, Burroughs Papers, box 2; and H. H. Cortelyou to Burroughs, Jan. 14, 1955, ISURF Papers, Burroughs General Correspondence from Dec. 28, 1954, folder. Also see Wilfred M. Witz to Burroughs, Nov. 4, 1954; and Burroughs to Witz, Nov. 19, 1954, ISURF Papers, Burroughs General Correspondence from Aug. 3, 1954, folder; and H. Ernest Bechtel to Burroughs, Mar. 10, 1955; Burroughs to Knodt, Mar. 28, 1955; and Knodt to Burroughs, Apr. 6, 1955, Burroughs Papers, box 2. Rumors continued in later 1955. See, for example, "Stilbestrol-Fed Cattle: How They're Selling Now," *Farm Journal*, 79 (Aug. 1955): 16; Chester Charles, "Stilbestrol," *Farm Quarterly*, 10 (Summer 1955): 48–49, 114–17; and "Successful Farming Report on Feed Stilbestrol," *Successful Farming*, 53 (July 1955): 36–37.

3. The quotation is from Leslie Johnson to Various Meat Packing Company Executives, c. Apr. 2, 1955, ISURF Papers, Burroughs General Correspondence from Feb. 1, 1954, folder. Also see Joseph Kastelic to Wise Burroughs, William Hale, C. Culbertson, and E. A. Kline, Jan. 6, 1955; Burroughs to Kastelic, Jan. 12, 1955; Kastelic to Burroughs, Jan. 17, 1955; Kastelic to L. W. McElroy, Jan. 17, 1955; and "Tentative Program—Meat Conference concerning Stilbestrol-Fed Cattle," ISURF Papers, Burroughs General Correspondence from Dec. 28, 1954, folder. Burroughs to H. Ernest Bechtel, May 4, 1955; and Bechtel to Burroughs, Apr. 26, 1955, Burroughs Papers, box 2; "Stilbestrol Continues to Look Good," *Farm Journal*, 79 (Apr. 1955): 15; Joseph Kastelic and Associates, "Beef Carcass Quality Unchanged When Stilbestrol Is Fed," *Flour and Feed*, 56 (June 1955): 26; "Stilbestrol Carcass Studies," *Nebraska Farmer*, 97 (Apr. 16, 1955): 72; "Effect of Stilbestrol on Cattle Is Small," *Cattleman*, 41 (May 1955): 148, 150; "Carcass Value Same with Stilbestrol," *Successful Farming*, 53 (June 1955): 4; Marion Simone, Floyd Carroll, Elly Hinreiner, and M. T. Clegg, "Effect of Corn, Barley, Stilbestrol, and Degree of Finish upon Quality of Beef," *Food Research*, 20 (1955): 512–29; U.S. Department of Agriculture, "Value of Stilbestrol for Beef-Cattle Feeding Confirmed by USDA Research, September 21, 1955," press release, ISURF Papers, Burroughs General Correspondence from Dec. 28, 1954, folder; "USDA Research Confirms Value of Stilbestrol Feeding," *Feedstuffs*, 27 (Oct. 1, 1955): 6; "Hormone Feeding Aspects Reviewed by USDA Scientist," ibid., 27 (Nov. 26, 1955): 25; and "Feeding Stilbestrol," *Agricultural Research*, 4 (Nov. 1955): 16.

4. Quincy Ayres to Horace Dawson, Dec. 10, 1954, ISURF Papers, Burroughs General Correspondence from Aug. 3, 1954, folder; Timothy Tilton to George A. Ordway, June 3 and 23, 1955; Tilton to Ayres, June 3, 1955; and Ordway to Tilton, June 16, 1955, ISURF Papers, Burroughs General Correspondence from Dec. 28, 1954, folder; and Ralph F. Kneeland, Jr., to F. N. Andrews, Nov. 3 and 18, 1954; and "Memorandum of Interviews—Use of Stilbestrol as Growth Promoters in Domestic Animals, May 18, 1955, and May 19, 1955," FDA Papers, Accession no. 88-82-57, box 21, vol. 280.

5. "Office Memorandum—FDA-NDA 9757 and 9770—Stilbestrol Premix for Steer Feeding, June 2, 1955," FDA Papers, Accession no. 88-82-57, box 21, vol. 280; and Quincy Ayres to Members of Iowa State College Research Foundation, Oct. 17, 1955; Timothy Tilton to J. Ordway, Aug. 24, 1955; and Tilton to Ayres, Dec. 14, 1955, ISURF Papers, Burroughs General Correspondence from Dec. 28, 1954, folder. Also see "Stilbestrol and Terramycin," *Flour and Feed*, 56 (July 1955): 20; John A. Rohlf, "Now . . . A Stilbestrol-Antibiotic Supplement," *Farm Journal*, 79 (July 1955): 32; "Lilly Report to the Feed Industry," *Feed Age*, 6 (Jan. 1956): 74–75; "Pfizer Stops Selling Straight Stilbestrol, Offers Combination," *Feedstuffs*, 28 (Sept. 15, 1956): 1, 5; Peter W. Janss to Ayres, Aug. 3, 1956; Ayres to Janss, Aug. 6, 1956; and E. A. Kelloway to Forrest Teel, Aug. 7, 1956, ISURF Papers, Burroughs General Correspondence from Dec. 28, 1954, folder; and "Seventeenth Annual Report of the Secretary-Manager of the Iowa State College Research Foundation, May 1, 1954–April 30, 1955," Burroughs Papers, box 3.

6. "Lilly Sales Hit New High; Increase in Use of Stilbosol Cited," *Feedstuffs*, 28 (Mar. 3, 1956):

69; "Lilly Quarterly Sales Surpass Previous Year," *Feed Age*, 7 (May 1957): 108; "Pfizer Sets New Record in '57 Sales, Earnings," *Feed Bag*, 34 (Mar. 1958): 71; "A Nation of Meat Eaters," *Nebraska Farmer*, 99 (Jan. 19, 1957): 6; E. W. Sheets, "Consumer Influence on Beef Types," *Shorthorn World*, 40 (Jan. 1, 1956): 5, 31–32; Kenneth Fulk, "Needed: New Terms to Describe the New Look in Beef," *Successful Farming*, 54 (Dec. 1956): 30; "Consumers' Preferences as Regards Beef," *Cattleman*, 43 (Jan. 1957): 118; Tom Adams, "Increased Cattle Feeding in South Texas," ibid., 42 (Feb. 1956): 34–35, 58; W. D. Farr, "Growth of Cattle Feeding in the West—Will More Big Feedlots Develop in the Midwest Also?" *Feedstuffs*, 29 (July 6, 1957): 26, 28, 30–31; "Industry Rated by U.S. Census," *Flour and Feed*, 57 (Dec. 1956): 3; "Geography of Meat Packing," *Journal of Agricultural and Food Chemistry*, 3 (Dec. 1955): 993; Earl L. Butz, "The New Look in Agriculture," *Feedstuffs*, 29 (May 18, 1957): 34, 38, 40, 53–56; "Beefmen Face Up to Consumer Whims," *Farm Journal*, 82 (Feb. 1958): 52; and Charles E. Ball, "Big Feedlots Come to the South," ibid., 83 (June 1959): 50-O, 54-P.

7. "Test-Tube Feeds," *Successful Farming*, 54 (Mar. 1956): 27–28, 172–73; Theodore P. Thery, "Magic Ingredients and the Feed Manufacturer," *Flour and Feed*, 57 (Oct. 1956): 4; Vernon Schneider, "Enzymes—The Newest Feed Additives," *Successful Farming*, 54 (Nov. 1956): 50–51, 80; "New Growth Factor," *Farm Journal*, 81 (Jan. 1957): 46; Philip J. Schaible, "Nitrofurans," *Successful Farming*, 55 (May 1957): 72; "Even Better Beef Hormones?" *Farm Journal*, 81 (May 1957): 39–40; "Brisk Business: Feed Additives," *Agricultural Research*, 5 (Oct. 1956): 10; "Another New Hormone for Fattening Steers," *Farm Journal*, 80 (Oct. 1956): 18; H. Ernest Bechtel, "Trends in Medicated Feeds," *Feedstuffs*, 28 (Mar. 17, 1956): 20, 22; "Hormone Implant Product for Steers Being Marketed," ibid., 28 (Sept. 22, 1956): 6; and Ralph McCall, "Ruminant Nutrition, New Frontier of Scientific Feeding Progress," *Feed Bag*, 33 (Dec. 1957): 7, 9, 95.

8. See, for example, "Stilbestrol Ear Implants," *Successful Farming*, 54 (Dec. 1956): 91; "More News about Stilbestrol Pellets," *Farm Journal*, 80 (Aug. 1956): 21; "New Mexico Beef Feeding Test Results Reported," *Feedstuffs*, 28 (Mar. 17, 1956): 23; "Reports of Cattle Fattening Tests at Colorado A and M College," ibid., 28 (Apr. 7, 1956): 48, 52; "Oklahoma Researchers Report on Livestock Feeding Experiments," ibid., 28 (June 2, 1956): 52–53; "Hormone and Antibiotic Boost Steer Gains in Arizona Experiment," ibid., 28 (June 14, 1956): 45; "Kentucky Researchers Report Results of Swine and Cattle Feeding Studies," ibid., 28 (Aug. 25, 1956): 20, 22–24; Charles W. Turner, "Biological Assay of Beef Steer Carcasses for Estrogenic Activity following the Feeding of Diethylstilbestrol at a Level of 10 Mg. per Day in the Ration," *JAS*, 15 (1956): 13–24; Raymond J. Douglas, Larkin H. Langford, and M. L. Buchanan, "Feeding Yearling Steers," *North Dakota Farm Research*, 20 (Nov.–Dec. 1957): 4–9; "Stilbestrol and Terramycin for Steers Being Fattened on Pasture," *Agricultural Research Report of the Agricultural Experiment Station of Virginia Polytechnic Institute*, 1957, 132–33; "Do Growth Boosters Pay on Calves?" *Farm Journal*, 81 (June 1957): 48; "Stilbestrol and Antibiotics in Rations for Yearling Steers," *Feed Bag*, 33 (Aug. 1957): 34; "Carcass Quality Not Hurt by Stilbestrol in Wyoming Test," *Feedstuffs*, 29 (Jan. 19, 1957): 17; "S.D. Stilbestrol Studies Reported," ibid., 29 (Feb. 23, 1957): 34; "Nutrition Research Findings Discussed at Nevada Meeting," ibid., 29 (Mar. 9, 1957): 42–44; O. D. Butler, "Livestock Feeding Research Reported at Meetings of Southern Agricultural Workers," ibid., 29 (Mar. 16, 1957): 42–44, 46–48; "Research on Stilbestrol for Ruminants," ibid., 29 (May 18, 1957): 18, 20, 22, 24; "Putting Research to Work," ibid., 29 (Oct. 5, 1957): 58, 60, 62; Marion Simone, M. T. Clegg, and Floyd Carroll, "Effect of Methods of Stilbestrol Administration on Quality Factors of Beef," *JAS*, 17 (1958): 834–40; "Stilbestrol," *South Dakota Farm and Home Research*, 9 (Feb. 1958): 23–29; J. P. Fontenot, R. F. Kelly, and J. A. Gaines, "Optimum Level of Stilbestrol Implant Was 12 Mg. for Grazing Beef Steers," *Agricultural Research Report of the Virginia Polytechnic Institute Agricultural Experiment Station for July 1, 1957–June 30, 1959*, 126–27; and "'New Thinking' Reviewed at Arizona Ruminant Meeting," *Feedstuffs*, 30 (July 5, 1958): 22; J. H. Jones, N. C. Fine, T. S. Neff, and W. L. Spangel, "Work with Cottonseed Hulls, Supplements, Stilbestrol in New Steer Nutrition Research," *Feed Bag*, 35 (Jan. 1959): 7, 9, 11, 18; "Idaho Stilbestrol Studies Reported," *Feedstuffs*, 31 (May 9, 1959): 25; and A. B. Nelson and L. S. Pope, "Subsequent Perfor-

mance of Stilbestrol-Implanted Cattle on Pasture and in the Feed Lot," *JAS*, 18 (1959): 1151.

9. See, for instance, "Stilbestrol in Fattening Bulls," *JAVMA*, 128 (1956): 161; "Implanting Stilbestrol Brings Good Results," *Successful Farming*, 54 (May 1956): 140; "New Test Results with Hormone Pellets," *Farm Journal*, 80 (Dec. 1956): 18; Earle W. Klosterman, "Relationship of Sex Hormones to Protein Levels for Fattening Cattle," *Feed Age*, 6 (July 1956): 32–35; and M. T. Clegg and F. D. Carroll, "Further Studies on the Anabolic Effect of Stilbestrol in Cattle as Indicated by Carcass Composition," *JAS*, 15 (1956): 37–47.

10. For Pfizer and the early commercialization of implants, see "OK Stilbestrol Pellets for Beef," *Farm Journal*, 80 (Jan. 1956): 22, 26; "Stilbestrol Pellets OK'ed for Feeder Cattle," *Successful Farming*, 54 (Mar. 1956): 4; "Pfizer to Market Stilbestrol Pellets, Implanting Device," *Feedstuffs*, 28 (Oct. 13, 1956): 1, 93; and "New Stilbestrol Implants Are Marketed by Pfizer," *Feed Bag*, 33 (May 1957): 84. For announcement of the awarding of the ISCRF patent, see "Iowa Foundation Granted Patent on Stilbestrol Use," *Feedstuffs*, 28 (May 12, 1956): 6; "Hormones for Cattle," *New York Times*, June 23, 1956, 26; "Iowa State Gets Patent on Diethylstilbestrol Feeds," *Journal of Agricultural and Food Chemistry*, 4 (1956): 496; and Iowa State College Information Service, "ISC Research Foundation Granted Patent on Livestock Feeding Discovery, May 8, 1956," ISURF Papers, Burroughs General Correspondence from Dec. 28, 1954, folder. For the Hale-Burroughs dispute, see "Deposition of William H. Hale, March 12, 1959," 229–31, 264–71, in *ISCRF* v. *Dawe's*, Burroughs Papers, box 2; and "Memo—Ruminant Nutrition Stuff from Wise Burroughs," Sept. 30, 1955, Burroughs Papers, box 2.

11. "Implanting Stilbestrol in Cattle," *JAVMA*, 129 (1956): 170; "DES Has Bacteriostatic Effect," *Journal of Agricultural and Food Chemistry*, 5 (July 1957): 481; James W. Miller, "375 Hear Nutrition Reports at Kansas Feed Conference," *Feedstuffs*, 28 (Jan. 14, 1956): 6, 73; "Kansas State Researchers Report on Livestock Feeding Experiments," *Feedstuffs*, 28 (June 23, 1956): 71–72; "Cattle Research Reported at Michigan Event," ibid., 28 (Sept. 29, 1956): 31; C. C. O'Mary and A. E. Cullison, "Effects of Low Level Implantation of Stilbestrol in Steers on Pasture," *JAS*, 15 (1956): 48–51; T. W. Dowe, J. Matsushima, V. H. Arthayd, and P. L. Jillson, "Diethylstilbestrol, Oral Administration vs. Implants for Fattening Cattle," *Forty-fifth Annual Feeders' Day— Nebraska Agricultural Experiment Station, April 26, 1957*, 18–20; "Illinois Cattle Tests Reported," *Feedstuffs*, 29 (Jan. 12, 1957): 62–63; "Carcasses Better If Stilbestrol Fed, Scientists Report," ibid., 29 (Feb. 23, 1957): 45; "Stilbestrol's Effect May Include Control of Harmful Organisms," ibid., 29 (July 6, 1957): 86; "Minnesota Station Reports Data on Cattle Studies," ibid., 29 (Oct. 5, 1957): 10; M. A. Madsen, R. J. Raleigh, and L. E. Harris, "Stilbestrol," *Utah Farm and Home Science*, 17 (Mar. 1956): 20, 25–26; M. T. Clegg, "Growth or Fattening with Stilbestrol Implants?" *Feedstuffs*, 29 (Aug. 3, 1957): 38, 45–46; "Hormones, Antibiotics, and a Chemobiotic for Cattle and Calves," ibid., 29 (Dec. 28, 1957): 24, 79; "Nutrition Experts Present Results of Their Cattle Feeding Research," ibid., 29 (Jan. 12, 1957): 41–45; "Stilbestrol Improves Summer Gains in Range Cattle," *Feed Age*, 8 (June 1958): 18; "Re-implanting Steers in Feedlots Speeds Gain," *Farm Journal*, 82 (Sept. 1958): 40; "More than One Stilbestrol Implant Pays," *Successful Farming*, 56 (Oct. 1958): 10; "Stilbestrol Boosts Gains on Pastured Steers," *North Dakota Farm Research*, 20 (Mar.–Apr. 1958): 4–7; "Pasture plus Stilbestrol," *Agricultural Research*, 6 (June 1958): 15; and "Do Pasture Implants Hurt Feedlot Gains?" *Farm Journal*, 83 (May 1959): 62.

12. For Purdue, see, for instance, W. M. Beeson, T. W. Perry, Martin Mohler, F. N. Andrews, and Martin Stob, "Combination of an Antibiotic and a Female Hormone for Fattening Steers," *JAS*, 16 (1957): 845–49; "Stilbestrol Can Increase Pasture Gains," *Successful Farming*, 55 (June 1957): 10; Martin Stob, T. W. Perry, F. N. Andrews, and W. M. Beeson, "Residual Estrogen in the Tissues of Cattle Treated Orally with Diethylstilbestrol, Dienestrol, Hexestrol, and Chlortetracycline," *JAS*, 15 (1956): 997–1102; "Putting Research to Work," *Feedstuffs*, 28 (May 12, 1956): 42–44; and "New Cattle Feeding Studies Reported at Purdue Event," ibid., 28 (May 5, 1956): 1, 73. For Iowa State, see Wise Burroughs, "Use of Hormones in Ruminant Nutrition," ibid., 29 (July 27, 1957): 30–32, 34–35; idem, "Stilbestrol and Hormone Feeding to Cattle and Other Livestock," ibid., 28 (July 28, 1956): 38, 40, 56–59; Roger Berglund, "Feed Men Hear Reports on New Iowa Research

Finding," ibid., 28 (Sept. 29, 1956): 10–12, 14–15; Joe Kastelic, Paul Homeyer, and E. A. Kline, "The Influence of the Oral Administration of Diethylstilbestrol on Certain Carcass Characteristics of Beef Cattle," *JAS*, 15 (1956): 689–700; "Summary of Stilbestrol Feeding," *Successful Farming*, 55 (Jan. 1957): 8; "New and Improved Methods of Using Stilbestrol Supplements in Cattle Fattening Rations," *Feedstuffs*, 30 (Mar. 22, 1958): 32–33, 76–77; "Higher Energy-Protein Ratio Ups Beef Grades," *Iowa Farm Science*, 13 (Mar. 1959): 13; Arthur W. Struempler and Wise Burroughs, "Stilbestrol Feeding and Growth Hormone Stimulation in Immature Ruminants," *JAS*, 18 (1959): 427–36; and "Carcass Grades on Stilbestrol Cattle above Live Grades," *Feedstuffs*, 31 (July 11, 1959): 36.

13. See, for example, Jack M. Curtis to R. S. Roe, July 15, 1955, FDA Papers, Accession no. 88-82-57, box 21, vol. 280; "Research Newsletter," *Journal of Agricultural and Food Chemistry*, 4 (Feb. 1956): 97; "Extra Stilbestrol Doesn't Hurt Cattle," *Farm Journal*, 80 (Apr. 1956): 62; "Stilbestrol Meat Safe for Consumption," *Science News Letter*, (Aug. 11, 1956): 85; "No Stilbestrol in Meat," *Journal of Agricultural and Food Chemistry*, 4 (1956): 659; "Stilbestrol-Fed Beef Found OK," *Cattleman*, 43 (Aug. 1956): 26; "No Hormone in Meat of Stilbestrol-Fed Steers," *Successful Farming*, 54 (Oct. 1956): 78; "No Hormone Evidence in Meat from Stilbestrol-Fed Steers," *Feed Age*, 7 (Apr. 1957): 32; and Ernest J. Umberger, Jack M. Curtis, and George H. Gass, "Failure to Detect Residual Estrogenic Activity in the Edible Tissues of Steers Fed Stilbestrol," *JAS*, 18 (1959): 221–26.

14. Committee on Animal Nutrition, Subcommittee on Hormones, *Hormonal Relationships and Applications in the Production of Meats, Milk, and Eggs* (Washington, D.C.: National Academy of Sciences and National Research Council, 1959), i; also see 11–17, 31–46.

15. The award citation is printed in "'Benefit to Man' Award Goes to I.S.C. Scientist," *Des Moines Tribune*, Nov. 18, 1958, 13. Also see "Wise Burroughs Receives John Scott Science Award," *Feed Age*, 8 (July 1958): 70.

16. Mrs. Robert Swenson to Wise Burroughs, Feb. 19, 1954, ISURF Papers, Burroughs General Correspondence from Apr. 28, 1953, folder.

17. Wise Burroughs to Mrs. Robert Swenson, Mar. 4, 1954, ISURF Papers, Burroughs General Correspondence from Feb. 1, 1954, folder; M. C. Goldman, "Poison by the Plateful," *Organic Gardening and Farming*, 1 (Mar. 1954): 47, 50–53.

18. C. N. Frey, "Chemicals in Food," *Science*, 120 (Sept. 9, 1954): 7-A.

19. For aspects of the failed food additive law, see "Food Additive Legislation," *Journal of Agricultural and Food Chemistry*, 2 (1954): 761; "Some Criticism of the FDA," ibid., 2 (1954): 809; and "Legislation and Regulation," ibid., 2 (1954): 1208. Also of interest are "100-Fold Margin of Safety," *Quarterly Bulletin of the Association of Food and Drug Officials*, 18 (Jan. 1954): 33–35; and "Food Industry Shifting from Chemical to Physical Technology," *Journal of Agricultural and Food Chemistry*, 2 (1954): 765.

20. The quotations are from "Protesting Food Additives," *Journal of Agricultural and Food Chemistry*, 3 (1955): 191–92; and "New Look at HEW," ibid., 3 (1955): 641. Also see, for example, "Food Additives Bills," ibid., 3 (1955): 292, 466; William C. Foster, "Food-Ingredient Legislation," *FDCLJ*, 10 (1955): 82–91; A. L. Miller, "Chemicals in Food," ibid., 10 (1955): 104–12; Vincent A. Kleinfeld, "Chemicals in Foods—A Legal Viewpoint," ibid., 9 (1954): 115–22; Roy S. Pruitt, "Chemical Additives to Food," ibid., 9 (1954): 178–84; George L. Maison," Aspects of the Scientific Functions of the Food and Drug Administration," ibid., 9 (1954): 620–24; and Bradshaw Mintener, "The Food Additive Problem," ibid., 11 (1956): 621–30.

21. John Cipperly, "Plans Set for Meeting on Medicated Feeds," *Feedstuffs*, 27 (Dec. 10, 1955): 1, 85.

22. AOAC procedures are discussed in J. Russell Couch, "New Nutrients Impose Problems upon Control Officials," *Flour and Feed*, 56 (Aug. 1955): 22–24. Also see "Scientific Investigations," in *Annual Report of the United States Department of Health, Education, and Welfare for the Year 1956*, 219–21. For the origin and early history of the organization, see Alan I Marcus, "Setting the Standard: Fertilizers, State Chemists, and Early National Commercial Regulation, 1880–1887,"

Agricultural History, 61 (Winter 1987): 47–73. Also see Kenneth Helrich, *The Great Collabora-tion: The First Hundred Years of the Association of Official Analytical Chemists* (Arlington, Va.: Association of Official Analytical Chemists, 1984).

23. "Best Wishes from Ike," *Feedstuffs*, 28 (Jan. 28, 1956): 1; and "Attention Focused on Ex-panding Use and Benefits of Medicated Feeds," ibid., 28 (Jan. 28, 1956): 6. The entire proceedings were published as Henry Welch and Felix Marti-Ibanez, eds., *Symposium on Medicated Feeds* (New York: Medical Encyclopedia, 1956); for Eisenhower's and Larricks's remarks, see i and 1–2.

24. The quotations are from Granville F. Knight, W. Coda Martin, Rigoberto Iglesias, and William E. Smith, "Possible Cancer Hazard Presented by Feeding Diethylstilbestrol to Cattle," in Welch and Marti-Ibanez, eds., *Symposium on Medicated Feeds*, 167–70. Also see "Government, Industry Experts Examine Regulation Problems," *Feedstuffs*, 28 (Jan. 28, 1956): 1, 77–78; "Scien-tists Discuss Use of Growth-Promoting Ingredients," ibid., 28 (Jan. 28, 1956), 6; "Feed Sym-posium Includes Talks on Production, Quality Control Problems," ibid., 28 (Jan. 28, 1956): 7; "Role of Feeds in Prevention, Treatment of Disease Outlined," ibid., 28 (Jan. 28, 1956): 7, 73; and Richard T. Claycomb, "F and DA Holds Meeting to Discuss Feed Industry Problems," *Feed Age*, 6 (Mar. 1956): 44–49. For Smith's previous carcinogenesis work, see William E. Smith, "The Neo-plastic Potentialities of Mouse Embryo Tissues: The Tumors Elicited from Gastric Epithelium," *Journal of Experimental Medicine*, 85 (1947): 459–78; idem, "The Neoplastic Potentialities of Mouse Embryo Tissues: The Tumors Elicited with Methylcholanthrene from Pulmonary Epi-thelium," ibid., 91 (1950): 87–103; and idem, "The Tissue Transplant Technic as a Means of Test-ing Materials for Carcinogenic Action," *Cancer Research*, 9 (1949): 712–23. For his institutional efforts, see William E. Smith, "Environmental Cancer Program at the Institute of Industrial Med-icine, New York University–Bellevue Medical Center," *Acta International Union against Cancer*, 9 (1953): 414–15.

25. The response to Smith's claims was captured in "Report Made on Health Significance of Drugs in Feeds," *Feedstuffs*, 28 (Jan. 28, 1956): 7, 78; Claycomb, "F and DA Holds Meeting"; and Welch and Marti-Ibanez, eds., *Symposium on Medicated Feeds*, 170–77; the quotation is from 177.

26. The quotations are from Waldemar Knempffert, "Hormone and Cancer," *New York Times*, Jan. 29, 1956, sec. 4, p. 9; and "Statement, Food Protection Committee, Food and Nutrition Board," *FDCLJ*, 12 (1957): 132, 192. Also see "Stilbestrol: A Case in Point," *Organic Gardening and Farming*, 3 (Jan. 1956): 39–41; "Feeds and Drugs," *Feedstuffs*, 28 (Feb. 25, 1956): 2; and "Food Additives Legislation," *Journal of Agricultural and Food Chemistry*, 4 (1956): 99, 191, 201.

27. The FDA-favored process is considered in John Cipperly, "Federal Control Developments May Present Trade Problems," *Feedstuffs*, 28 (Feb. 25, 1956): 1, 79. Also see "The Head of the Food and Drug Administration Bends Administrative Policy in Approach to Food Legislation," *Journal of Agricultural and Food Chemistry*, 4 (1956): 572; "FDA Is Asked for Tighter Controls on Medicated Feeds," *Feedstuffs*, 28 (Apr. 21, 1956): 1; "Major Action Taken to Simplify Labeling of Medicated Feeds," ibid., 28 (July 14, 1956): 1, 85; John H. Collins, "Importance of Medicated Feeds to the U.S. Food and Drug Administration," ibid., 28 (Sept. 29, 1956): 44, 45, 49; "Off the Editor's Chest," *Consumers' Research Bulletin*, 38 (Nov. 1956): 2, 28; and Fred Bailey, "Food Chemicals," *Nebraska Farmer*, 98 (Nov. 17, 1956): 12, 46.

28. Cipperly, "Federal Control Developments," 79.

29. "Memo from the Editors," *Feed Age*, 6 (Apr. 1956): 15.

30. "Preface," *Selected Proceedings of the Council on Consumer Information*, 3 (1957): 1–2; George P. Larrick, "The Consumer Looks at Chemicals in Our Food," ibid., 3 (1957): 18–22; "Food Additives Legislation," *Journal of Agricultural and Food Chemistry*, 5 (1957): 560–61; "Physicians Urged to Protect Foods," *New York Times*, May 31, 1957, 21; "Chemical Testing," *Nebraska Farmer*, 99 (Jan. 19, 1957): 52; John H. Collins, "Functions and Policies of the Food and Drug Administration," *Feedstuffs*, 29 (June 15, 1957): 34–36, 38; "Food and Drug Administra-tion," in *Annual Report of the United States Department of Health, Education, and Welfare*, 1957, 207; and John L. Harvey, "Report from the Food and Drug Administration," *FDCLJ*, 12 (1957): 71–80; Also see, for instance, Bernard L. Oser, "Food-Additives Legislation Drags On," ibid., 12

(1957): 199–208; Max Leva, "The Food-Additives Amendment," ibid., 12 (1957): 236–41; John A. Gosnell, "Food-Additive Matter—A Review of Fundamentals," ibid., 12 (1957): 369–74; Edward Brown Williams, "The Food-Additive Matter—Comments on Proposed Legislation," ibid., 12 (1957): 553–65; and George A. Michael, "Chemical Additives—Public Health Problem No. 1," ibid., 12 (1957): 639–48, 656. On consumers' perspectives, see Colston Warne, "Outline of Topics for Discussion of Federal and State Food, Drug, and Cosmetic Laws and Their Enforcement," *Quarterly Bulletin of the Association of Food and Drug Officials*, 23 (Jan. 1959): 29–33.

31. The quotations are from *Acta Unio Internationalis contra Cancrum*, 13 (1957): 187–88; "U.S. Denies Charge of Food Content," *New York Times*, Feb. 23, 1957, 10; and "Food Additives Arouse Dispute," ibid., June 30, 1957, 34. Also see "Food Legislation Suggestion," ibid., Jan. 24, 1957, 21. For Smith and the Union, see *Congressional Record*, 85th Cong., 1st sess., A1353; and *Acta Unio Internationalis contra Cancrum*, 11 (1955): 72–76; 13 (1957): 77–85, 156–63, 179–81. For the Union's Rome meeting and the American press, see John Lear, "Food and Cancer," *Saturday Review*, 39 (Oct. 6, 1956): 57–62; "Science and the Citizen," *Scientific American*, 195 (Oct. 1956): 68–71; and Arnaldo Cortesi, "Cancer Is Traced to Feed Additives," *New York Times*, Aug. 21, 1956, 31.

32. *Congressional Record*, 85th Cong., 1st sess., A1351.

33. The quotations from Smith's letter which appear in this paragraph and the two that follow are from ibid., A1352–54. For a different interpretation of Smith, the Union, and the relationship with Delaney, see Phyllis Anderson Meyer, "The Last Per Se: The Delaney Cancer Clause in United States Food Regulation" (Ph.D. diss., U of Wisconsin, 1983), 125–206.

34. The quotations are from "To Avoid the Things That Cause Cancer," *Consumer Bulletin Annual*, 1958, 133. Also see *Congressional Record*, 85th Cong., 2d sess., 3630–32, 7783, A1377. For Hueper's arguments, see W. C. Hueper, "Recent Developments in Environmental Cancer," *A.M.A. Archives of Pathology*, 58 (1954): 475–523; and idem, "Potential Role of Non-nutritive Food Additives and Contaminants as Environmental Carcinogens," ibid., 62 (1956): 218–49. The "per se" argument was derived from the 1938 Food, Drug, and Cosmetics Act and its General Safety Clause dealing with individual substances "generally recognized as safe."

35. The quotation is from *Congressional Record*, 85th Cong., 2d sess., 16304. Also see "To Strengthen a Good Law," *New York Times*, Sept. 29, 1957, sec. 4, p. 10; "Scientists and Food Additives," *Journal of Agricultural and Food Chemistry*, 5 (1957): 640, 642; and "Chemical-Additives Bill Recommended," *FDCLJ*, 12 (1957): 197–98, 256.

36. "New York City Considers Action on Hormones in Meat," *Feedstuffs*, 30 (Aug. 2, 1958): 7.

37. Ibid.

38. "New Hope for Food Additives Law," *Journal of Agricultural and Food Chemistry*, 6 (1958): 563.

39. *Congressional Record*, 85th Cong., 2nd sess., 16304–6, 17412–24, 19358–59, 19641; "Bill on Food Additives Gains," *New York Times*, Aug. 14, 1958, 19; "Flemming Hails Drug Law," ibid., Sept. 10, 1958, 7; "After 10 Years—A Food Additives Law," *Journal of Agricultural and Food Chemistry*, 6 (1958): 635; "Firmer Additive Control Seen in New FDA Bill," *Feedstuffs*, 30 (Aug. 30, 1958): 1, 75; and "New Food-Additives Law Enacted," *FDCLJ*, 13 (1958): 549–50, 602–5.

40. "An Act to Protect the Public Health by Amending the Federal Food, Drug, and Cosmetic Act to Prohibit the Use in Food of Additives Which Have Not Been Adequately Tested to Establish Their Safety" (P.L. 85-929), 71 *U.S. Statutes* (1958): 1784–89. Also see "An Act to Prohibit the Movement in Interstate Commerce of Adulterated and Misbranded Food, Drugs, Devices, and Cosmetics, and for Other Purposes," 52 *U.S. Statutes* (1938): 1040–59. For a history of the passage of the 1938 act, see Charles O. Jackson, *Food and Drug Legislation in the New Deal* (Princeton: Princeton UP, 1970), especially 175–200.

41. *Congressional Record*, 85th Cong., 2d sess., 17414–15.

42. *Congressional Record*, 85th Cong., 2d sess., 17422; and "New Additives Law," *Journal of Agricultural and Food Chemistry*, 6 (1958): 725. Also see "The Scientists' Forum," *FDCLJ*, 13

(1958): 394–400, 477–79, 592–601. For critiques of the "per se" argument, see Frank A. Vorhes, Jr., "Food Additives," *Journal of the Association of the Official Analytical Chemists* (hereafter, *JAOAC*), 42 (1959): 109–12; John A. Osmundsen, "New Law Likely to Improve Foods," *New York Times*, Feb. 24, 1959, 1, 22; and "Food Additives," ibid., Feb. 27, 1959, 24.

43. The quotations are from Graham DuShane, "Locking the Barn Door First," *Science*, 129 (Jan. 23, 1959): 177. Also see "Firmer Additive Control Seen in New FDA Bill," *Feedstuffs*, 30 (Aug. 30, 1958): 1, 75; "FDA Issues Statement on New Additive Amendment," ibid., 30 (Sept. 13, 1958): 4; John Cipperly, "Capital Comment," ibid., 30 (Sept. 13, 1958): 4; "FDA Details Provisions of New Additive Amendment," ibid., 30 (Sept. 20, 1958): 92–93; George P. Larrick, "The New Food-Additives Law," *FDCLJ*, 13 (1958): 634–48; and Fredus N. Peters, Jr., "The Scientific Significance of the Amendment to the Food-Manufacturing Industry," ibid., 13 (1958): 800–4. At least some consumer groups were not satisfied with the new law. See, for example, "The 1958 Amendment to the Food and Drug Act," *Consumers' Research Magazine*, 41 (Nov. 1, 1958): 11–13.

44. John Cipperly, "Proposed Rules under Additives Law Announced," *Feedstuffs*, 30 (Nov. 29, 1958): 1, 77; "Food Additives," *Federal Register*, Dec. 9, 1958, 9511–17; and "New Additives Regulations," *Journal of Agricultural and Food Chemistry*, 7 (1959): 12, 14.

CHAPTER 3. REGULATION AS OPPOSITION

1. Peripheral consumerist periodicals continued to wave the red flag. See, for instance, Thomas Powell, "New Poisons Imperil Our Meat," *Organic Gardening and Farming*, 6 (Mar. 1959): 99–103. The FDA proved willing to give feed manufacturers additional time to consider the new law's ramifications. See "Move Time on Additives Regulations," *Journal of Agricultural and Food Chemistry*, 7 (1959): 3; and "FDA Doings," ibid., 7 (1959): 226.

2. The program is outlined in Carla S. Williams, "The FDA's Consumer Consultant Program," *FDCLJ*, 15 (1960): 748–54. Also see Orville L. Freeman, "Consumer Representation in Government," *Selected Proceedings of the Council on Consumer Information*, 6 (1960): 2–12.

3. John Cipperly, "Bans New Uses of Substances in Feed, Drinking Water," *Feedstuffs*, 31 (May 30, 1959): 1, 77; "FDA Issues Statement Clarifying Drug Status," ibid., 31 (May 30, 1959): 1, 77; Charles G. Durbin, "Animal Feeds and Feed Additives under Federal Law," *Proceedings of the Cornell Nutrition Conference for Feed Manufacturers*, 1959, 83–90; Arthur A. Checchi, "Food-Additive Procedures and Policies," *FDCLJ*, 14 (1959): 591–96; "New Crackdown Coming on Feed Additives," *Successful Farming*, 57 (June 1959): 8; "Chemical Feed Additives," *Cattleman*, 45 (May 1959): 23; "Bans Wider Stilbestrol Use in Feeds," *Des Moines Register*, July 9, 1959, 7; and *Federal Register* (May 30, 1959): 4376.

4. "Food Additives Amendment . . . Its Effect on Feed Manufacturing," *Feed Age*, 9 (June 1959): 96; Eldon H. Roesler, "Dangers in the Delaney Amendment," *Feed Bag*, 35 (Aug. 1959): 15; John Cipperly, "More Observations on FDA Policy Statement Outlined," *Feedstuffs*, 31 (June 6, 1959): 1, 8; "'Common Meeting Ground' on Drug Problem Seen," ibid., 31 (June 20, 1959): 1, 101; John Cipperly, "'Area of Compromise' between FDA and Feed Trade Still Held Possible," ibid., 31 (June 27, 1959): 1, 89; idem, "Period of Consultation on Drugs Seen; New Concept under Law Indicated," ibid., 31 (July 4, 1959): 1, 89; "Trade Efforts Continue on Drug Problem," ibid., 31 (July 4, 1959): 1, 89; John Cipperly, "Further Consultation with FDA Seen," ibid., 31 (July 11, 1959): 1, 113; "FDA Clamps Freeze on Feed Additive Use," *Nebraska Farmer*, 101 (July 4, 1959): 11; "Delaney Amendment FDA Headache," *Feed Bag*, 35 (July 1959): 23; "Pulse of the Industry," ibid., 35 (Sept. 1959): 16; and Bernard L. Oser, "Current Problems Posed by the Food-Additives Amendment," *FDCLJ*, 14 (1959): 574–83.

5. "AFMA Maps Program on Drug Problem; Recommendations Listed," *Feedstuffs*, 31 (July 11, 1959): 1, 112, 113; "AFMA Outlines Current Drug Situation under Delaney Clause," ibid., 31 (Nov. 7, 1959): 1, 105; and Drew Pearson, "Cancer Food Problems," story reproduced from the *Clinton Herald*, Nov. 18, 1959, Burroughs Papers, box 2.

6. AFMA's position is explained in "AFMA Issues Statement on Additive Law," *Feedstuffs*, 31

(June 6, 1959): 1, 105; and "AFMA Questions FDA Interpretation of Additive Law," ibid., 31 (June 20, 1959): 1, 105. Also see "Industry Spokesmen Pursue Action on Drug Problems," ibid., 31 (June 27, 1959): 1, 89; John Cipperly, "Foundation for Hope on Delaney Amendment Seen," ibid., 31 (Aug. 8, 1959): 1, 97; "Pulse of the Industry," *Feed Bag*, 35 (July 1959): 16; "AFMA Statement," *Feed Age*, 9 (Aug. 1959): 64; and Pearson, "Cancer Food Problems."

7. The AHI agenda is reported in "Results of Conference with FDA on Arsenicals Outlined by Drug Group," *Feedstuffs*, 31 (July 11, 1959): 1, 113; "Say New Technique May Clear Up Stilbestrol Problem," ibid., 31 (July 18, 1959): 4; and George L. Gates, "Ask for Reasonable Approach to Drug Dilemma," ibid., 31 (Oct. 17, 1959): 1, 4, 6. Also see John Cipperly, "Equitable Solution to Trade-FDA Drug Ruling Problems Expected," ibid., 31 (July 18, 1959): 1, 105; Wise Burroughs to D. L. Bruner, Aug. 24, 1959, Burroughs Papers, box 2; and D. L. Bruner, "Interpretation and Enforcement of Additive Regulations," *Feedstuffs*, 31 (Nov. 14, 1959): 66–67, 69.

8. John Cipperly, "Outlook for Early Change in Drug Policy Reported Dim," *Feedstuffs*, 31 (Oct. 24, 1959): 1, 4; Samuel Alfred, "Feeds and the Food Additives Amendment," ibid., 31 (Sept. 26, 1959): 90–92; and "AFMA Outlines Current Drug Situation under Delaney Clause," ibid., 31 (Nov. 7, 1959): 1, 105.

9. Lilly's contentions are made known in "Lilly Cites Information Showing Safety of Stilbestrol Use," *Feedstuffs*, 31 (July 4, 1959): 7, 90; and "Lilly Issues Statement on FDA Letter; Asserts Safety of Stilbestrol," ibid., 31 (July 25, 1959): 1, 4. Also see "Eli Lilly and Co.," *Feed Age*, 9 (July 1959): 58, 60.

10. Minute Books of the Iowa State Board of Regents, June 19, 1959, 504, Special Collections Department, Parks Library, Iowa State University.

11. The quotations are from Wise Burroughs to James Hilton, July 6, 13, and 15, 1959, ISURF Papers, Burroughs General Correspondence from Dec. 28, 1954, folder. Also see "Don't Worry about Your Beefsteak," *Des Moines Register*, July 12, 1959, 16-G; and "Plan Talks on Stilbestrol," ibid., July 15, 1959, 19.

12. Wise Burroughs to James Hilton, July 15, 1959, *ISURF* Papers, Burroughs General Correspondence from Dec. 28, 1954, folder.

13. Arthur Kirschbaum, "The Role of Hormones in Cancer: Laboratory Animals," *Cancer Research*, 17 (1957): 432–39; W. U. Gardner, "Experimental Induction of Uterine Cervical and Vaginal Cancer in Mice," *Cancer Research*, 19 (1959): 170–76; Lloyd W. Hazleton, "Biological Evaluation of Agricultural and Food Chemicals," *Journal of Agricultural and Food Chemistry*, 5 (1957): 336–39; and G. E. Mitchell, Jr., A. L. Neumann, and H. H. Draper, "Metabolism of Tritium-Labeled Diethylstilbestrol by Steers," ibid., 7 (1959): 509–12.

14. The quotations are from John Cipperly, "Health Dept. Opposes Cancer Clause Removal," *Feedstuffs*, 31 (Nov. 14, 1959): 1, 121; idem, "Connotations of Cranberry Incident for Feed, Other Industries Cited," ibid., 31 (Nov. 21, 1959): 1, 109; and "Government by Publicity," *Chemical Week*, 85 (Nov. 21, 1959): 12. Also see Carroll P. Streeter, "What Follows That Cranberry Crackdown?" *Farm Journal*, 83 (Dec. 1959): 33; "Chemicals and Food," *Wallaces Farmer*, 84 (Nov. 21, 1959): 12; "The Health, Education, and Welfare Dept. Is More Militant," *Chemical Week*, 85 (Nov. 28, 1959): 51–52; Pearson, "Cancer Food Problems"; Jack Anderson, "Berry Tiff Only a Breeze," published in *Clinton Herald*, Nov. 30, 1959, Burroughs Papers, box 2; Helen Thomas, "Hormones," *Ames Daily Tribune*, Dec. 2, 1959, 6; "Stilbestrol in Poultry May Be Banned," *Journal of Agricultural and Food Chemistry*, 7 (1959): 804; "Stilbestrol Study Set: Cancer Link," *Des Moines Register*, Dec. 4, 1959: 5; "Health Study Set on Meat Hormone," *New York Times*, Dec. 4, 1959, 25; "Secretary's Press Conference on Stilbestrol Delayed," *Feedstuffs*, 31 (Dec. 5, 1959): 1, 101; "A New Crackdown on Stilbestrol May Be Ahead," *Chemical Week*, 85 (Dec. 12, 1959): 48; and "Halt Sale of Chickens Treated with Stilbestrol," *Feedstuffs*, 31 (Dec. 12, 1959): 1, 96.

15. Ernest J. Umberger to R. S. Roe, Sept. 30, 1957; and Umberger and George H. Gass to Roe, Feb. 3, 1958, FDA Papers, Accession no. 88-82-57, box 22, vol. 281; and Ernest J. Umberger, Jack M. Curtis, and George H. Gass, "Failure to Detect Residual Estrogenic Activity in the Edible Tissues of Steers Fed Stilbestrol," *JAS*, 18 (1959): 221–26.

16. The rejoinder to Flemming is from John Cipperly, "Flemming Statement," *Feedstuffs*, 31 (Dec. 12, 1959): 5. Also see Bess Furman, "Poultry Treated by Drug Banned," *New York Times*, Dec. 11, 1959, 1, 26; "Hormones and Chickens," *Time*, 74 (Dec. 21, 1959): 32; "Text of Secretary's Statement," *Feedstuffs*, 31 (Dec. 12, 1959): 1, 93; "Washington Newsletter," *Chemical Week*, 85 (Dec. 19, 1959): 29–30; John Cipperly, "Suggests Way to End Stilbestrol-Feed Problem," *Feedstuffs*, 31 (Dec. 19, 1959): 1, 89; "Flemming Seeks Food Law Change," *New York Times*, Dec. 24, 1959, 20; "Row over Pure Food," *U.S. News and World Report*, 47 (Dec. 28, 1959): 34–35; Jay Richter, "Washington Letter," *Nebraska Farmer*, 102 (Jan. 2, 1960): 10; and "Effects of Government Moves under FDA Ruling Are Far Reaching," *Feed Age*, 10 (Jan. 1960): 45–46.

17. William M. Blair, "Flemming Action Called Political," *New York Times*, Dec. 14, 1959, 41; Fred E. Tunks, "Broiler Demand Affected Little, Apparently, by Caponette Ban," *Feedstuffs*, 31 (Dec. 19, 1959): 1, 8; and John Cipperly, "U.S. Said to Be Accelerating Drive on Chemical Misuse," ibid., 32 (Jan. 2, 1960): 1, 90.

18. The quotations are from William M. Blair, "Crop-Spray Fight Going to Congress," *New York Times*, Jan. 20, 1960, 17; "Pesticide Problem Studied by Cabinet," ibid., Jan. 23, 1960, 43; "Special USDA Statement Endorses 'Safe' Chemicals in Production of Foods," *Feedstuffs*, 32 (Jan. 16, 1960): 1, 105; and "Farm Groups Ask Ike for Study of Chemical Laws," ibid., 32 (Feb. 6, 1960): 24.

19. The quotations are from "Eisenhower Asks Additives Report," *New York Times*, Feb. 13, 1960, 21; and "Wisconsin Committee's Named to Develop Policy on Chemicals," *Feedstuffs*, 32 (Feb. 13, 1960): 83. Also see "Another Dangerous Food Additive Banned," *Consumers' Research Magazine*, 43 (Feb. 1960): 29–30; "The Cranberry Fuss," ibid., 43 (Mar. 1960): 4; James L. Trawick to Robert K. Enders, Jan. 27, 1960, FDA Papers, Accession no. 88-82-57, box 22, vol. 283; "President Directs Study of Additives," *JAMA*, 172 (Feb. 27, 1960): 25; "Food and Drug Administration," *Wallaces Farmer*, 85 (Apr. 16, 1960): 6; "United States to Pay Indemnity for Cranberry Losses," *Science*, 131 (1960): 1033–34; "U.S. Sets Payment for Berry Losses," *New York Times*, May 5, 1960, 23. The AMA sided with the USDA over HEW. See, for instance, Harold M. Schmeck, Jr., "A.M.A. Aide Scores Warnings on Food," *New York Times*, Jan. 2, 1960, 1, 4; John H. Talbolt, "Cranberries, Charcoal, and Chickens," *JAMA*, 172 (Jan. 2, 1960): 62; and Bess Furman, "Flemming Scores A.M.A. Article Critical of Crackdown on Drugs," *New York Times*, Jan. 8, 1960, 17.

20. The quotations are from Adrian L. Ringuette, "Medicated Animal Feed under the Food-Additives Amendment of 1958: A Case Study," *FDCLJ*, 15 (1960): 339; Edward J. Matson, "Scientific Judgment in Law and Regulation," *FDCLJ*, 15 (1960): 75; and John Cipperly, "Scientists Air Varied Views on Cancer Clause," *Feedstuffs*, 32 (Apr. 9, 1960): 1, 93. Also see idem, "House Unit Gets Proposals for Delaney Clause Changes," ibid., 32 (Jan. 30, 1960): 1, 97; idem, "Chemical Industry Spokesmen Voice Views Opposing FDA's 'Broad Powers,'" ibid., 32 (Feb. 13, 1960): 4, 119; Bess Furman, "Drug Industry Asks U.S. to Ease Clause on Color Additives," *New York Times*, Feb. 12, 1960, 31; Thomas P. Carney, "Scientific Judgment Ignored by Delaney Amendment Because of Its Rigid Nature," *Feed Bag*, 36 (Mar. 1960): 7; and Graham DuShane, "Zero Tolerance," *Science*, 131 (1960): 697.

21. See, for example, *Congressional Record*, 86th Cong., 2d sess., 1827, 2252–53, 14609, A341–43, A906, A1014–15, A1039, A1373–74, A2492–94.

22. The quotations are from George L. Gates, "Delaney Clause Sharply Rapped at GFDNA Event," *Feedstuffs*, 32 (Mar. 12, 1960): 1, 7; "Trade Urged to Contact Congress on Additive Bill," ibid., 32 (Mar. 26, 1960): 1, 107; and Fred E. Tunks, "Ag Industries Tackle Food 'Scare' Problems," ibid., 32 (Apr. 9, 1960): 1, 94. Also see "Need for a Realistic Approach," *Feed Bag*, 36 (Apr. 1960): 15; "Organization Meets to Develop Policy on Color and Food Additives Issue," *Feed Age*, 10 (May 1960): 98; "Industry Interest in Congressional Action on Additives Is Mounting," *Feedstuffs*, 32 (Apr. 2, 1960): 1, 97; "Wanted: Sound Judgment, Facts," ibid., 32 (Apr. 9, 1960): 1, 92; "Zero Tolerance Panned," ibid., 32 (Apr. 9, 1960): 1, 7; Eldon H. Roesler, "New Storage Pact, Food and Drug Relations National's Topic," *Feed Bag*, 36 (Apr. 1960): 78, 82–83, 86–87; and Earl

L. Butz, "The Impact of FDA Regulations on Research and Living Standards," *Feedstuffs*, 32 (Mar. 26, 1960): 10–12.

23. Butz, "Impact"; "Delaney Clause Hit in Abbott Report," *Feedstuffs*, 32 (Mar. 19, 1960): 4; "Says Research on Feed Additives Cut by Food, Drug Law," ibid., 32 (Mar. 26, 1960): 4; "Experts Discuss Role of Agricultural Chemicals in Production of Food," ibid., 32 (Apr. 9, 1960): 10; "Food Additives Law Reported to Be Curtailing Research," *Science*, 131 (1960): 975–76; and "Delaney Clause Said to 'Threaten Future Progress,'" *Feedstuffs*, 32 (Apr. 23, 1960): 1, 96.

24. William Longgood, *The Poisons in Your Food* (New York: Simon and Schuster, 1960, 92–164, 222–52, especially 129–64.

25. William J. Darby, review of *The Poisons in Your Food* by William Longgood, *Science*, 131 (1960): 979. For a similar perspective, see, for example, John A. Osmundsen, "Chemicals for Dinner," *New York Times*, May 1, 1960, sec. 7, p. 22.

26. "Poisons in Your Food: A Review and Commentary," *Consumers' Research Magazine*, 43 (May 1960): 12–14; "Poisons," *New York Times*, May 29, 1960, sec. 7, p. 14; "Food Additives," *Science*, 132 (1960): 156–58; and "Stilbestrol," ibid., 132 (1960): 1680, 1683. Hueper's speech was printed in *Congressional Record*, 86th Cong., 2d sess., A3307–8. Hueper was a German emigré; his first name was sometimes given as Wilhelm.

27. The quotations are from "Food Unit Urged in Cancer Fight," *New York Times*, May 15, 1960, 60; and "Additive Report Released by Science Advisory Committee," *Science*, 131 (1960): 1596–98. Also see Graham DuShane, "Risk and Benefit," ibid., 131 (1960): 1581; "USDA Opposes Delaney Clause," *Feedstuffs*, 32 (May 28, 1960): 4; "Report of Panel on Food Additives," *FDCLJ*, 15 (1960): 421–22, 447–54; and "Wisconsin Group Suggests Delaney Clause Amendment," *Feedstuffs*, 32 (June 4, 1960): 101.

28. "Presidential Panel Report Seen as 'Turning Point' in Additive Problems," *Feedstuffs*, 32 (May 21, 1960): 1, 4; "Pulse of the Industry," *Feed Bag*, 36 (June 1960): 16; "President's Panel Recommends Policies on Food Additives Amendment Issues," *Feed Age*, 10 (June 1960): 45; and "The Committee Can Only Recommend," ibid., 10 (July 1960): 90.

29. For the "Delaney phobia" quotations, see John Cipperly, "Delaney Clause Not Changed in Bill," *Feedstuffs*, 32 (June 4, 1960): 1, 101. Also see idem, "No Proposals Made to Implement Report on Delaney Clause," ibid., 32 (May 28, 1960): 1, 99; "Feed Men Urged to Contact Senate on Delaney Clause," ibid., 32 (June 18, 1960): 1, 111; John Cipperly, "Time Running Out for Additive Bill," ibid., 32 (June 25, 1960): 1, 97; idem, "Congress Approves Additive Measure with Cancer Clause," ibid., 32 (July 9, 1960): 1, 97; "The Whole 'Additive' Battle," *Cattleman*, 47 (Aug. 1960): 22; and *Congressional Record*, 86th Cong., 2d sess., A4450–51.

30. Thomas E. Letch, "Effect of FDA Regulation Main Topic at Nebraska Conference," *Feedstuffs*, 32 (Oct 8, 1960): 1, 104; John Cipperly, "Farm, FDA Policies," ibid., 32 (Nov. 12, 1960): 1, 100; H. L. Schilt, "The Effect of FDA Regulations on the Feed Industry," ibid., 32 (Nov. 19, 1960): 18–19, 22–24; Martin E. Newell, "Delaney Clause Pros, Cons Debated," ibid., 32 (Nov. 26, 1960): 1, 101; "Lilly's Dr. Carney Asks Delaney Clause Change," *Feed Bag*, 36 (Dec. 1960): 13; "ACS to Hold Symposium on Delaney Amendment," *Feed Age*, 10 (Nov. 1960): 49; Bess Furman, "Additive Curbs on Food Scored," *New York Times*, Dec. 9, 1960, 64; "Hess and Clark Dedicates Feed Additives Lab," *Feed Bag*, 36 (Sept. 1960): 78; T. W. Perry, "Changes Coming in Beef Production," *Successful Farming*, 58 (Mar. 1960): 52, 121, 124–25; "What's New," ibid., 58 (Apr. 1960): 6; K. B. Kerr, "What's Ahead for Feed Additives?" *Feedstuffs*, 32 (Apr. 16, 1960): 18, 19, 100; "Outlook," *Successful Farming*, 58 (Dec. 1960): 10; "Stilbestrol Nets $125,000 a Year for Iowa Foundation," *Feedstuffs*, 32 (July 2, 1960): 8, 89; "Washington Views," *Journal of the American Pharmaceutical Association: Practical Edition*, 21 (1960): 401, 689, 759; and "Report of the Food and Drug Administration," *Annual Report—U.S. Department of Health, Education, and Welfare*, 1960, 232, 234, 236–39, 252–55, 257–59.

31. Umberger's quotations are from Ernest J. Umberger to Robert S. Roe, Oct. 3, 1960, FDA Papers, Accession no. 88-82-57, box 22, vol. 283. Also see Shelbey T. Greg to George P. Larrick, June 13, 1960; and Paul L. Day to Roe, Oct. 6, 1960, FDA Papers, Accession 88-82-57, box 22,

vol. 283. The study that led Umberger to question the metabolites of DES was published as a short communication two years later. It was extremely preliminary, used massive single doses of DES, and never followed up. See A. M. Gawienowski, H. W. Knoche, and H. C. Moser, "The Metabolism of $_{14}$C- and $_3$H- Labeled Diethylstilbestrol in the Rat," *Biochimica et Biophysica Acta*, 65 (1962): 150–52. At least one group would call for a DES ban until the metabolite question was investigated and resolved. See "Harmful Drug in Meats?" *Consumers' Research Magazine*, 45 (May 1962): 29, 39.

32. Ernest Umberger to A. J. Lehman, Jan. 5, 1961, FDA Papers, Accession no. 88-82-57, box 22, vol. 284.

33. "Stilbestrol Rating Still High—But," *Cattleman*, 46 (Mar. 1960): 12; "Warn Cattle Feeds on Stilbestrol," *Successful Farming*, 58 (Apr. 1960): 6; J. Kenneth Kirk, "Additives and the FDA," *FDCLJ*, 16 (1961): 361–65; "Annual Report of the Food and Drug Administration," in *Annual Report—U.S. Department of Health, Education, and Welfare*, 1961, 313–22, 348–53; Roy A. Keller, "Gas Chromatography," *Scientific American*, (Oct. 1961): 58–67; George P. Larrick, "Applications of Science in Assuring Safety of the Food Supply," in Food and Nutrition Board, Food Protection Committee, *Science and Food: Today and Tomorrow* (Washington: National Academy of Sciences and National Research Council, 1961), 57–62; Nathan Mantel and W. Ray Bryan, "'Safety' Testing of Carcinogenic Agents," *Journal of the National Cancer Institute*, 27 (1961): 455–70; "Food Additives," *Science*, 133 (1961): 947–48; Falvia L. Richardson and Georgia Hall, "Mammary Tumors and Mammary-Gland Development in Hybrid Mice Treated with Diethylstilbestrol for Varying Periods," *Journal of the National Cancer Institute*, 25 (1960): 1023–33; Howard B. Andervont, Michael B. Shimkin, and Harry Y. Canter, "Susceptibility of Seven Inbred Strains and the F 1 Hybrids to Estrogen-Induced Testicular Tumors and Occurrence of Spontaneous Testicular Tumors in Strain BALB/c Mice," ibid., 25 (1960): 1069–80; Howard B. Andervont, Michael B. Shimkin, and Harry Y. Canter, "Some Factors Involved in the Induction or Growth of Testicular Tumors in BALB/c Mice," ibid., 25 (1960): 1083–96; George H. Gass to E. J. Umberger, Oct. 10 and Dec. 5, 1961; and Umberger to Gass, Dec. 18, 1961, FDA Papers, Accession no. 88-82-57, box 22, vol. 284; Umberger to Leo Friedman, Apr. 12, 1962, FDA Papers, Accession no. 88-82-57, box 22, vol. 285; and "Stilbosol," *Successful Farming*, 60 (Dec. 1962): 11.

34. *Congressional Record*, 87th Cong., 1st sess., 3921–26, 4874–75; "Kefauver Hearings," *Journal of the American Pharmaceutical Association: Practical Edition*, n.s. 1 (1961): 145; "Kefauver Submits Bill on Drug Curbs," *New York Times*, Apr. 13, 1961, 22; Tom Wicker, "Drug Pricing Bill Opposed by A.M.A.," ibid., July 6, 1961, 31; Marjorie Hunter, "Ribicoff Favors Drug-Curb Bill and Predicts It Will Cut Prices," ibid., Sept. 14, 1961, 17; "Safer, Cheaper Drugs," ibid., Sept. 18, 1961, 28; and Alvin Shuster, "Drug Makers Accept U.S. Tests on Effectiveness of Medications," ibid., Dec. 8, 1961, 27. The Kefauver hearings have had substantial ex post facto analysis. See, for instance, Richard Harris, *The Real Voice* (New York: Macmillan, 1964), 1–136; and Peter Temin, *Taking Your Medicine: Drug Regulation in the United States* (Cambridge: Harvard UP, 1980), 120–23.

35. The quotations are from "Industry Pushes for Delaney Clause Change; Results Still Not Clear," *Feedstuffs*, 33 (July 1, 1961): 1, 8; R. Blackwell Smith, "The Meaning of Safety as Regards Food Additives," in Food and Nutrition Board, Food Protection Committee, *Science and Food*, 49–50; "Critics of Food Adulteration Given the Brush-Off," *Consumers' Research Magazine*, 44 (Apr. 1961): 24–28; "Excess Sugar, Chemical Additives in Our Diet, and the Food 'Experts,'" ibid., 44 (Aug. 1961): 26–27; and "Food and Nutrition," *Consumer Bulletin Annual*, 1961–62, 176–77. Also see "National News," *Feed Age*, 11 (Jan. 1961): 43; "Chemical Spokesman Says 'Atmosphere of Understanding' Needed," *Feedstuffs*, 33 (May 6, 1961): 6, 184; "Pity the Layman," *Feed Age*, 11 (Dec. 1961): 82; and Thomas K. Jukes, "Food Additives," *Science*, 134 (1961): 798, 869.

36. Kennedy's quotations are from "Text of Kennedy's Message to Congress on Protections for Consumers," *New York Times*, Mar. 16, 1962, 16. Also see "Kennedy Asks for Broad New Authority to Reduce Tariffs," *New York Times*, Jan. 12, 1962, 13; "President's Message on Consumer

Protection," *FDCLJ*, 17 (1962): 224; Walter Sullivan, "Nutritionists Bid U.S. Expand Supervision of Food Additives," *New York Times*, Apr. 8, 1962, 1, 82; "Annual Report of the Food and Drug Administration," in *Annual Report of the U.S. Department of Health, Education, and Welfare*, 1962, 312–17; and "Drug Law Changes Pushed," *Journal of the American Pharmaceutical Association: Practical Edition*, n.S. 2 (1962): 203.

37. "Not Much Hope for Delaney Exemption," *Feedstuffs*, 34 (Mar. 3, 1962): 55; "Legislation to Exempt No-Residue Additives Reported Not Likely," ibid., 34 (Mar. 24, 1962): 4; "FDA Advisory Group Report Expected to Be Given by June 30," ibid., 34 (Apr. 14, 1962): 8, 87; "To Seek Legislation for Separate Controls on Veterinary Drugs," ibid., 34 (June 2, 1962): 1, 78; "Feed Men Urged to Support New Bill Which Would Ease Drug Problems," ibid., 34 (July 14, 1962): 1, 75; and *Congressional Record*, 87th Cong., 2d sess., 12713.

38. Robert E. Eckardt, "Environmental Carcinogenesis: Guest Editorial," *Cancer Research*, 22 (1962): 395–97. The essay to which Eckard was responding was W. C. Hueper, "Environmental Carcinogenesis and Cancers," ibid., 21 (1961): 842–57. Dubos's comments first appeared in René J. Dubos, "Man vs. Environment," *Industrial Medicine and Surgery*, 30 (1961): 369–73. Also see *Congressional Record*, 87th Cong., 2d sess., 13515; Wilbur J. Cohen to Hubert H. Humphrey, Apr. 11, 1962, FDA Papers, Accession no. 88-82-57, box 22, vol. 285; "Senate Gets Bill to Ease Industry Drug Problems," *Feedstufs*, 34 (July 21, 1962): 4; Morris Fishbein, "Control of Drugs," *New York Times*, May 12, 1962, 22; Edward G. Feldmann, "Dual Role of the FDA," *Journal of Pharmaceutical Sciences*, 51, (May 1962): I; and "American Chemical Society Symposium," *FDCLJ*, 17 (1962): 102–3.

39. For thalidomide, see, for example, Robert K. Plumb, "Deformed Babies Traced to a Drug," *New York Times*, Apr. 12, 1962, 37, 58; "Doctor's Action Bars Birth Defects," ibid., July 16, 1962, 24; "Used by Some in U.S.," ibid., July 20, 1962, 12; "Uncertainty on Drugs," ibid., July 27, 1962, 24; Marjorie Hunter, "U.S. Will Tighten Test Drug Rules," ibid., Aug. 3, 1962, 12; William M. Freeman, "Sales of Tranquilizers Decrease in Wariness over Thalidomide," ibid., Aug. 3, 1962, 12; "A Medal for Dr. Kelsey," *Chemical Week*, 91 (Aug. 11, 1962): 5; "The Thalidomide Lesson," *Science*, 137 (Aug. 17, 1962): 497; and "U.S. Again Warns on Thalidomide," *New York Times*, Aug. 24, 1962, 12. For a brief history of the event, see Ralph Adam Fine, *The Great Drug Deception (New York: Stein and Day,* 1972), 167–81. For *Silent Spring*, see Rachel Carson, *Silent Spring* (Boston: Houghton Mifflin, 1962); "You're Accused of Poisoning Food," *Farm Journal*, 86 (Sept. 1962): 29, 52; "Needed . . . Cool Minds," *Feed Age*, 12 (Sept. 15 1962): 98; "Lining Up on Pesticides," *Chemical Week*, 91 (Sept. 15, 1962): 41; "An Anti-Ag Chemical Mood Is to Be Reckoned With," *Feed Age*, 12 (Oct. 1962): 53–54; and "The Power of the Half Truth," ibid., 12 (Nov. 1962): 88.

40. See, for instance, "U.S. Investigates Testing of Drugs," *New York Times*, Aug. 1, 1962, 19; "Inadequate Test Charged," ibid., Aug. 1, 1962, 19; "Dr. Kelsey's Predecessor Told Senators of Drug Perils in 1960," ibid., Aug. 10, 1962, 1, 6; and "The Dangerous Environment of Man," *Consumers' Research Magazine*, 45 (Aug. 1962): 23–24.

41. "Drug Control Plan Advances in Senate," *New York Times*, July 13, 1962, 8; "Dr. Kelsey Calls for Tighter Restrictions on Drugs as Senate Hearings Open," ibid., Aug. 7, 1962, 15; Marjorie Hunter, "Senate Approves Drug Safety Bill; Vote Unanimous," ibid., Aug. 24, 1962, 1, 12; John A. Osmunden, "Scientists Fear New Laws May Curb Drug Research," ibid., Aug. 26, 1962, 1, 60; Marjorie Hunter, "Administration Is Urging House to Accept Senate's Drug Bill," ibid., Sept. 19, 1962, 22; "Unlock the Drug Bill," ibid., Sept. 20, 1962, 32; Marjorie Hunter, "House Votes Bill on Drug Control," ibid., Sept. 28, 1962, 1, 19; idem, "Strong Drug Bill Finally Emerges," ibid., Oct. 3, 1962, 23; "Drug Reform Bill Is Signed at White House, with Dr. Kelsey Present," ibid., Oct. 11, 1962, 31; "AVMA Supports FDA Control of Animal Feed Additives and Drugs," *JAVMA*, 141 (1962): 541; "Tougher Drug Laws," *Cattleman*, 49 (Sept. 1962): 31; "A.V.M.A. Urges F.D.A. Control over Additives and Drugs," ibid., 49 (Oct. 1962): 142; Fred Bailey, "Washington Report," *Feed Age*, 12 (Aug. 1962): 51; 12 (Sept. 1962): 55; "What's New in Washington," *Successful Farming*, 60 (Sept. 1962): 15; 60 (Oct. 1962): 15; "Washington Newsletter," *Chemical Week*, 91 (July 21, 1962):

55; "Riding a Scare," ibid., 91 (Aug. 11, 1962): 23; "Swift Action on Drugs," ibid., 91 (Aug. 18, 1962): 52; "Bitter Pill for Drugmakers," ibid., 91 (Sept. 1, 1962): 17; "Sweeter Taste for Proposed New Drug Laws," ibid., 91 (Sept. 15, 1962): 41–42; "New FDA Rules Will Affect You," *Wallaces Farmer*, 87 (Sept. 1, 1962): 8; "First Action on Drug Law Changes Taken," *Feedstuffs*, 34 (Aug. 11, 1962): 1, 87; John Cipperly, "House-Approved Drug Bill Would Ease Trade Snags," ibid., 34 (Sept. 29, 1962): 1, 71; "Hurdle in Drug Legislation Cleared," ibid., 34 (Oct. 6, 1962): 1, 95; "Drug Bill," *Journal of the American Pharmaceutical Association: Practical Edition*, n.s. 2 (1962): 517; and John L. Harvey, "The Omnibus Bill," *FDCLJ*, 17 (1962): 553–76.

42. The quotations are from 87 *U.S. Statutes* (1958): 570–71, 781. Also see *Congressional Record*, 87th Cong., 2d sess., 11250–53, 15717–18, 17289–96, 17364–92, 17395–422, 21049–100, 21242–46, 22037–57, 22315–25. The Kefauver hearings in the late 1950s and early 1960s helped marshal sentiment for the bill, which also required manufacturers to prove that substances used in products were generally considered to be efficacious; merely to demonstrate that a chemical was generally accepted as harmless was no longer sufficient.

43. *Congressional Record*, 87th Cong., 2d sess., 22053–57; Marjorie Hunter, "Food and Drug Agency Faces Two Investigations in Congress," *New York Times*, Oct. 14, 1962, 82; "Shakeup Ahead for FDA?" *Chemical Week*, 91 (Oct. 20, 1962): 25; "Fuss over FDA," *Feedstuffs*, 34 (Nov. 3, 1962): 4, 79; "FDA Criticism," ibid., 34 (Dec. 8, 1962): 7; Fred Bailey, Jr., "Washington Report," *Feed Age*, 12 (Nov. 1962): 47; and "Feed Control Officials Discuss New Drug Bill," ibid., 12 (Nov. 1962): 51.

44. Marjorie Hunter, "Shake-Up Asked for Drug Agency," *New York Times*, Oct. 26, 1962, 8.

45. Ibid. Also see "New Pressures on the Food and Drug Administration," *Chemical Week*, 91 (Nov. 3, 1962): 27; Boisfeuillet Jones, "Consumer Protection Activities," *FDCLJ*, 17 (1962): 808–12; and Marjorie Hunter, "Shifts Planned in Drug Agency," *New York Times*, Nov. 27, 1962, 27.

46. John Cipperly, "FDA Expected to Examine New Applications Closely," *Feedstuffs*, 34 (Oct. 13, 1962): 1, 75; "Stilbestrol Petitions Filed by FDA," ibid., 34 (Dec. 22, 1962): 59; and George P. Larrick to Ralph G. Smith, Oct. 16, 1962; and Ernest J. Umberger to Alan Spiher, Nov. 19, 1962, FDA Papers, Accession no. 88-82-57, box 22, vol. 287.

CHAPTER 4. MONITORS AND MONITORING

1. For a useful overview of the 1960s, see, for example, William O'Neill, *Coming Apart: An Informal History of the 1960s* (Chicago: Quadrangle Books, 1971); and Morris L. Dickstein, *Gates of Eden: American Culture in the 1960s* (New York: Oxford UP, 1987). Also see Tom Shachtman, *Decade of Shocks: Dallas to Watergate, 1963–1974* (New York: Poseidon, 1983); and Christopher Lasch, *The Culture of Narcissism: American Life in an Age of Diminishing Expectations* (New York: Norton, 1978).

2. For interpretations of the new crusades, see, for example, Erma Angevine, ed., *Consumer Activists: They Made a Difference* (New York: Consumers' Union Foundation, 1982); Robert N. Mayer, *The Consumer Movement: Guardians of the Marketplace* (Boston: Twayne, 1989), 34–58; and Mark V. Nadel, *The Politics of Consumer Protection* (Indianapolis: Bobbs-Merrill, 1971), 31–44, 101–29, and 155–218.

3. "Stilbestrol Regulation Expected 'within Matter of Weeks,'" *Feedstuffs*, 35 (Feb. 9, 1963): 1, 83; "FDA Begins Action on Requests for New Stilbestrol Use," ibid., 35 (Feb. 23, 1963): 1, 87; "New Clearances for Diethylstilbestrol," *Feed Age*, 13 (Mar. 1963): 88; "Tighten Regulations on Beef Cattle Feed Drug," *Cattleman*, 49 (Mar. 1963): 141; Don Muhn, "New Ruling on Stilbestrol Use in Feed," *Des Moines Register*, Mar. 6, 1963, 1; "FDA Issues Requirements on Diethylstilbestrol," *Feed Age*, 13 (Apr. 1963): 26; and "Requirements Issued for Safe Use of Diethylstilbestrol," *JAVMA*, 142 (1963): 792–93. The only letter received which protested the new DES rule was R. R. Stevens to HEW Hearing Clerk, Mar. 15, 1963, FDA Papers, Accession no. 88-82-57, box 22, vol. 297.

4. The quotations are from "HEW Letter on Animal Drug Amendments Bill (H.R. 7247),

September 16, 1963," *Food Drug Cosmetic Law Reporter: Developments*, 1963–67, 80125–27; "FDA Statement on Animal Drug Amendments Bill (H.R. 7247), September 17, 1963," ibid., 1963–67, 80127–29; and "If Present Regulations Had Existed Earlier, Would Not Now Have Growth Promotants," *Feedstuffs*, 35 (Sept. 28, 1963): 2, 78. Also see "Streamlined FDA Clearance Rules," *Feed Age*, 13 (Aug. 1963): 56; "Keep the Trends Moving," ibid., 13 (Dec. 1963): 66; "Drug Bill Hearings Set; Larrick Opposes Animal Drug Law," *New York Times*, Sept. 18, 1963, 21; "Strong Support for Drug Legislation," *Feedstuffs*, 35 (Sept. 21, 1963): 1, 92–94; and "Industry Problems Illustrate Need for Changes in FDA Law, Procedures," ibid., 35 (Sept. 28, 1963): 2, 77.

5. If Present Regulations, 78.

6. For the trade papers' and the Agricultural Research Institute's reaction to Burroughs's platform, see "Tight Controls," *Wallaces Farmer*, 88 (Oct. 5, 1963): 52; "Rapid National Growth—The Hard Way," *Feed Age*, 13 (Nov. 1963): 92; "Report of Projects and Proposals Committee," *Proceedings of the Agricultural Research Institute*, 12 (1963): 269; "Report of the Subcommittee on Feed Additives," ibid., 14 (1965): 224; and "Report of the Feed Adjuvant Regulations Committee," ibid., 15 (1966): 247–48. For the reactions of the FDA, HEW, and Congress, see Arthur L. Davis, "Confidential Draft of Animal Drug Bill under Study by FDA," *Feedstuffs*, 35 (Dec. 14, 1963): 4; and "No Action on Drug Bill; Humphrey Backs Position of FDA," ibid., 35 (Dec. 21, 1963): 1, 72–73.

7. Edward G. Feldmann, "Selection of Advisors for FDA," *Journal of Pharmaceutical Sciences*, 52 (Sept. 1963): i; "FDA Reorganization to Be Initiated," *Feedstuffs*, 35 (Nov. 2, 1963): 1, 4; "FDA Is Reorganized, Putting Added Emphasis on Its Role in Research," *Wall Street Journal*, Nov. 14, 1963, 15; "FDA Reorganization to Require Six Months," *Feedstuffs*, 35 (Nov. 9, 1963): 73; and Elinor Langer, "FDA: Drug Agency Answers Critics by Attempting to Step Up Science, but Many Critical Problems Remain," *Science*, 142 (Dec. 4, 1963): 1280–82.

8. The quotation on the one-hundred-to-one ratio is from "100-Fold Margin of Safety," *Quarterly Bulletin of the Association of Food and Drug Officials*, 18 (Jan. 1954): 33–35. Also see Bernard L. Oser, "Recent Developments on the Food-Additives Front," *FDCLJ*, 14 (1959): 260–61.

9. The cautionary quotations are from Food and Nutrition Board, Food Protection Committee, *Problems in the Evaluation of Carcinogenic Hazard from Use of Food Additives* (Washington, D.C.: National Academy of Sciences and National Research Council, 1960), 12, 18. Also see Morton L. Levin, Irwin D. J. Bross, and Paul R. Sheehan, "Food Additives," *Science*, 133 (Mar. 24, 1961): 947–48.

10. Food and Nutrition Board, Food Protection Committee, *Problems in the Evaluation*, 31–32.

11. Nathan Mantel and W. Ray Bryan, "'Safety' Testing of Carcinogenic Agents," *Journal of the National Cancer Institute*, 27 (1961): 457, 458, 468.

12. George H. Gass, Don Coats, and Nora Graham, "Carcinogenic Dose-Response Curve to Oral Diethylstilbestrol," *Journal of the National Cancer Institute*, 33 (1964): 971–77. Also of interest is Howard B. Andervont and Thelma B. Dunn, "Occurrence of Mammary Tumors in Castrated Agent-Free Male Mice after Limited or Repeated Exposure to Diethylstilbestrol," ibid., 33 (1964): 143–47.

13. See, for example, "Drug Agreement Progress Seen," *Feedstuffs*, 36 (Feb. 15, 1964): 1, 82; "Animal Drug Bill Runs into Delays," ibid., 36 (May 9, 1964): 2; "AHI Hears Roberts on Bill Compromises," *Feed Age*, 14 (June 1964): 77; "House Committee Change Noted; AHI Continues Work on New Legislation," *Feedstuffs*, 37 (July 31, 1965): 2; "AFMA Joins AHI in Supporting New FDA Lgislation," ibid., 37 (Oct. 23, 1965): 1, 67; "AHI Accepts Most of FDA Proposals on Its Drug Bill," ibid., 38 (July 2, 1966): 1, 5; and "Animal Drug Bill Apparently Killed by Late Opposition," ibid., 38 (Oct. 15, 1966): 1, 84.

14. Mintz's and Galbraith's quotations were on the dust jacket of Morton Mintz, *The Therapeutic Nightmare* (Boston: Houghton Mifflin, 1965). Also see Douglas W. Cray, "New Book Upsets U.S. Drug Makers," *New York Times*, Nov. 28, 1965, sec. 3, pp. 1, 12; Louis Lasagna, "The Medicines We Take," ibid., Nov. 28, 1965, sec. 7, p. 32. For contemporary defenses of the New

Journalism, see either Tom Wolfe, *The New Journalism* (New York: Harper and Row, 1973); or Everette E. Dennis and William L. Rivers, *Other Voices: The New Journalism in America* (San Francisco: Canfield, 1974).

15. "Report of the FDA Commissioner," in *Annual Report of the U.S. Department of Health, Education, and Welfare*, 1965, 323–24.

16. Larrick's quotation is from "Larrick to Retire from FDA; Reappraisal of Agency Set," *Feedstuffs*, 37 (Nov. 20, 1965): 1, 77–78. Also see "FDA Official Warns Feeders against Mis-Use of Additives," ibid., 37 (Feb. 27, 1965): 1, 8; "FDA Committee to Study Antibiotic Use," ibid., 37 (May 8, 1965): 2; "FDA Calls for Tissue Residue Data on Drugs," ibid., 37 (June 5, 1965): 6; "FDA Animal Drug Activities Upgraded into New Bureau," ibid., 37 (Oct. 2, 1965): 1, 65; "FDA's Advisory Council Discusses Feed Regulation," ibid., 37 (Nov. 27, 1965): 2, 76; "Washington Reports," *Feed Age*, 15 (Nov. 1965): 39; and William H. Summerson, "The Role of Scientific Research in the Food and Drug Administration," *FDCLJ*, 20 (1965): 427–32. The FDA also received criticism from consumers and others. See, for instance, "Group's 'Watchdog' Functions Reassigned," *Feedstuffs*, 37 (Jan. 30, 1965): 7; "Rep. Harris Orders Study of FDA," ibid., 37 (Apr. 17, 1965): 4; "FDA Public Relations Are at Low Ebb," *Feed Age*, 15 (Feb. 1965): 33; "It's an Unfamiliar Role for FDA," ibid., 15 (June 1965): 33; and "U.S. Laws Termed Bar to New Drugs," *New York Times*, Mar. 8, 1965, 34.

17. The quotations are from "USDA Is 'Department of Consumers,'" *Wallaces Farmer*, 93 (Mar. 23, 1968): 59; and Edward G. Feldmann, "Zero Tolerance Intolerable," *Journal of Pharmaceutical Sciences*, 54 (Sept. 1965): 1. Also see "Industry Groups Oppose USDA Moves," *Feedstuffs*, 37 (Jan. 9, 1965): 2, 78; "New Drive to Head Off Meat Residues," *Farm Journal*, 89 (Apr. 1965): 44; "Reorganization Move Begun by Freeman," *New York Times*, Feb. 9, 1965, 57; "Washington Roundup," *Cattleman*, 51 (May 1965): 136; "A Ray of Hope," *Feed Age*, 15 (Aug. 1965): 68; Bernard L. Oser, "The Mathematical, Legal, and Chemical Concepts of Zero," *FDCLJ*, 20 (1965): 597–604; and "'No Residue' and 'Zero Tolerance,'" ibid., 20 (1965): 608–22.

18. John N. S. White, "The Dangers of Feeding Stilbestrol to Beef Cattle," *Natural Food and Farming* (Feb. 1966): 35–38. Also see Trienah Meyers to Thomas M. Pelly, Feb. 9, 1966; Anthony M. Gawienowski to James L. Goddard, Feb. 24, 1966; Bruce Rothwell to Abraham Ribicoff, Mar. 26, 1966; Gawienowski to Wallace S. Brammell, Mar. 30, 1966; "Memorandum of Telephone Conversation between Mrs. Lane and Alexander Downs, Jr., April 7, 1966"; H. M. Dry to Ralph Yarborough, Apr. 18, 1966; Josephine Arliss to Clifford P. Case, June 6, 1966, W. N. Swain to Case, June 20, 1966; J. B. Thomas to George Murphy, June 21, 1966, FDA Papers, Accession no. 88-82-57, box 22, vol. 291; Rank and Filers of Local Union no. 10, Rocky Mount, N.C., to the USDA, May 9, 1966, "Memorandum from William H. Summerson to M. D. Kinslow, Nov. 16, 1966," FDA Papers, Accession no. 88-82-57, box 22, vol. 292; H. H. Roswell, "Publisher's Report," *National Police Gazette*, 171 (May 1966): 2; "Parade's Special Intelligence Report," June 20, 1966, 3; John N. S. White to Augustus Hawkins, Oct. 20, 1966, FDA Papers, Accession no. 88-82-57, box 22, vol. 291; and George McGrath, "Meat—America's No. 1 Health Hazard . . . Why?" *National Police Gazette*, 171 (Dec. 1966): 28–29.

19. Goddard's indictment is from "Drug Manufacturers Accused by FDA Chief of 'Irresponsibility,'" *Wall Street Journal*, Apr. 7, 1966, 2. Also see Jonathan Spivak, "New FDA Chief Faces Task of Increasing Unit's Scientific, Administrative Capacity," ibid., Jan. 12, 1966, 4; John Walsh, "Food and Drug Administration: Test for Leadership Vaccine," *Science*, 151 (1966): 801–3; "FDA Has Huge Growth," *JAVMA*, 148 (1966): 236; "The President's Message on Consumer Interests," *FDCLJ*, 21 (1966): 180; "Food and Drug Agency's New Head Takes Over," *Feedstuffs*, 38 (Jan. 22, 1966): 1, 72; and Edward G. Feldmann, "New Era at FDA," *Journal of Pharmaceutical Sciences*, 55 (Apr. 1966): 1.

20. Goddard explained his motivation in "Committee on Veterinary Medicine to Advise FDA," *JAVMA*, 148 (1966): 730–31. Also see "Washington Report," *Feed Age*, 16 (Apr. 1966): 43; 16 (June 1966): 35–36; "FDA Announces Advising Committee," *Feedstuffs*, 38 (Feb. 12, 1966): 6, 68; "The Advisory Committee," *Feed Age*, 16 (Mar. 1966): 7; "Dr. C. D. Van Houweling Expected to Head

FDA Veterinary Unit," *Feedstuffs*, 38 (Oct. 22, 1966): 1, 68; "New FDA Magazine Explained," ibid., 38 (Nov. 12, 1966): 8, 89; and Jack Kiesner, "FDA Vet Bureau's Growth Traced," ibid., 38 (Dec. 3, 1966): 2, 74. Goddard also appointed former Senator Maurine Neuberger as consultant on consumer relations and approved launching a "new four color, slick paper FDA magazine" to press the agency's public case. Called the *FDA Papers*, it claimed "not [to] be a news magazine," but to "define agency policies in a clear, concise manner" to "reflect the official FDA position." (These quotations are from "New FDA Magazine Explained.") The new magazine was soon renamed *FDA Consumer*. To reduce confusion, the periodical will be referred to by the latter designation throughout this book.

21. The quotations from the committee report are from Jane E. Brody, "Study Urges Care in Antibiotic Use," *New York Times*, Aug. 22, 1966, 1, 29. Also see "Drugs Approved before '62 Face Rescreening," *New York Times*, Mar. 19, 1966, 1, 30; "Academy to Study Efficacy of Drugs," ibid., June 25, 1966, 31; Jack Kiesner, "FDA Expected to Study Antibiotics," *Feedstuffs*, 38 (Aug. 13, 1966): 1, 8; "New York Times Questions Feed Use of Antibiotics, Asks Official Inquiry," ibid., 38 (Aug. 13, 1966): 8; and "Big Brother and Antibiotics," *Feed Management*, 17 (Oct. 1966): 80.

22. Jane E. Brody, "F.D.A. Seeks Curb on Drugs in Food," *New York Times*, Aug. 23, 1966, 36; "Government Plans High Level Meeting on Use of Antibiotics," *Feedstuffs*, 38 (Sept. 3, 1966): 1, 74; "ARS Begins Study of Drug Resistance," ibid., 38 (Sept. 24, 1966): 4; "Medicated Feed Symposium Set for June 5–7," ibid., 38 (Oct. 22, 1966): 4; "Responsibility of FDA Veterinary Branch Probably Will Increase," ibid., 38 (Nov. 26, 1966): 14; "Resolutions," *Proceedings of the Agricultural Research Institute*, 16 (1967): 277; "Medicated Feeds," *Feed Management*, 18 (Apr. 1967): 8; Food and Nutrition Board, Food Protection Committee, *Toxicants Occurring Naturally in Foods*, Publication no. 1354 (Washington, D.C.: National Academy of Sciences and National Research Council, 1966); and Committee on Animal Nutrition, Subcommittee on Hormones, *Hormonal Relationships and Applications in the Production of Meats, Milk, and Eggs*, Publication no. 1415 (Washington, D.C.: National Academy of Sciences and National Research Council, 1966).

23. The quotation is from Jack Kiesner, "FDA Official Calls Antibiotic Use 'Potential Hazard,'" *Feedstuffs*, 39 (July 15, 1967): 2, 88. Also see "Scientific and Public Health Issues of Medicated Feeds to Be Discussed at Symposium," *JAVMA*, 150 (1967): 110–11; "The Barn Door Will Stay Unlocked," *Feed Management*, 18 (Jan. 1967): 80; "Plans Complete for Symposium on Drug Usage in Feeds," *Feedstuffs*, 39 (Apr. 1, 1967): 9, 72; "Technical Advisory Committee Report," *Proceedings of the Nutrition Council of the AFMA*, May 1967, 7; "No Sure Human Hazard Shown at FDA's Vet Drug Symposium," *Feedstuffs*, 39 (June 10, 1967): 1, 69; "Independent Agency Research Guidelines Asked at Symposium," ibid., 39 (June 10, 1967): 1, 72; "Dr. Goddard Asks 'Better Handle' on Drug Usage," ibid., 39 (June 10, 1967): 1, 73; "Drug Resistant Bacteria Issue Reviewed," ibid., 39 (June 17, 1967): 3, 75; Harold M. Schmeck, Jr., "More Data Urged on Animal Drugs," *New York Times*, June 6, 1967, 96; idem, "Scientists Study Feed Antibiotics," ibid., June 11, 1967, 54; "Washington's Impact," *Feed Management*, 18 (July 1967): 6; "National Academy of Sciences," *JAVMA*, 151 (1967): 405; "FDA Head Comments on Additive Safety and Further Work," ibid., 39 (Aug. 26, 1967): 4, 65; and "Feed Adjuvants Committee Report," *Proceedings of the Agricultural Research Institute*, 16 (1967): 264–65. Proceedings of the medicated feeds conference were published as National Academy of Sciences, *The Use of Drugs in Animal Feeds: Proceedings of a Symposium*, Publication no. 1679 (Washington, D.C.: National Academy of Sciences, 1969).

24. Some personalized the charges, likening Goddard to "a kind of Rachel Carson crusader who would trade adequate food supplies, and perhaps even starvation, for the right to hear the birds sing every morning." Others pointed to the movie *In like Flint* to argue that Goddard's characteristic smirk, mannerisms, and "flair for the dramatic" reminded them of this James Bond spoof's title character. The personalized charges, as well as the quotations in the text, are from William C. Coleman, "Antibiotics—at the Crossroads," *Feed Management*, 18 (Aug. 1967): 28, 30; "No Case Was Made—We Demand Facts," ibid., 18 (Aug. 1967): 29, 31; Louis Rothschild, Jr., "The New FDA," *Quarterly Bulletin of the Association of Food and Drug Officials of the United States*, 31 (Oct. 1967): 187–96; Wallace Werble, "The New FDA," ibid., 31 (Oct. 1967): 197–201; "F.D.A.

Attacked by A.M.A.'s Chief," *New York Times*, Apr. 30, 1966, 34; and Jane Brody, "Ex-Drug Official Scores Controls," ibid., Apr. 20, 1966, 30.

25. "Memorandum of a Conference between the Consumer and Marketing Service of the USDA and the Bureau of Veterinary Medicine of the FDA, February 3, 1967, and February 24, 1967," FDA Papers, Accession no. 88-82-57, box 22, vol. 293; Fred J. Kingma, "Establishing and Monitoring Drug Residue Levels," *FDA Consumer*, 1 (July 1967): 8–9, 31–33; and C. D. Van Houweling, "Drugs in Animal Feeds? A Question without an Answer," ibid., 1 (Sept. 1967): 11–15.

26. "Its Best to Do as Told," *Feedlot*, 9 (Aug. 1967): 26; "Deputy Rankin Speaks to AAFCO on FDA's Worry about Drugs in Feeds," *Feed Management*, 18 (Sept. 1967): 91; "Drugs for Livestock," *Successful Farming*, 65 (Sept. 1967): 11; "Residue Problems Compounded by Big Feed lots," *JAVMA*, 151 (Dec. 2, 1967): 2, 69; "FDA Reports More Frequent Detection of Stilbestrol Residues," *Feedstuffs*, 39 (Dec. 16, 1967): 7; and "Memorandum of Conference between the CMS, USDA, and BVM, FDA, March 24, 1968," FDA Papers, Accession no. 88-82-57, box 22, vol. 294.

27. The FDA wanted to avoid the morass of legal proceedings in seeking compliance. It expressed profound concern over the apparent disregard for regulations and warned cattlemen that they faced severe penalties. Farm publications hoped the specter of a stilbestrol-free beef industry would sober up producers. See, for example, "FDA Reports More Frequent Detection of Stilbestrol Residues," 7; Director, Kansas City District, to Director, Bureau of Veterinary Medicine, Mar. 24, 1968, FDA Papers, Accession no. 88-82-57, box 22, vol. 294; "Problems Exist, but Progress Made," *Feedstuffs*, 40 (Jan. 20, 1968): 1, 65; Fred J. Delmore, "Voluntary Compliance," *FDA Consumer*, 2 (Feb. 1968): 12–15; Edwin I. Goldenthal, "Current Views on Safety Evaluation of Drugs," ibid., 2 (May 1968): 14–18; and "Warning on Drugs," *Successful Farming*, 66 (June 1968): 9.

28. Luther G. Ensminger to George Schwartzman, Mar. 1, 1967; Ensminger to Carl Ponder, Apr. 11, 1967; and Ponder to Schwartzman, Aug. 22, 1967, FDA Papers, Accession no. 88-82-57, box 22, vol. 293; Schwartzman to William Rader, Mar. 25, 1968, Ponder to Schwartzman, July 5, 1968; Director, Kansas City District, to Director, Bureau of Medicine, Aug. 6, 1968; and Ernest J. Umberger to Helen L. Reynolds, Feb. 14, 1969, FDA Papers, Accession no. 88-82-57, box 22, vol. 294; Carl Ponder, "Diethylstilbestrol in Eggs, Animal Tissue, and Milk," Jan. 15, 1969, FDA Papers, Accession no. 88-82-57, box 22, vol. 294; George Schwartzman, "Report on Drug Residues in Animal Tissues," pts. 1 and 2, *JAOAC*, 51 (1968): 270–71; 52 (1969): 226; and Louis N. Jones, Murray Seidman, and B. C. Southworth, "Quantitative Determination of Diethylstilbestrol by Thin-Layer Chromatography," *Journal of Pharmaceutical Sciences*, 57 (1968): 646–49. Also see William Horwitz, "The Role of the AOAC in Food and Drug Control," *Quarterly Bulletin of the Association of Food and Drug Officials*, 30 (Apr. 1966): 81–89.

29. Goddard is quoted in "FDA Proposes to Restrict Some High Level Oral Antibiotics," *Feedstuffs*, 40 (Apr. 20, 1968): 1, 73. Also see "Goddard Resigns as FDA Commissioner," *Feedstuffs*, 40 (May 25, 1968): 1, 46; "Dr. Herbert Ley to Be New FDA Head," ibid., 40 (June 8, 1968): 1; Edward G. Feldmann, "Changing the Guard at FDA," *Journal of Pharmaceutical Sciences*, 57 (July 1968): 1; and "New FDA Commissioner Talks about Government Issues Affecting Agribusiness," *Feedstuffs*, 40 (Oct. 5, 1968): 2, 78.

30. The quotations are from Ron Lutz, "Groups Cite Opposition to FDA Drug Ban Talk," *Wallaces Farmer*, 93 (June 22, 1968): 5; "Sees No Evidence to Link Animal Antibiotics to Human Disease," *Feedstuffs*, 40 (June 22, 1968): 7, 58; and "Opponents of Antibiotic Restrictions Continue to File Comments," ibid., 40 (July 20, 1968): 7, 58. Also see "Antibiotics—Food Producing Animals—Restricted Use," *Food Drug Cosmetic Law Reporter: Developments*, 1967–70, 40307–8; William E. Brennan, "Administrative Hearings in FDA," *FDA Consumer*, 2 (July–Aug. 1968): 14–17, 25–26; "FDA Proposes to Restrict Oral Antibiotics," 1, 73; "Washington's Impact," *Feed Management*, 19 (May 1968): 8; "Washington Report," *Successful Farming*, 66 (Sept. 1968): 7; 66 (Nov. 1968): 9; "Animal Scientists, Others Question Antibiotic Policy," *Feedstuffs*, 40 (Aug. 10, 1968): 4, 57; and "Brouse Tells of AHI Program to Study Effects of Antibiotics," ibid., 40 (Oct. 12, 1968): 7, 85. Also see William R. Pendergast, "Have the FDA Hearing Regulations Failed Us,"

FDCLJ, 23 (1968): 524–31; and "The Diagnosis and Treatment of FDA Hearings," *Quarterly Bulletin of the Association of Food and Drug Officials*, 34 (1970): 23–28.

31. "Animal Drug Bill Passes House, Now Awaits Action in Senate," *Feedstuffs*, 38 (Oct. 8, 1966): 1, 78; "Animal Drug Bill Introduced; Approval Said to Be Likely," ibid., 39 (Jan. 28, 1967): 1, 66; "House Approval Expected on Animal Drug, Feeds Bill," ibid., 39 (Oct. 7, 1967): 1, 8; "Senate May Not Consider Animal Drug Bill This Year," ibid., 39 (Oct. 28, 1967): 1, 54; "House Approves Animal Drug Bill; Hearing in Senate Put Off until 1968," ibid., 39 (Nov. 11, 1967): 2, 74; "Washington's Impact," *Feed Management*, 18 (June 1967): 6; "Excerpts from House Report No. 875 Accompanying H.R. 3639, October 31, 1967," *Food Drug Cosmetic Law Reporter: Developments*, 1967–70, 40743–46; "Washington's Impact," *Feed Management*, 19 (Feb. 1968): 8; 19 (Sept. 1968): 8; "Vote on Animal Drug Amendment by Senate Foreseen," *Feedstuffs*, 40 (May 25, 1968): 1, 147; "FDA Amendments Added to Drug Bill by Senate Unit," ibid., 40 (June 15, 1968): 1, 68; "Animal Drug Bill Cleared, Goes to President," ibid., 40 (July 6, 1968): 1; and "The Job Goes On," ibid., 40 (July 20, 1968): 10.

32. The optimistic sentiments are from "Report of the FDA Commissioner," in *Annual Report of the United States Department of Health, Education, and Welfare for 1968*, 291, 315; and Colston E. Warne, "The Consumer Movement, a New and Significant Force in the American Economy," *Quarterly Bulletin of the Association of Food and Drug Officials*, 32 (1968): 11–12. Also see "Washington's Impact," *Feed Management*, 19 (Aug. 1968): 6; and Charles C. Johnson, Jr., "The Future of Consumer Protection," *FDCLJ*, 23 (1968): 608–13.

33. The quotations of the Committee of Sixteen are from "Industry Group Calls for Stronger FDA, with Voluntary Compliance Program," *Feedstuffs*, 41 (Feb. 1, 1969): 7, 51. Also see "Drug Men Welcome F.D.A. Changes," *New York Times*, Jan. 5, 1969, sec. 3, p. 38; William P. Woods, "The Murky Crystal Ball—Future Drug Legislation," *FDCLJ*, 24 (1969): 119–25; and Thomas H. Jukes, "Observations on Safety of Feeding Antibiotics to Animals," ibid., 41 (Mar. 1, 1969): 12.

34. Gordon E. Bivens, "Editorial Note," *Journal of Consumer Affairs*, 1 (1967): 5–6; and Dan M. Burr, *Abuse of Trust: A Report on Ralph Nader's Network* (Chicago: Regnery Gateway, 1982), 25–33, 39–46, 51–71, 177–79, 184–89, 208–9. Also of use are Thomas R. Dunlap, *DDT: Scientists, Citizens, and Public Policy* (Princeton: Princeton UP, 1981), 142–51; Charles McCarry, *Citizen Nader* (New York: Saturday Review, 1972), 181–237; Nadel, *Politics of Consumer Protection*, 176–218; and Michael Pertschuk, *Revolt against Regulation: The Rise and Pause of the Consumer Movement* (Berkeley and Los Angeles: U of California P, 1982), 13–45.

35. Nader is quoted in "Nader Starts Drive on Food Industry," *New York Times*, Apr. 18, 1969, 17. Also see "FDA Eases Proposed Rules on Antibiotics in Food Animals, Modifies Ad Restrictions," *Wall Street Journal*, May 19, 1969, 6; John D. Morris, "Nader Recruits Eighty New Student 'Raiders' to Investigate the Operations of a Dozen Federal Agencies," *New York Times*, June 1, 1969, 43; and "FDA Official Reviews Current Status of Antibiotic Use in Food Animals," *Feedstuffs*, 41 (July 12, 1969): 8.

36. Nelson is quoted in "U.S. Drug Testing Termed Inadequate at Senate Inquiry," *New York Times*, Apr. 24, 1968, 48; and Walter Rugaber, "Senator Seeks a Federal Center to Take Over Testing of Drugs," ibid., July 30, 1969, 1, 43. Also see Alan T. Spiker, Jr., "Premarketing Clearance: A Bulwark of Consumer Protection," *FDA Consumer*, 3 (July–Aug. 1969): 4–7; and "Senator Nelson Would Have FDA Test All New Drugs," *Feedstuffs*, 41 (Aug. 2, 1969): 6.

37. The Ley committee quotations are found in Harold M. Schmeck, Jr., "Food-Drug Board Found Too Limited," *New York Times*, Aug. 8, 1969, 1, 17; and Jack Kiesner, "Report Criticizes, Advises FDA on Feed, Drug Work," *Feedstuffs*, 41 (Aug. 9, 1969): 1, 8. For an early discourse on risk-benefit analysis in regulating matters, see Chauncey Starr, "Social Benefit versus Technological Risk," *Science*, 165 (Sept. 19, 1969): 1232–38.

38. "Agribusiness Relations with USDA, FDA at Favorable Point," *Feedstuffs*, 41 (Sept. 6, 1969): 1, 216; "Concentrates," *Chemical and Engineering News*, Sept. 8, 1969, 39; "Government Officially Announces Cyclamate Sweeteners Will Be Taken off Market Early Next Year," *New York Times*, Oct. 19, 1969, 58; *Food Drug Cosmetics Law Reporter: Developments*, 1967–70,

40934–35, 40959–61; Richard D. Lyons, "Congressman Says Actress's Speech Helped Bar Cycla-mates," *New York Times*, Oct. 22, 1969, 26; and "A Decision on Cyclamates," *FDA Consumer*, 3 (Nov. 1969): 6–7.

39. See, for example, "Pressure for FDA Reorganization Increases," *Feedstuffs*, 41 (Oct. 18, 1969): 6; "How U.S. Keeps Tabs on Food Additives," *National Observer*, Nov. 3, 1969, 7; Phillip H. Abelson, "Chemicals and Cancer," *Science*, 166 (Nov. 7, 1969): 693; Joel R. Kramer, "Yester-day Cyclamates, Today 2, 4, 5-T, Tomorrow DDT," *Science*, 166 (Nov. 7, 1969): 724; Sandra Blakeslee, "Food Safety a Worry in Era of Additives," *New York Times*, Nov. 9, 1969, 1, 74; James J. Nagle, "Rep. Hall Backs Food Plan Change," ibid., Nov. 11, 1969, 75; and "On Cyclamates," *Chemical and Engineering News*, 47 (Dec. 1, 1969): 6–8.

40. The quotations are from "Nixon Asks Congress for Thorough FDA Review," *Feedstuffs*, 41 (Nov. 1, 1969): 1, 67; and "Finch Wants Softer Law Setting Cyclamate Levels," *New York Times*, Nov. 6, 1969, 5. Also see "A.M.A. Assails Law Barring Cyclamate," ibid., Nov. 8, 1969, 25; "Delaney Clause: AMA Asks for Change," *Chemical and Engineering News*, 47 (Nov. 17, 1969): 14; "Finch Removing Ley as FDA Head in Major Shakeup," *Wall Street Journal*, Dec. 11, 1969, 2; "Veterinary Unit Survives FDA Re-shuffle—Ley Doesn't," *Feedstuffs*, 41 (Dec. 13, 1969): 1, 8; and "FDA Reorganization," ibid., 41 (Dec. 20, 1969): 10.

41. Ley's indictment is reported in Richard D. Lyons, "Ousted F.D.A. Chief Charges 'Pressure' from Drug Industry," *New York Times*, Dec. 31, 1969, 1, 13. Also see "Ley to Resign Today as HEW Unit Chief; Edwards to Take Job," *Wall Street Journal*, Dec. 12, 1969, 26; "Dr. Ley Leaving U.S. Service Today," *New York Times*, Dec. 12, 1969, 33; "Trouble over Drugs on the Market," ibid., Jan. 4, 1970, sec. 4, p. 5; and "Department of Health, Education, and Welfare Reorganiza-tion Order," *Food Drug Cosmetic Law Reporter: Developments*, 1967–70, 40971–73.

42. *Congressional Record*, 91st Cong., 1st sess., 40615–16. Also see "FDA Explains Policy regarding Stilbestrol Feeding, Implants," *Feedstuffs*, 40 (Nov. 23, 1968): 4, 64; "Results with New Carrier for DES Implants Reported," ibid., 40 (Dec. 7, 1968): 2, 61; "Washington's Impact," *Feed Management*, 20 (Jan. 1969): 8; Director, Kansas City District, to Director, Bureau of Veterinary Medicine, May 22, 1969, FDA Papers, Accession no. 88-82-57, box 22, vol. 295; "FDA Notes Better Compliance with Withdrawal Rules," *Feedstuffs*, 41 (July 12, 1969): 1, 55; "Stilbestrol Use," ibid., 41 (Aug. 2, 1969): 10; "DES Feed-Implant Combination Sought by Hess and Clark," ibid., 41 (Oct. 11, 1969): 4; and "Ernest J. Umberger to Lee Stelling, July 10, 1969, FDA Papers, Accession no. 88-82-57, box 22, vol. 295.

43. Pharmacological Testing Department of Eli Lilly and Company, "Biological Assay for the Estimation of Diethylstilbestrol Residue in Cattle Tissue following a 20-Mg. Oral Treatment;" and Ernest J. Umberger to Ann Holt, June 20, 1969, FDA Papers, Accession no. 88-82-57, box 22, vol. 295; "FDA Asked to OK Double Stilbestrol Limit," *Farm Journal*, 93 (Aug. 1969): B13; "Elanco Announces More Active Form of Diethylstilbestrol, *Feedstuffs*, 41 (Nov. 22, 1969): 4; "Scientists Explain Chemistry Involved in DES Finding," ibid., 41 (Nov. 29, 1969): 4, 52; "Two New Gain Boosters for Beef Feeders," *Farm Journal*, 94 (Jan. 1970): 30; "Elanco Perfects Improved Form of Diethylstilbestrol," *Feed Management*, 21 (Jan. 1970): 50–51; "Discovery Leads to 'Tune-Up' for DES," *Feedlot Management*, 12 (Jan. 1970): 58; "Now, a New Kind of Stilbestrol," *Successful Farming*, 68 (Jan. 1970): 38; and R. S. Rutherford, "Gas Chromatographic Determination of CIS-and Trans-Diethylstilbestrol in Feed Premix," *JAOAC*, 53 (1970): 1242–43.

CHAPTER 5. HEARTS, NOT MINDS

1. Erma Angevine, "The Right to Excellence," *FDCLJ*, 25 (1970): 46.

2. James S. Turner, *The Chemical Feast: The Ralph Nader Study Group Report on Food Protec-tion and the Food and Drug Administration* (New York: Grossman, 1970), 101, 116, 192, 248–49. Also see Philip M. Boffey, "Nader's Raiders on the FDA: Science and Scientists 'Misused,'" *Sci-ence*, 168 (Apr. 17, 1970): 349–52. Congress and the FDA investigated charges of scientific falsi-fication and distortion within the agency but found nothing improper. See, for example, "Manipula-tion of Laboratory Tests Is Laid to Aides in Federal Drug Agency," *New York Times*, May 1, 1970,

21; "F.D.A. Probe Needed," ibid., May 4, 1970, 36; "Federal Food and Drug Agency Demotes a Critic," ibid., May 27, 1970, 27; and *Hearing on the Regulatory Policies of the Food and Drug Administration before a Subcommittee of the Committee on Government Operations, House of Representatives, Ninety-first Congress, Second Session, June 9, 1970* (Washington, D.C.: GPO, 1970).

3. For a sampling of the response, see, for example, David E. Rosenbaum, "F.D.A. Called Tool of Food Industry," *New York Times*, Apr. 9, 1970, 17; "'Nader's Raiders' Castigate FDA, Charging Failure in Enforcement of Food Regulations," *Wall Street Journal*, Apr. 9, 1970, 10; "FDA Influenced Too Much by Food Trade, Nader Report Charges," *Feedstuffs*, 42 (Apr. 11, 1970): 1, 8; "Nader's Report Calls for Review of Adding Antibiotics to Animal Feeds," ibid., 42 (Apr. 18, 1970): 2; "Nader Raiders Attack Recommendations on Veterinary Drugs," *JAVMA*, 157 (1970): 277–78; and John Leonard, "Nader's Raiders Ride Again," *New York Times*, July 24, 1970, 29.

4. Turner, *Feast*, 174–75, 178, 229–34, 251.

5. Roger Berglund, "Communicating with Consumer Advocates," *Feedstuffs*, 42 (Apr. 18, 1970): 10. Also see "Nader Report Challenged," ibid., 42 (May 30, 1970): 10; Neil Tietz, "Public Must Be Better Informed of Medicated Feed Benefits: Affleck," ibid., 42 (June 6, 1970): 7; and "Too Often Politics Influence Action by the Regulators: Patton," ibid., 42 (May 9, 1970): 6. For Nader's intentions, see "FDA Influenced Too Much," 8; "Nader Raiders Attack," 278; and John D. Morris, "Nader Plans Expanded Summer Investigations of Government and Busines," *New York Times*, June 21, 1970, 51.

6. Edwards set out the agency's position in George Gates, "FDA Moving Ahead on Feed Additive Issues," *Feedstuffs*, 42 (May 2, 1970): 1, 51; "More Attention to Be Given to Safeguarding Food Supply: Edwards," ibid., 42 (May 23, 1970): 26; and "Edwards Outlines Role of FDA, Sees Emphasis on 'Scientific Judgment,'" ibid., 42 (July 4, 1970): 2.

7. John S. Lang, "Cancer-Inciting Hormone Found in U.S. Beef Supply," *Des Moines Register*, June 24, 1970, 1. On Edwards's actions at the FDA, see "FDA's Edwards Names Task Force to Study Use of Antibiotics in Feeds," *Feedstuffs*, 42 (May 9, 1970): 1, 50; "More Representative Task Force Needed," ibid., 42 (May 23, 1970): 14; "F.D.A. Managers Are Enthusiastic about Changes They Are Making," *New York Times*, May 31, 1970, 46; "U.S. Aims to Erase Law Banning Food Additives That Cause Cancer," *Wall Street Journal*, June 10, 1970, 17; Jonathan Spivak, "Total Ban on Cyclamates in Food Hinted as FDA Plans for Session of Experts," ibid., June 23, 1970, 4; and "AHI Requests Representation of Trade on Antibiotic Task Force," *Feedstuffs*, 42 (July 18, 1970): 2.

8. Lang, "Cancer-Inciting Hormone," 1.

9. See chapter 3, above, for the history and purpose of the proviso.

10. Van Houweling is quoted in "Van Houweling Notes Increased Attention to Animal Drug Matters," *Feedstuffs*, 42 (Aug. 8, 1970): 4. Trenkle's remarks are captured in "Animal Scientists Defend Stilbestrol against Cancer Charge," ibid., 42 (July 4, 1970): 2, 55. Also see "Government Officials Claim DES Reports in Press 'Misleading,'" ibid., 42 (July 11, 1970): 8; "DES' Record of Safety, Usefulness Cited in Statement," ibid., 42 (June 27, 1970): 4; and "Tired of Consumerism," *Feedlot Management*, 12 (Sept. 1970): 11; Ralph Sanders, "Food Residues," *Successful Farming*, 68 (Sept. 1970): 32–33; and "Federal Pressure Certain to Build as Consumerism Emphasis Continues," *Feed Management*, 21 (Oct. 1970): 18–20.

11. Ron Lutz, "Crackdown on Residues in Meats," *Wallaces Farmer*, 95 (Aug. 8, 1970): 16 and 16-A; "FDA Warns Cattlemen of DES Hormone Ban If It Persists in Beef," *Wall Street Journal*, Aug. 21, 1970, 16; "Cattle Industry Warned on the Use of a Hormone," *New York Times*, Aug. 23, 1970, 51; and "Heed Regulations for Drugs in Feeds," *Wallaces Farmer*, 95 (Sept. 12, 1970): 5.

12. "Educational Program on DES Use Planned," *Feedstuffs*, 42 (Sept. 12, 1970): 4, "DES Cleared at 5–20 Mg. Range," ibid., 42 (Sept. 19, 1970): 6; "Double Dose to Cattle of Cancer-Linked Hormone," *Des Moines Register*, Sept. 17, 1970, 1, 3; "New Level for Stilbestrol Is Approved," *Wallaces Farmer*, 95 (Oct. 10, 1970): 13; "Use Additive Properly," *Feedstuffs*, 42 (Oct. 10, 1970): 10; "Stabilized Stilbestrol for Hormone Feeding," *Feedlot Management*, 12 (Nov. 1970): 38; "New Level for DES," ibid., 12 (Nov. 1970): 16–17; "DES Approved for Broader Beef Use,"

Feed Management, 21 (Nov. 1970): 45; "Higher Stilbestrol Level Ok'd," *Farm Journal*, 94 (Nov. 1970): B12; and "Withdraw Stilbestrol before You Sell Cattle," *Wallaces Farmer*, 95 (Nov. 28, 1970): 22.

13. "FDA to Prosecute Feeders If Drug Residues Found; Voluntary Cattle Program Planned," *Feedstuffs*, 42 (Oct. 10, 1970): 1, 65; Don Muhm, "Cattlemen Back Stilbestrol Step," *Des Moines Register*, Oct. 18, 1970, 1-F and 2-F; "Drug Withdrawal," *Feedstuffs*, 42 (Nov. 14, 1970): 10; Greg Lauser, "Public Concern May Force Curtailment of Drug Use in Feeds: AFMA's Boyd," ibid., 42 (Dec. 19, 1970): 1, 57; "New Pressure on Livestock Feeders," *Farm Journal*, 94 (Dec. 1970): 21; "New Residue Control Plan Needs Total Cooperation," *Feedlot Management*, 12 (Dec. 1970): 9; "DES Certificates," ibid., 12 (Dec. 1970): 62; and "Cattlemen Assume Lead in Regulatory DES Use," *JAVMA*, 158 (1971): 157–58.

14. The CMS position is outlined in "Cut Check on Residue in Meats," *Des Moines Register*, Nov. 22, 1970, 12-G. Also see "USDA Reduces Meat Sampling for DES, Antibiotic Residues," *Feedstuffs*, 42 (Nov. 28, 1970): 4.

15. Jack Kiesner, "C and MS to Sharply Increase Sampling of Beef Carcasses for DES Residues," *Feedstuffs*, 42 (Dec. 12, 1970): 7.

16. George D. Lakata, "Methods Evaluation in the Chemical Analysis of Drugs in Animal Tissues," *JAOAC*, 53 (1970): 225–26; Anthony J. Malanoski, "Regulatory Control of Hormone Residues by Chemical Analysis," ibid., 53 (1970): 226–28; and "BVM Reviews Activities in Areas of Antibiotics, Liquid Supplements," *Feedstuffs*, 43 (Jan. 30, 1971): 2.

17. On the eve of these congressional hearings, the FDA cited its first DES-feeding violator. See "U.S. Moving to Sue Farmers for Trace of Drugs in Cattle," *New York Times*, Feb. 9, 1971, 41; "FDA to Set Up Enforcement on Medicated Feeds," *Feedstuffs*, 43 (Feb. 27, 1971): 6; "Industrywide Group to Launch Drug Certification Program March 1," ibid., 43 (Feb. 27, 1971): 6; John A. Rohlf, "Your Beef Business," *Farm Journal*, 95 (Apr. 1971): B20; "Animal Drug Withdrawal Certificate," *Farm Quarterly*, 26 (May–June 1971): 55; "Stilbestrol Use Certificates," *Successful Farming*, 69 (Jan. 1971): 21; "Drug Residues," ibid., 69 (Mar. 1971): 7; S. L. McHenry, "The Animal Drug Certification Program," *FDA Consumer*, 5 (Apr. 1971): 18–20; "Drug Certification Program," *IBPA Journal*, 1 (May 1971): 13; "NADCC Starts Drug Certification Program," *JAVMA*, 158 (1971): 1510–11; "Washington's Impact," *Feed Management*, 22 (May 1971): 6; and "Drug Withdrawal Certificate Clarified," *Wallaces Farmer*, 96 (June 26, 1971): 3.

18. "House Hearing to Investigate Human Health Implications of Feed Additives," *Feedstuffs*, 43 (Feb. 6, 1971): 2; *Congressional Record*, 92d Cong., 1st sess., 3984–92; and "House Subcommittee to Hear Testimony on Food, Feed Additives," *Feedstuffs*, 43 (Mar. 13, 1971): 5. The quotation in the text defines the subject of the Senate committee hearing, as given in the *Congressional Record*.

19. *Regulation of Food Additives and Medicated Animal Feeds: Hearings before a Subcommittee of the Committee on Government Operations, House of Representatives, Ninety-second Congress, First Session, March 16, 17, 18, 29, and 30, 1971* (Washington, D.C.: GPO, 1971): 167, 172–78, 190–95, 228–33, 324–33, 369–92, 408–68, 504–16, 534–40, 568–93; Fountain's concluding remarks are found on the last page. Also see, on the hearing, "FDA's Edwards Reports Interim Conclusions of Antibiotics Task Force," *Feedstuffs*, 43 (Mar. 20, 1971): 2, 81; "News from Washington," *JAVMA*, 158 (1971): 1619; "Officials Asked about Programs for Control of Stilbestrol Use," *Feedstuffs*, 43 (Mar. 27, 1971): 8; James Risser, "U.S. Probe: Stilbestrol Beef Residue," *Des Moines Register*, Mar. 28, 1971, 1, 5; and "Congressional Hearing May Cause Step-up in Drug Residue Activity by FDA, USDA," *Feedstuffs*, 43 (Apr. 3, 1971): 1, 52. For the AOAC and DES residues in 1971, see George D. Lakata, "Report on Drug Residues in Animal Tissues," *JAOAC*, 54 (1971): 281–82.

20. *Chemicals and the Future of Man: Hearings before the Subcommittee on Executive Reorganization and Government Research of the Committee on Government Operations, United States Senate, Ninety-second Congress, First Session, April 6 and 7 1971* (Washington, D.C.: GPO,

1971), 6–7, 12–14, 17–20; and "Stilbestrol Use, USDA Sampling Methods Hit during Senate Hearing," *Feedstuffs*, 43 (Apr. 10, 1971): 2, 52. See also "Sweden Bans Importing Beef Fed Hormones—Cites DES," *Des Moines Register*, Oct. 8, 1970, 1.

21. See, for example, "Offers Rebutal on DES," *Drovers' Journal*, Dec. 24, 1970, 8; "Compleat Consumer," *National Observer*, Mar. 1, 1971, 8; "Dose-Exposure Concept," *Feedstuffs*, 43 (Mar. 6, 1971): 10; "Drug Legislation," *JAVMA*, 158 (1971): 1156; Robert Steffen, "Why We Must Grow Good Beef without Hormones or High-Energy Rations," *Organic Gardening and Farming*, 18 (Apr. 1971): 61–63; "Restrictions," *Cattleman*, 57 (Apr. 1971): 24; "For Consumers of Food," *Consumer Bulletin*, 54 (June 1971): 4; and Melvin A. Bernarde, *The Chemicals We Eat* (New York: American Heritage, 1971).

22. Arthur L. Herbst and Robert E. Scully, "Adenocarcinoma of the Vagina in Adolescence," *Cancer*, 25 (1970): 745–57; and Arthur L. Herbst, Howard Ulfelder, and David C. Poskanzer, "Adenocarcinoma of the Vagina," *New England Journal of Medicine*, 284 (Apr. 22, 1971): 878–81.

23. The quotations are from Alexander D. Langmuir, "New Environmental Factor in Congenital Disease," *New England Journal of Medicine*, 284 (Apr. 22, 1971): 912–13; and "FDA Denies Connection between DES Feeding, Cancer of Women," *Feedstuffs*, 43 (Apr. 24, 1971): 8. Also see Brian Sullivan, "Girls' Cancers Laid to Mothers' Drug," *Washington Post*, Apr. 23, 1971, A3.

24. Burroughs is quoted in Arlo Jacobson, "Sees No Link in Use for Livestock," *Des Moines Register*, Apr. 23, 1971, 2; Editorial, *Wallaces Farmer*, 96 (June 12, 1971): 18; "Says Alarm over DES Unfounded," *Ames Tribune*, June 23, 1971, 1; and "Says Stilbestrol-Fed Beef Thoroughly Tested," *Des Moines Register*, June 24, 1971, 8. For some examples of the persistent alarm, see, for instance, "Hormones Linked to Cancer Cases," *New York Times*, June 20, 1971, 50; and Judah Folkman, "Transplacental Carcinogenesis by Stilbestrol," *New England Journal of Medicine*, 285 (Aug. 12, 1971): 404–5.

25. Proxmire is quoted in "Proxmire Urges Ban on Cattle Hormone," *New York Times*, May 26, 1971, 49; and "Proxmire Seeks DES Feeding Ban; Farr, Cattlemen's Head Retorts," *Feedstuffs*, 43 (May 29, 1971): 2, 86. Also see "Ribicoff Questions Safety of Feed Additives, Effectiveness of Monitoring," ibid., 43 (May 1, 1971): 5, 36; "Demands Immediate Ban on DES," *Des Moines Tribune*, May 25, 1971, 12; and "Washington Newsletter," *Chemical Week*, 108 (June 23, 1971): 19.

26. Lyng's remarks are reported in "Lyng Says No Meat with DES Residues Reaches Consumers," *Feedstuffs*, 43 (May 29, 1971): 86; and "Department of Agriculture Denies Beef Contamination," *New York Times*, May 30, 1971, 35. Also see "FDA's Edwards Defends DES Monitoring Program," *Feedstuffs*, 43 (June 5, 1971): 6, 139; "Hardin's Comments," ibid., 43 (June 12, 1971, 10; "Predict Ban on Hormone Would Raise Beef Costs," *Des Moines Register*, June 29, 1971, 2; "FDA Official Tells Feeders to Protect Beef Reputation," *Feedstuffs*, 43 (July 24, 1971): 5, 34; "DES Ban Would Boost Beef Prices," *Feed Management*, 22 (Aug. 1971): 44; and "USDA Puts Price Tag on DES Ban," *Farm Journal*, 95 (Aug. 1971): B2.

27. "Feed Men Have Obligation to Avoid Drug Residue Problem, Texans Told," *Feedstuffs*, 43 (Apr. 24, 1971): 2, 36; George Gates, "Industry's Drug Certificate Program Getting Word to Producers, AHI Told," ibid., 43 (May 15, 1971): 1, 53; "Public Not Informed of Benefits of Medicated Feeds, Glennon Tells AHI," ibid., 43 (May 22, 1971): 4; Don Muhm, "Support Feed Additive Use," *Des Moines Register*, May 28, 1971, 17; "Political Approach, Too," *Feedstuffs*, 43 (May 29, 1971): 10; "Proxmire Seeks DES Feeding Ban; Farr, Cattlemen's Head Retorts," ibid., 43 (May 29, 1971): 2, 86; Kendrick McHenry, "Washington Roundup," *Cattleman*, 58 (Aug. 1971): 26–27; and William C. Coleman, "Drugs Are Vital to the Food Supply," *Feed Management*, 22 (Oct. 1971): 14–16.

28. The following articles contain the quotations listed in the text: Jonathan Spivak, "FDA Drug Policy Running into More Flak from Makers, Users, Faces Key Hearings," *Wall Street Journal*, Apr. 30, 1971, 3; James G. Discoll, "FDA Watchdog Doesn't Bite," *National Observer*, May 17, 1971, 1, 7; Jonathan Spivak, "Why Does the FDA Fall So Short?" *Wall Street Journal*, May 20, 1971, 20; Victor Cohn, "Panel on FDA Favors Wider Inspection Role," *Washington Post*, May 28,

1971, A2; Harold M. Schmeck, Jr., "Report Criticizes F.D.A. over Its Scientific Effort," *New York Times*, May 28, 1971, 1, 10; and "HEW Proposes Expanding FDA into Consumer Agency; Nixon OK's Idea," *Feedstuffs*, 43 (July 24, 1971): 2, 40.

29. The book is summarized and Wellford is quoted in "Nader Group to Issue Report on Ag, Poultry Industries," *Feedstuffs*, 43 (May 1, 1971): 1, 43; and "Nader Report Faults FDA, USDA for Roles on Drug Use," ibid., 43 (July 17, 1971): 6, 47.

30. These quotations are from Harrison Wellford, *Sowing the Wind: A Report from Ralph Nader's Center for Study of Responsive Law on Food Safety and the Chemical Harvest* (New York: Grossman, 1972), 127–28, 156–57.

31. See, for instance, "In Response to Nader Report, USDA Defends Role in Food Safety," *Feedstuffs*, 43 (July 24, 1971): 1, 41; Dorothy Cottrell, "The Price of Beef," *Environment*, 13 (July–Aug. 1971): 44–51; "Ralph Nader vs. the USDA," *Farm Journal*, 95 (Sept. 1971): 42; Jerry Carlson, "Tougher Meat Inspection," ibid., 95 (Sept. 1971): 13, 23–24, 41; Elias Colomboto to USDA Secretary, June 20, 1971; Mara Hisiger to USDA Secretary, June 20, 1971; and "Maude Caruso to USDA Secretary, Aug. 15, 1971, FDA Papers, Accession no. 88-82-57, box 74, folder labeled "Folder 121, 131, 135C, 144—Diethylstilbestrol Proposal to Increase."

32. David G. Hawkins to Clifford M. Hardin, Oct. 8, 1971, FDA Papers, Accession no. 88-78-36, box 74, folder labeled "Folder 121, 131, 135C, 144—Diethylstilbestrol Proposal to Increase.

33. The FDA's new strictures are reported in "U.S. to Require Seven-Day Withdrawal Period, Mandatory Certification for Stilbestrol," *Feedstuffs*, 43 (Oct. 16, 1971): 1, 64. Also see James Risser, "Find Prohibited Residues of the Drug DES in Beef," *Des Moines Register*, Oct. 9, 1971, 1; William B. Mead, "Cancer Chemical Is Found in Beef," *Washington Post*, Oct. 9, 1971, A2; "Press Release—USDA and FDA Announce Intentions to Tighten Controls on Diethylstilbestrol, October 12, 1971," FDA Papers, Accession no. 88-88-07, box 1, DES Liquid Premixes folder; "Fallout from the DES Report," *Feedstuffs*, 43 (Oct. 16, 1971): 10; Morton Mintz, "Agriculture Admits Error in Cattle Hormone Report," *Washington Post*, Oct. 17, 1971, A4; "Rapid Implementation Seen for New DES Controls," *Feedstuffs*, 43 (Oct. 23, 1971): 4, 38; *Federal Register*, 36 (Oct. 23, 1971): 20534; "DES's Last Chance," *Feed Age*, 22 (Dec. 1971): 5; and "USDA and FDA to Tighten Controls on DES," *JAVMA*, 159 (1971): 1738.

34. Stein is quoted in "Natural Resources Defense Council, Inc., Environmental Defense Fund, Inc., National Welfare Rights Organization, Federation of Homemakers, and Maryann Stein v. Elliot Richardson, Secretary of Health Education and Welfare, Charles C. Edwards, Commissioner of Food and Drugs, Sam D. Fine, Associate Commissioner for Compliance, Food and Drug Administration, C. D. Van Houweling, Director, Bureau of Veterinary Medicine, Food and Drug Administration . . . ," 3–5, in David G. Hawkins to Hearing Clerk, Nov. 4, 1971, FDA Papers, Accession no. 88-78-36, box 74, folder labeled "Folder 121, 131, 135C, 144." Also see C. W. McMillan to Hearing Clerk, Oct. 28 1971, FDA Papers, Accession no. 88-88-07, box 1, DES Liquid Premixes folder: Indianapolis Stock Yards Marketing Institute to Clifford Hardin, Oct. 21, 1971; William A. Dowden to Hardin, Oct. 21, 1971; R. D. Hempstead to Hearing Clerk, Nov. 4, 1971; and Lee H. Boyd to Hearing Clerk, Dec. 8, 1971, FDA Papers, Accession no. 88-82-57, box 74, folder labeled "Folder 121, 131, 135C, 144—Diethylstilbestrol Proposal to Increase"; "Washington Newsletter," *Chemical Week*, 109 (Oct. 20, 1971): 27; "AFMA Supports DES Withdrawal Rules, Certification," *Feedstuffs*, 43 (Nov. 6, 1971): 6; "New Stilbestrol Regulations," *IBPA Journal*, 1 (Nov. 1971): 5; "Washington Roundup," *Cattleman*, 58 (Nov. 1971): 140; and *Congressional Record*, 92d Cong., 1st sess., 36259.

35. Hawkins outlined his points in "Natural Resources Defense Council v. Richardson," 3–5, 8–11, 15–20, 24–27. Also see "U.S. Is Accused of Failing to Shield Public from Misuse of Synthetic Hormone in Feed," *Wall Street Journal*, Oct. 29, 1971, 14; Philip A. McCombs, "Suit Filed on Cattle Hormone," *Washington Post*, Oct. 19, 1971, A2; James Risser, "Court Asked to Ban DES in Animal Feed," *Des Moines Register*, Oct. 29, 1971, 1, 15; "Court Asked to Ban Disputed

Hormone," *New York Times*, Oct. 29, 1971, 25; and Jane E. Brody, "Disturbing Hints of a Possible Link to Cancer," ibid., Oct. 31, 1971, sec. 4, p. 14.

36. Ironically, Hawkins's suit came on the heels of a study championing DES as an effective morning-after pill in human pregnancies. See "Controversial Drug Is Found Effective against Pregnancies," *Wall Street Journal*, Oct. 26, 1971, 19; and Jane E. Brody, "Success Reported for Next-Day Pill," *New York Times*, Oct. 26, 1971; 19.

37. Proxmire's charges are reported in "Proxmire Asks Senate to Ban Use of DES," *Feedstuffs*, 43 (Nov. 6, 1971); 2; "Proxmire Bill Would Ban Hormone Given to Cattle," *New York Times*, Nov. 9, 1971, 33; and Morton Mintz, "Proxmire Bill Would Impose Ban on Cancer-Linked Cattle Hormone," *Washington Post*, Nov. 9, 1971, A3. Also see *Congressional Record*, 92d Cong., 1st sess., 39734–35, 40092, 40261, 42312–15; and "FDA Acts to Curb Use of Disputed DES Due to Cancer Link; Hearings on Drug Set," *Wall Street Journal*, Nov. 10, 1971, 12.

38. *Regulation of Diethylstilbestrol (DES): Its Use as a Drug for Humans in Animal Feeds (Part I). Hearings before a Subcommittee of the Committee on Government Operations, House of Representatives, Ninety-second Congress, First Session, November 11, 1971* (Washington, D.C.: GPO, 1972), 33, 35, 48. On the FDA's warning regarding the medical use of DES, see "FDA Warns on Hormone in Pregnancy," *New York Times*, Nov. 10, 1971, 21; "FDA Warning on DES," *Washington Post*, Nov. 10, 1971, B4; and "FDA Advice: Don't Use DES during Pregnancy," *Des Moines Tribune*, Nov. 10, 1971, 13.

39. *Regulation of Diethylstilbestrol (Part I)*, 35.

40. Ibid., 59, 68, 72.

41. Ibid., 93, 96.

42. Ibid., 105–6; "FDA Head Hopeful New DES Controls Will End Residues; Otherwise, Ban Likely," *Feedstuffs*, 43 (Nov. 13, 1971): 4, 60; and Morton Mintz, "Cancer Expert Attacks Use of Cattle Fattener," *Washington Post*, Nov. 12, 1971, A2. For a more sober view, see James Risser, "Argue Cancer Threat from Use of DES," *Des Moines Register*, Nov. 14, 1971, 1, 5.

43. The hearing was related to the Democratic-inspired, Nixon-supported crusade to wipe out cancer. The House and the Senate had passed different versions of an anticancer bill, and the measure's passage seemed to await only the formation of a joint conference committee. Just four days after the Fountain hearing, Congressman Cornelius Gallagher (D-N.J.) invited the Soviet Union's premier, Aleksey Kosygin, to join the United States in a ten-year race to cure cancer. The relationship of American cattle-raising methods to environmental quality also seemed problematic. Wastes generated by feedlot animals posed potential pollution problems. A steer excreted approximately one cubic foot of nitrogenous and highly toxic solid and liquid waste daily, which raised groundwater and runoff concerns. In Kansas alone, environmentalists traced fifteen major fish kills in one year to feedlot runoffs. See "Reaction Cool to Cancer Bill," *Science and Government Report*, 1 (Apr. 1, 1971): 2; "Cancer Research Legislation," *JAVMA*, 159 (1971): 1188–89; "Kosygin Rejects Cancer Race," *Science and Government Report*, 1 (Nov. 25, 1971): 2; and "Cancer Bill Compromise in Works," ibid., 1 (Nov. 25, 1971): 2–4. For feedlots, see "Now It's Beef and Mutton," *Newsweek*, 78 (Nov. 8, 1971): 85; and David W. Harker, "Does Beef-Fattener Leave Cancer Seeds in Meat," *National Observer*, Nov. 20, 1971, 10. For a rejoinder, see James A. Affleck, "Medicated Feeds and Ecology," *Shorthorn World*, 56 (Aug. 1971): 142–43.

44. "Consumer Attitudes," *Successful Farming*, 69 (Apr. 1971): 16; and Letters to the Editor, *Washington Post*, Dec. 8, 1971; A23.

45. Hutt already had confronted Congress. When Senator Moss demanded that Hutt reveal his client list before assuming office, Hutt had maintained that the FDA post did not require Senate confirmation and that Moss's order was totally out of line. John D. Morris, "F.D.A. Appointee to Face Hearing," *New York Times*, Sept. 12, 1971, 59; and idem, "Assurances Given by F.D.A. Counsel," ibid., Sept. 18, 1971, 32.

46. *Regulation of Diethylstilbestrol (DES): Its Use as a Drug for Humans in Animal Feeds, Part 2. Hearings before a Subcommittee of the Committee on Government Operations, House of Repre-*

sentatives, Ninety-second Congress, First Session, December 13, 1971 (Washington, D.C.: GPO, 1971), 113–18, 182–91.

47. Ibid., 194–210. Of use on the hearings are *Federal Register*, 36 (Dec. 8, 1971): 23292; Morton Mintz, "Rules Seen Decreasing Chemical in Beef, Lamb," *Washington Post*, Dec. 14, 1971, A4; James Risser, "Still Find DES in Beef, but Problem to End Soon: FDA," *Des Moines Register*, Dec. 14, 1971, 4; and "Natural Estrogens Found in Some Foods, but DES Ban Likely If Residues Persist," *Feedstuffs*, 43 (Dec. 18, 1971): 4, 43. Predictably, *Feedstuffs* and the *Washington Post* highlighted entirely different aspects of the hearing.

48. *Regulation of Diethylstilbestrol, Part* 2, 225, 251–52.

49. Ibid., 253–76, 317–26, 337–46: quotations on 253–54. See also "Memorandum of Telephone Calls to A. J. Malanoski and Robert Munns, December 9, 1971," FDA Papers, Accession no. 88-82-57, box 1, vol. 12.

50. *Regulation of Diethylstilbestrol, Part* 2, 114, 136–38, 140, 142. Also see J. Richard Crout to Henry E. Simmons, Dec. 9, 1971; Leo Friedman to George A. Brubaker, Mar. 10, 1972; C. D. Van Houweling to Charles Edwards, Dec. 21, 1971; and Richard P. Lehmann to Van Houweling, Dec. 21, 1971, FDA Papers, Accession no. 88-82-57, box 1, vol. 9.

51. M. Adrian Gross to Nathan Mantel, Feb. 11, 1972; and Mantel to Gross, Feb. 27, 1972, FDA Papers, Accession no. 88-82-57, box 1, vol. 9. Also see Morton Mintz, "DES Question: New Controversy," *Washington Post*, Feb. 14, 1972, B1, B5. Gross's connection with Mantel predated the imbroglio. See M. Adrian Gross, O. Garth Fitzhugh, and Nathan Mantel, "Evaluation of Safety for Food Additives: An Illustration Involving the Influence of Methyl Salicylate on Rat Reproduction," *Biometrics*, 26 (1970): 181–94.

52. M. Adrian Gross to William W. Wright, Mar. 13, 1972, FDA Papers, Accession no. 88-82-57, box 1, vol. 9. *Regulation of Diethylstilbestrol, Part* 2, 173–77. Also see Leo Friedman to V. O. Wodicka, Jan. 20, 1972; A. J. Kowalk and R. L. Gillespie to Friedman, Feb. 11, 1972, and "Memo of Meeting—Leo Friedman, A. J. Kowalk, and R. L. Gillespie, February 14, 1972," FDA Papers, Accession no. 88-82-57, box 1, vol. 9; Kowalk to Friedman, Mar. 21, 1972, FDA Papers, Accession no. 88-82-57, box 1, vol. 11.

53. Morton Mintz, "FDA Scientists Propose Ban on Cattle Growth Stimulant," *Washington Post*, Mar. 23, 1972, G7.

54. *Congressional Record*, 92d Cong., 2d sess., 2967, 3493, 4356, 5059, 5491, 8129–30, 8774–75; "Legislation against DES," *JAVMA*, 160 (1972); 916; Morton Mintz, "Benefits vs. Hazards of a Fattening Agent for Cattle," *Washington Post*, Mar. 11, 1972, A16; "Fountain Committee Centering Too Much on FDA, Minority Member Charges," *Feedstuffs*, 44 (Mar. 6, 1972): 5, 43; "Edwards Bites Watchdog," *Science*, 175 (Mar. 24, 1972): 1347; "Delaney Joins Sponsors of Bills to Ban DES Use," *Feedstuffs*, 44 (Feb. 14, 1972): 5, 53; "DES Found in Animal Livers," *Des Moines Register*, Feb. 24, 1972; 1; and "Two Positive DES Samples, *JAVMA*, 160 (1972): 1369. For the more immediate concerns, see "Task Force Recommends Limits on Certain Antibiotics in Feed Unless Safety Data Established," *Feedstuffs*, 44 (Jan. 31, 1972): 4, 56; "FDA Says Drugs in Animal Feeds Require New Tests to Assess Hazards to Humans," *Wall Street Journal*, Feb. 1, 1972, 9; Victor Cohn, "Petition Seeks Meat-Additive Ban," *Washington Post*, Feb. 10, 1972, C1, C14; idem, "Red Food Dye Unsafe?" ibid., Feb. 11, 1972, B1, B4; "Task Force Minority Objects to Health Hazard Statement," *Feedstuffs*, 44 (Feb. 7, 1972): 4, 56; G. O. Kermode, "Food Additives," *Scientific American*, 226 (Mar. 1972): 15–21; Jean Mayer, "Additives: What's in the Food We Put in Our Mouths," *Washington Post*, Mar. 9, 1972; C2; and Morton Mintz, "Hill Asked for Act to Repay Cyclamate Ban Losses," ibid., Mar. 26, 1972, E3.

55. The FDA statement is quoted in "Cross-Contamination Cases with DES, Liquid Feeds Reported," *Feedstuffs*, 44 (Mar. 6, 1972): 2, 50. Also see "HEW News, March 10, 1972" FDA Papers, Accession no. 88-88-07, box 1, DES Liquid Premixes Folder; *Federal Register*, 37 (Mar. 11, 1972): 5264; "Propose Ban on Liquid DES Used in Livestock Feeds," *Des Moines Register*, Mar. 11, 1972, 16; "Hormone Ban Proposed," *New York Times*, Mar. 11, 1972, 59; and "FDA to Propose Ban on Use of DES in Liquid Feeds, *Feedstuffs*, 44 (Mar. 13, 1972): 2, 53.

56. Mrs. Melvin A. Danielson to Leslie Aspin, Mar. 24, 1972; and Ida Honorof to Hearing Clerk, Mar. 19, 1972, FDA Papers, Accession no. 88-88-07, box 1, DES Liquid Premixes folder. Also see Ruth H. Harmer to Hearing Clerk, Mar. 19, 1972; John N. White to Hearing Clerk, Mar. 20, 1972, "Granville F. Knight to Hearing Officer, Mar. 21, 1972; and Warren Hutchinson to Carl T. Curtis, Mar. 29 1972, FDA Papers, Accession no. 88-88-07, box 1, DES Liquid Premixes folder.

57. "Telegram—Barkley Allied Chemical Corporation to Bob Price, March 14, 1972"; John Tower to Charles C. Edwards, Mar. 17, 1972; Roy Dinsdale to Carl T. Curtis, Mar. 20, 1972; Larry J. Kennedy to Thomas Railsback, Mar. 21, 1972; Morris Brock to Jay Linderman, Mar. 21, 1972; William Glennon and Marvin L. Vinsand to Hearing Clerk, Mar. 29, 1972; Clarence W. Blase to Curtis, Mar. 30, 1972; LeRoy Werner to Curtis, Apr. 3, 1972; Z. A. McCasland and Robert H. Schiller to Manuel Lujan, Apr. 3, 1972; Norman Luthy to Curtis, Apr. 3, 1972; Frank C. Lathrop to Hearing Clerk, Mar. 27, 1972; M. Appelbaum to Hearing Clerk, Mar. 27, 1972; Dean Simonsen to Hearing Clerk, Mar. 31, 1972; C. S. Pruitt to Hearing Clerk, Mar. 31, 1972; Wayne K. Hill to Hearing Clerk, Apr. 5, 1972; E. J. Korbel to Hearing Clerk, Apr. 6, 1972; Frank N. Rawlings to Hearing Clerk, Apr. 7, 1972, FDA Papers, Accession no. 88-88-07, box 1, DES Liquid Premixes folder; Randy Bridson, "Liquid Feed Men Irked by Proposed DES Ban," *Feedstuffs*, 44 (Mar. 20, 1972): 1, 88; "AFMA Strategy on Liquid DES Ban Outlined," ibid., 44 (Apr. 3, 1972): 2; and "AFMA, NFIA Join Forces to Ask Hearing on Liquid DES Ban," ibid., 44 (Apr. 10, 1972): 2, 48.

58. The Hawkins quotations appear in Morton Mintz, "Cancer-Causing Chemical Still Found in Meat," *Washington Post*, Mar. 19, 1972, E2. and "Frank L. Joe, Jr., to William Price, Apr. 20, 1972, FDA Papers, Accession no. 88-82-57, box 1, vol. 12. Also see "USDA Finds DES Residues in Two More Cases," *Feedstuffs*, 44 (Mar. 13, 1972): 5, 45; "Another Stilbestrol Residue Found," ibid., 44 (Apr. 3, 1972): 4; "FDA Still Unsure about Stilbestrol Ban," ibid., 44 (Apr. 24, 1972): 1, 37–38; and James Risser, "Three Cases of DES in Beef; Iowa Infraction Cited," *Des Moines Register*, Apr. 21, 1972; 4. For the FDA, see *Congressional Record*, 92d Cong. 2d sess., 12637, 13164.

59. "Report of the Special Committee on Feed Additives," *JAS*, 35 (1972): 303–5.

CHAPTER 6. ENTITLE AND VICTIMIZE

1. Jim Hightower, *Hard Tomatoes, Hard Times: A Report of the Agribusiness Accountability Project on the Failure of America's Land Grant College Complex* (Cambridge, Mass.: Schenkman, 1973), 49–51. Also see "HEW Chief Promises to Strengthen FDA in Bid to Counter Moves for New Agency," *Wall Street Journal*, May 4, 1972, 23; "Suit Is Filed to Bar Two Meat Additives," *New York Times*, May 4, 1972, 48; Victor Cohn, "Aide of Nader Sues to Ban Color Additives for Meat," *Washington Post*, May 4, 1972, D28; "HEW Secretary Edwards Argues against Making Food and Drug an Independent Agency," *Feedstuffs*, 44 (May 8, 1972): 6, 42; Joan Arehart-Treichel, "Antibiotics in Animal Feeds: Threat to Human Health," *Science News*, 101 (May 27, 1972): 348–49; and "Relationship of Land Grant Colleges, Agribusiness Attacked," *Feedstuffs*, 44 (June 5, 1972): 4, 45.

2. See, for example, "New DES Residue Detectors Investigated," *Feedstuffs*, 44 (Mar. 6, 1972): 5; "FDA Looking into New Assay Method for Detecting Drugs in Meat Tissues," *FDA Consumer*, 6 (Mar. 1972): 26; Virgil O. Wodicka to William G. Drury, May 10, 1972, FDA Papers, Accession no. 88-82-57, box 1, vol. 12; "Washington Roundup," *Cattleman*, 58 (May 1972): 24; "Illegal Hormone Residues in Cattle Found Up Sharply," *New York Times*, May 7, 1972, 39; "Illegal DES Use Doubles, U.S. Reports," *Washington Post*, May 19, 1972, A25; "USDA Finds Eight More Livers with DES Residues," *Feedstuffs*, 44 (May 8, 1972): 4, 42; "New Product Helps DES Withdrawal," *Farm Journal*, 96 (May 1972): B2; "New Withdrawal Feed," *Feedlot Management*, 14 (May 1972): 14; and Jeff Cox, "What You Should Know about Buying Beef," *Organic Gardening and Farming*, 19 (May 1972): 73–79.

3. For Cardon's Statements, see Daryl Natz, "New Interpretation of Drug Regulation Called for; Other Feed Industry Problems Cited," *Feedstuffs*, 44 (May 22, 1972): 1, 56. For Scherle, see "Vetoing the Law of Nature," *IBPA Journal*, 2 (July 1972): 6. Dawe's Laboratory, a Chicago-based animal biologics company, planned to recreate the Gass study, but to use liver-conjugated stilbes-

trol in place of the pure crystalline variety. It sought funding from a consortium of industry interests for a three-year-long investigation to demonstrate either that Gass's 6.25-ppb figure was inaccurate or that conjugated DES lost its possible carcinogenic properties. The biologics firm lured Gass to design and head the project, the significance of which grew when the company submitted the research protocols to the USDA, the FDA, the National Cancer Institute, and the American Society of Animal Science for review and approval. For example, see "FDA Test Method May Be Confusing DES, Molasses," *Feedstuffs*, 44 (May 1, 1972): 2, 42–43; "Industry Support Sought for New DES Research," ibid., 44 (May 8, 1972): 2, 49; Dawe's Laboratory, "Frontiers in Nutrition," 247 (June–July 1972): 959, FDA Papers, Accession no. 88-88-07, box 1, Stilbestrol folder; "Interpretation Makes a Big Difference," *Feedlot Management*, 14 (July 1972): 7; "Iowan Introduces Bill to Modify Delaney Clause," *Feedstuffs*, 44 (June 26, 1972): 2, 45; "Butz in Wyoming, Iowa Representative on House Floor Defend Stilbestrol Use," ibid., 44 (July 3, 1972): 1, 38; and "Washington Roundup," *Cattleman*, 59 (July 1972): 22–23.

4. The quotations are from "DES Ban," *Feed Management*, 23 (June 1972): 4; George A. Montgomery, "Longer Feeding Period on the Way," *Feedlot Management*, 14 (May 1972): 26; Fred Bailey, Jr., "From DES to Fertilizer—We're in a Fight for Survival," *Successful Farming*, 70 (May 1972): 7; Ralph Saunders, "Feed Additives; Will We Lose Them?" ibid., 70 (May 1972): 19; "Expect Ban of DES If Residues Continue," *Wallaces Farmer*, 97 (May 13, 1972): 18; and "You Don't Want the Alternatives," *Feedlot Management*, 14 (June 1972): 7, 54.

5. Edwards's sentiments are contained in "Diethylstilbestrol; Notice of Opportunity for Hearing on Proposal to Withdraw Approval of New Animal Drug Applications," *Federal Register*, 37 (June 21, 1972): 12251–52. Also see "DES traces Found in Ore., Calif. Steers," *Washington Post*, June 10, 1972, A7; "DES Residues," *Feedlot Management*, 14 (June 1972): 18, 26; "USDA News Release—USDA Reports Fifteen DES-Positive Meat Samples in Eight States, June 16, 1972," ISURF Papers, Burroughs Uncatalogued folder; "HEW News Release 'Diethylstilbestrol, June 16, 1972,'" FDA Papers, Accession no. 88-82-57, box 27, vol. 247; "Another Look at DES," *Chemical Week*, 110 (June 28, 1972): 14; "FDA Considering Ways to Curb DES Residues in Cattle, Sheep Livers," *Wall Street Journal*, June 19, 1972, 8; Harold M. Schmeck, Jr., "F.D.A. Weighs Ban on a Cattle Feed," *New York Times*, June 17, 1972, 32; and "FDA Issues Statement on Diethylstilbestrol," *JAVMA*, 161 (1972): 249–50.

6. Morton Mintz, "FDA Sets Hearing on DES Drug Use," *Washington Post*, June 17, 1972, A3; "Another Delay on DES," ibid., June 23, 1972, A18; "Rep. Fountain Attacks Proposed DES Hearing," *Feedstuffs*, June 26, 1972, 2, 50; and *Congressional Record*, 92d Cong., 2d sess., 21517–18, 23814.

7. "'Concerned' FDA Calls for Hearing on Stilbestrol," *Feedstuffs*, June 19, 1972, 1, 56–57; and *Congressional Record*, 92d Cong., 2d sess., 22171–72, 23002. For the Iowa State University situation, see Arne Paulsen to Marvin A. Anderson, June 23, 1972; and Arne Paulsen and Bob Hazel, "Summary of DES Situation, June 23, 1972," ISURF Papers, Burroughs Uncatalogued folder; ISURF, "Annual Report of the Secretary-Manager of the ISURF, April 1, 1971–March 31, 1972," Burroughs Papers, box 3; "Tighten Stilbestrol Withdrawal Rules," *Wallaces Farmer*, 96 (Nov. 13, 1971): 24; Wil Groves and Bob Dunaway, "Stilbestrol Threatened by Illegal Residues," ibid., 96 (Nov. 27, 1971): 15; and "Defends U.S. Meat Supply," *IBPA Journal*, 1 (Dec. 1971): 23. ISURF's stilbestrol patents were set to expire on June 19, 1973, and the university gradually withdrew from active participation in stilbestrol's defense. It produced only an internal memo outlining a DES ban's consequences to the cattle industry.

8. Paul E. Corneliussen, comp., "Report of the Analytical Workshop on Diethylstilbestrol in Feeds, June 26 thru 28, 1972, Washington, D.C.," FDA Papers, Accession no. 88-82-57, box 1, vol. 12. Despite the title, the symposium also covered DES residues.

9. Morton Mintz, "Cancer Institute Head Urges FDA Ban DES," *Washington Post*, July 10, 1972, A2. Also see "U.S. Tells Big Rise in DES in Livestock," *Des Moines Tribune*, June 29, 1972, 1; Morton Mintz, "Cancer Agent Rises in Meat," *Washington Post*, July 4, 1972, A3; "Inspectors Report Increase in Cattle Hormone Residues," *New York Times*, July 8, 1972, 12; "More Beef Found

Faulty," ibid., July 23, 1972, 32; and "Four More DES Residues Found," *Feedstuffs*, 44 (July 24, 1972): 2.

10. The quotations are from Florence Ellis to James A. Burke, July 1, 1972; and Jesse E. O. Berry to Joel T. Broyhill, July 4, 1972, FDA Papers, Accession no. 88-88-07, box 1, Diethylstilbestrol folder; Marian Snuggs to Charles Edwards, July 5, 1972, FDA Papers, Accession no. 88-88-07, box 1, DES-FDC-D-494 folder; Mr. and Mrs. Nye Adams to Edwards, July 26, 1972, FDA Papers, Accession no. 88-82-57, box 27, vol. 248; Mrs. Joseph T. Hansford to Hearing Clerk, July 29, 1972, FDA Papers, Accession no. 88-88-07, box 1, Diethylstilbestrol folder; Yvonne L. Zorick to Edwards, June 25, 1972. FDA Papers, Accession no. 88-88-07, box 1, DES-FDC-D-494 folder; and Mrs. C. Mosher to Roy A. Taylor, June 27, 1972, FDA Papers, Accession no. 88-88-07, box 1, Diethylstilbestrol folder. Also see, for example, Victor Kerlias to Edwards, July 4, 1972; and Ronni Snyder to Edwards, July 12, 1972, FDA Papers, Accession no. 88-82-57, box 1, Diethylstilbestrol FDC-D-494 folder; Celia M. Nakano to HEW, July 5, 1972; Mabel R. Simmons to Hearing Clerk, July 7, 1972; and Andrew Skapura to Hearing Examiner, July 10, 1972, FDA Papers, Accession no. 88-88-07, box 1, Stilbestrol folder; and Jessie M. Corpening to Edwards, June 17, 1972; Elizabeth H. Frisco to Edwards, June 18, 1972; Sara Campbell Rockwell to Edwards, June 29, 1972; Don Ternes to Edwards, July 4, 1972; and Gordon W. Newell to Edwards, July 14, 1972, FDA Papers, Accession no. 88-88-07, box 1, DES-FDC-D-494 folder.

11. David G. Hawkins to Elliot L. Richardson, July 12, 1972; Susan Barlow to Hearing Clerk, July 8, 1972; and R. Collins Vallee to Hearing Clerk, July 28, 1972, FDA Papers, Accession no. 88-88-07, box 1, Diethylstilbestrol folder; and G. E. Cornelius to FDA, July 24, 1972; and Alice Shabecoff to Charles Edwards, July 19, 1972, FDA Papers, Accession no. 88-82-57, box 27, vol. 249.

12. The quotation is from O. D. Butler to Hearing Clerk, July 21, 1972, FDA Papers, Accession no. 88-88-07, box 1, Stilbestrol folder. Also see Oakley M. Ray to Hearing Clerk, July 6, 1972; C. W. McMillan to Hearing Clerk, July 18, 1972; and Billy Ray Gowdy to Tom Steed, July 5, 1972, FDA Papers, Accession no. 88-88-07, box 1, Stilbestrol folder; Ed Biaggini, Jr., to Charles C. Edwards, June 29, 1972, FDA Papers, Accession no. 88-88-07, box 1, DES-FDC-D-494 folder; William C. Donnell to Hearing Clerk, July 13, 1972, FDA Papers, Accession no. 88-88-07, box 1, DES Liquid Premixes folder; and C. B. Christensen to C. D. Van Houweling, July 28, 1972, FDA Papers, Accession no. 88-82-57, box 27, vol. 248.

13. The quotations are from Roman Hruska, Paul J. Fannin, Quentin N. Burick, Peter H. Dominick, Carl T. Curtis, Wallace F. Bennett, John Tower, Gale W. McGee, Clifford P. Hansen, Milton R. Young, Clinton P. Anderson, James O. Eastland, and Henry Bellmon to Charles C. Edwards, June 20, 1972, FDA Papers, Accession no. 88-88-07, box 1, DES-FDC-D-494 folder; "Cattle Feeders' Day Program Announced," *IBPA Journal*, 2 (July 1972): 23; "Ask 'Realistic' DES Residue Level," *Ames Tribune*, July 14, 1972, 1. Also see Don Muhm, "Cattlemen Urge Allowing 'Realistic' DES Amounts," *Des Moines Register*, July 14, 1972, 1.

14. The quotation is from Eugene I. Lambert to Hearing Clerk, July 31, 1972, FDA Papers, Accession no. 88-88-07, box 1, Diethylstilbestrol folder. An example of a response to the FDA's proposed action on implants is "Pfizer, Inc., to FDA Commissioner in the Matter of the Proposed Withdrawal of Pfizer NADA Nos. . . . 9783, 9770, 11356, and 9757, July 1972," copy provided to me by Pfizer, Inc. For protests on other aspects of the proposed DES ban, see M. Appelbaum to Hearing Clerk, July 10, 1972; Ian D. Smith to Hearing Clerk, July 11, 1972; Harold J. Daub, Jr., to Hearing Clerk, July 19, 1972; Howard R. Harrison to Hearing Clerk, July 19, 1972; R. O. Clutter to Hearing Clerk, July 19, 1972; Wayne K. Hill to Hearing Clerk, July 20, 1972; Vincent DeFelice and Eugene I. Lambert to Hearing Clerk, July 20, 1972; Ralph F. Elliott to Hearing Clerk, July 21, 1972; John R. Stafford to Hearing Clerk, July 21, 1972; and James L. Kaler to Hearing Clerk, July 21, 1972, FDA Papers, Accession no. 88-88-07, box 1, Diethylstilbestrol folder, "Elanco Reports Hearing on Use of DES in Feeds," *Feedstuffs*, 44 (July 24, 1972): 2, 47; and "Dawe's Laboratories, Sixteen Other Firms, Ask DES Hearing," ibid., 44 (July 31, 1972): 2, 43.

15. *Regulation of Diethylstilbestrol (DES), 1972: A Hearing before the Subcommittee on Health*

of the Committee on Labor and Public Welfare, United States Senate, Ninety-second Congress, Second Session, on S. 2818, July 20, 1972 (Washington, D.C.: GPO, 1972), 6–16. Also see "Kennedy Sets Hill Hearing on Continued Use of DES," *Washington Post*, July 14, 1972, A3; "Senate Subcommittee Calls Hearing on Use of DES; No Residue Found in Week," *Feedstuffs*, 44 (July 17, 1972): 6; "Proxmire and Anti-Stilbestrol Forces Line Up against Beef Spokesmen at Senate Hearing," ibid., 44 (July 24, 1972): 2, 48; and David W. Hacker, "Cancer Agent in Meat Makes Political Stew," *National Observer*, July 29, 1972, 4.

16. *Regulation of Diethylstilbestrol (DES)*, 1972, 16–26, quotation on 19.

17. William Proxmire to Charles C. Edwards, July 28, 1972, FDA Papers, Accession no. 88-82-57, box 27, vol. 247; and "Transcript of Proceedings—Food and Drug Administration, National Advisory Drug Committee, Special Meeting on Diethylstilbestrol, July 25, 1972," 110–127, copy provided to me by Pfizer, Inc., and in my possession.

18. Nicholas Wade, "DES: A Case Study of Regulatory Abdication," *Science*, 177 (July 28, 1972): 335–37. Wade's piece generated heated rejoinders. See "Letters," ibid., 178 (Oct. 13, 1972): 117–18.

19. "Diethylstilbestrol—Order Denying Hearing and Withdrawing Approval of New Animal Drug Applications for Liquid and Dry Premixes, and Deferring Ruling on Implants," *Federal Register*, 37 (Aug. 4, 1972): 15747–50.

20. Ibid. Also see "Farmers: 'We'll Use DES,'" *Des Moines Tribune*, Aug. 3, 1972; 1–3; "FDA Bans Use of Hormone DES in Feeds for Animals; Residue Found to Persist," *Wall Street Journal*, Aug. 3, 1972; 6; Clark Mollenhoff, "Ban Doesn't Bar Implant of Hormone," *Des Moines Register*, Aug. 3, 1972; 2; Stuart Auerbach, "U.S. Bans Production of Cattle Fattener," *Washington Post*, Aug. 3, 1972, A1 and A4; Richard D. Lyons, "F.D.A. Bans Drug Linked to Cancer," *New York Times*, Aug. 3, 1972, 1, 53; and "HEW News—Diethylstilbestrol, August 2, 1972," FDA Papers, Accession no. 88-82-57, box 27, vol. 254. The Beltsville research was later published as T. S. Rumsey, R. R. Oltjen, F. L. Daniels, and A. S. Kozak, "Depletion Patterns of Radioactivity and Tissue Residues in Beef Cattle after the Withdrawal of Oral 14C-Diethylstilbestrol," *JAS*, 40 (1975): 539–49.

21. The quotations are from Mrs. A. Brewster Lawrence, Jr., to Charles Edwards, Aug. 3, 1972, FDA Papers, Accession no. 88-82-57, box 27, vol. 250; and Blanche E. Sherrington to Edwards, Sept. 1, 1972, FDA Papers, Accession no. 88-82-57, box 27, vol. 253. Also see Irene D. Genco to Edwards, Sept. 15, 1972, FDA Papers, Accession no. 88-82-57, box 27, vol. 253; H. Holman to Secretary of Agriculture, Aug. 3, 1972, and P. Kleinhaus to FDA, Sept. 4, 1972, FDA Papers, Accession no. 88-82-57, box 27, vol. 252; "Fattening Profits," *Nation*, 215 (Sept. 18, 1972): 196–97; August Gribbon, "Government Bans Cattle-Feed Hormone," *National Observer*, Aug. 12, 1972, 7; "The Hormone Ban Spreads Confusion," *Business Week*, Aug. 12, 1972, 29; "How Much Is Too Much," *New Republic*, 167 (Aug. 19 and 26, 1972): 7; Mrs. Gordon Lamont to Edwards, Aug. 8, 1972, FDA Papers, Accession no. 88-88-07, box 1, Diethylstilbestrol folder.

22. The quotations are from Edward M. Kennedy to Charles Edwards, Aug. 4, 1972, FDA Papers, Accession no. 88-82-57, box 27, vol. 250; and Bob Price to Edwards, Aug. 16, 1972, FDA Papers, Accession no. 88-82-57, box 27, vol. 252. Also see *Congressional Record*, 92d Cong., 2d sess., 26449–51, 26553–55, 26563, 27197, 28615, 28726, 29447, 29452, 29700, 30595–97, 31537–50, 31853, 32865; Peter Barton Hutt to Edwards, Sept. 25, 1972, FDA Papers, Accession no. 88-82-57, box 27, vol. 254; "Senate Unit Votes Total, Immediate Ban on DES," *Ames Tribune*, Aug. 4, 1972, 1, 12; "Senate Subpanel Backs Bill to Ban Use of DES in Any Form on Animals," *Wall Street Journal*, Aug. 4, 1972, 20; "Canada to Halt DES Use in Animal Feeds, Jan. 1," ibid., Aug. 5, 1972; 21; "Subcommittee Votes Complete Ban on DES," *Feedstuffs*, 44 (Aug. 7, 1972): 1; "FDA Extension of DES Permit Denounced by Rep. Fountain," *Washington Post*, Aug. 7, 1972, A2; Nicholas Wade, "FDA Invents More Tales about DES," *Science*, 177 (Aug. 11, 1972): 503; "Senate Panel Approves Ban on Additive in Cattle Feeds," *New York Times*, Aug. 16, 1972, 40; "DES Ban in Cattle Feed Voted by Senate Unit," *Wall Street Journal*, Aug. 16, 1972, 4; Tim O'Brien, "Senate Unit Votes an Immediate Ban of Cattle-Fattener," *Washington Post*, Aug. 16,

1972, A2; B. M. Keller to L. H. Fountain, Sept. 11, 1972, and Fountain to Edwards, Sept. 12, 1972, FDA Papers, Accession no. 88-82-57, box 27, vol. 252; James Risser, "Official Says Phase Out of DES by FDA Illegal," *Des Moines Register*, Sept. 13, 1972, 1; Morton Mintz, "Incomplete Ban on DES Held Illegal," *Washington Post*, Sept. 13, 1972, A24; "Senate Approves Three Health Measures," ibid., Sept. 21, 1972, A3; "Senate Passes Measure for Immediate Banning of DES in Cattle Feed," *Wall Street Journal*, Sept. 21, 1972, 15; "Medical Aid Bill Voted by Senate," *New York Times*, Sept. 21, 1972, 36; "Senate Approves Ban of All DES Forms; Similar House Action Not Yet Scheduled," *Feedstuffs*, 44 (Sept. 25, 1972): 5; and "Second Report Defends Phase-Out of Stilbestrol," ibid., 44 (Sept. 25, 1972): 4, 46. A later opinion sided with Fountain. See "Another Opinion Says DES Phaseout Was Illegal," ibid., 44 (Nov. 13, 1972): 5, 55

23. The quotations are from George Anthan, "Farm State Congressmen Fight DES Ban," *Des Moines Register*, Aug. 4, 1972, 1, 5; "Butz Regrets Necessity of Stilbestrol Ban," *Feedstuffs*, 44 (Aug. 7, 1972): 4; "The Fight to Modify the Delaney Amendment Is Beginning," *Chemical Week*, 111 (Aug. 9, 1972): 10; Earl L. Butz, "Meat Prices," *Vital Speeches of the Day*, 38 (Aug. 15, 1972): 647–49; "Cancer Clause Changes," *Chemical Week*, 111 (Aug. 16, 1972): 22; Nicholas Wade, "Delaney Anti-Cancer Clause: Scientists Debate on Article of Faith," *Science*, 177 (Aug. 18, 1972): 588–91; "FDA Defends DES Phaseout Period; Kennedy Charges Politics, Favoritism to Industry," *Feedstuffs*, 44 (Aug. 21, 1972): 2, 84; "A Gradual Phase-out for Diethylstilbestrol (DES)," *Chemical Week*, 111 (Aug. 23, 1972): 19. Also see "Legislation to Modify the Delaney Clause," *JAVMA*, 161 (1972): 477; Clark Mollenhoff, "Dilemma on Food Additives," *Des Moines Register*, Sept. 17, 1972, 1A, 5A; Richard D. Lyons, "Administration Seeks Relaxation of Ban on Cancer-Causing Food Additives," *New York Times*, Sept. 21, 1972, 35; "FDA's Edwards Points to Problems of Delaney Clause," *Feedstuffs*, 44 (Sept. 25, 1972): 7, 41; and *Congressional Record*, 92d Cong., 2d Session, 27592. Debates over the Delaney clause found their way into the famous Senate Hearings on Nutrition and Human Needs. See *Hearings before the Select Committee on Nutrition and Human Needs of the United States Senate, Ninety-second Congress, Second Session on Nutrition and Human Needs* (Washington, D.C.: GPO, 1973), Pts. 4A–C (Sept. 19–21, 1972), pp. 807–15, 857–59, 869–71, 1213–19, 1279–1301, 1314–18, 1669, 1671–74, 1679–81, 1721–26.

24. Leo Friedman, Director of the FDA's Division of Toxicology, joined his administrative colleagues in criticizing the Delaney clause, but proposed a concrete alternative when he deplored the "scientific attitude which accepts everything that is conceivable as possible." That perspective, "together with the logic of probability, leads inevitably" to an intellectually indefensible belief that "only the complete absence of a potential cause can assure the absolute absence of any risk." Instead of zero tolerance, Friedman pressed for a "biologically inactive level," the largest dosage at which no biological response occurred. The maximum allowable concentration of substances in doubt should be limited to no more than "one molecule per cell." See "'Threshold Level' for Response from Additives Urged," *Feedstuffs*, 44 (Aug. 28, 1972): 40.

25. Cattlemen made their sentiments known in "ANCA, AFMA Are Disappointed with DES Decision," *Feedstuffs*, 44 (Aug. 7, 1972): 1, 45; and Wayne Anderson, "Feed Men, Others See Big Switch to DES Implants in Wake of Ban in Feeds," ibid., 44 (Aug. 14, 1972): 1, 60; and "DES—The First Shoe," *Feed Management*, 23 (Sept. 1972): 56. Also see G. P. Lofgreen to Charles C. Edwards, Aug. 5, 1972, FDA Papers, Accession no. 88-82-57, box 27, vol. 253; "Several Replacements for Stilbestrol in Feeds," *Wallaces Farmer*, 97 (Aug. 26, 1972): 8; B. W. Beadle to Edwards, Sept. 7, 1972, FDA Papers. Accession no. 88-82-57, box 27, vol. 252; John A. Rohlf, "Your Beef Business," *Farm Journal*, 96 (Sept. 1972): B20; "Oral DES Ban Stuns Industry," *Feedlot Management*, 14 (Sept. 1972): 56–68; "They Win on DES Too," *Successful Farming*, 70 (Sept. 1972): 7; "DES: Chance for Survival Exists . . . If," *Feedlot Management*, 14 (Oct. 1972): 6; "DES Substitutes," ibid., 14 (Oct. 1972): 90; "Stilbestrol Alternatives," *Farm Quarterly*, 27 (Fall 1972): 62; Morton Mintz, "Meat Additive Risk Laid to FDA," *Washington Post*, Aug. 20, 1973, F2; and Thomas H. Jukes, "Antibiotics in Animal Feeds and Animal Production," *Bioscience*, 22 (1972): 526–34. Iowa State University defended DES feeding and attacked implanting as dangerous and difficult. See Don Muhm, "Problems in Using DES as Implant," *Des Moines Register*, Aug. 4,

1972; I, 5; and Jim Hildebrand, "Feeding of Animal Growth Drug Is Banned," *Iowa State Daily*, Aug. 10, 1972, 1–2. The testimony of consumerists before the Senate Committee on Nutrition and Human Needs provided cattlemen no solace. See *Hearings before the Committee on Nutrition and Human Needs*, 1549–85, 1667.

26. Wayne Hill to Commissioner of Food and Drugs, Aug. 28, 1972, FDA Papers, Accession no. 88-88-07, box I, Diethylstilbestrol folder; "Feed Mix Order Protested," *New York Times*, Sept. 17, 1972, 8; and "Washington Newsletter," *Chemical Week*, III (Sept. 27, 1972): 10.

27. Edward O. Haenni to C. D. Van Houweling, July 31, 1972, FDA Papers, Accession no. 88-82-57, box 27, vol. 247; George Lakata to William Price, Aug. 4, 1972; C. A. Armstrong to Regional Food and Drug Directors, Aug. 4, 1972, FDA Papers, Accession no. 88-82-57, box 27, vol. 248; Herbert Friedlander to John C. Evans, Sept. 7, 1972, FDA Papers, Accession no. 88-82-57, box 27, vol. 252; "FDA Obtains First Conviction in DES Case; More Residues Found in Liver Samples," *Feedstuffs*, 44 (Sept. 18, 1972): 7; Anne Amsie to John W. Howard, Oct. 13, 1972; Howard to Price, Oct. 16, 1972; and Frank L. Joe, Jr., to Howard, Oct. 16, 1972," FDA Papers, Accession no. 88-82-57, box 27, vol. 355; and Price to J. C. Evans, Nov. 17, 1972; and Evans to Friedlander, Nov. 30, 1972, FDA Papers, Accession no. 88-82-57, box 27, vol. 257; Albert C. Kolbye, Jr., to Van Houweling, Jan. 15, 1973, FDA Papers, Accession no. 88-82-57, box I, vol. 12; Wilbert Shimoda to N. H. Booth, Feb. 5, 1973, FDA Papers, Accession no. 88-82-57, box 27, vol. 360; Edward H. Allen to Price, Apr. 10, 1973, FDA Papers, Accession no. 88-82-57, box 28, vol. 363; Shimoda and David Batson to H. Dwight Mercer and Five Others, Apr. 16, 1973, FDA Papers, Accession no. 88-82-57, box 27, vol. 362.

28. The quotations are from "FDA Tells Firms It Won't Rescind Ban on DES in Feed on Basis of New Data," *Feedstuffs*, 44 (Nov. 6, 1972): 2, 60. Also see Edward J. Thacker to T. W. Edminster, Sept. 5, 1972, FDA Papers, Accession no. 88-82-57, box 27, vol. 252; "USDA Releases Progress Reports on DES Feeding Research Underway," *Feedstuffs*, 44 (Oct. 23, 1972): 4; Edminster to C. D. Van Houweling, Nov. 21, 1972, FDA Papers, Accession no. 88-82-57, box 27, vol. 257; "No DES Detected in Tissues Fourteen Days After Withdrawal," *Feedstuffs*, 44 (Dec. 4, 1972): 4, 51; "DES Implants Clear First Tests," ibid., 44 (Oct. 16, 1972): 70; Walter E. Byerley to Charles C. Edwards, Oct. 20, 1972; and "Memorandum of Conference—Dawe's Laboratories, Chemetron Corporation, and FDA, October 25, 1972," FDA Papers, Accession no. 88-82-57, box 27, vol. 355; and Sam D. Fine to Walter E. Byerley, Oct. 24, 1972, FDA Papers, Accession no. 88-88-07, box I, DES Liquid Premixes folder. These studies were published as P. W. Aschbacher and E. J. Thacker, "Metabolic Fate of Oral Diethylstilbestrol in Steers," *JAS*, 39 (1974): 1185–92.

29. The fact that the new National Center for Toxicological Research was nearly completed may have contributed to the agency's decision. The project, created by executive order in January 1971, was jointly administered by the FDA and the Environmental Protection Agency and was set at the former biological warfare laboratories in Pine Bluff, Arkansas. Its mission included investigation into the consequences of long-term, low-dosage exposure to toxicants; the relationship of toxicants to basic metabolic processes in animals; improved methodologies for determining chemical safety; and means of extrapolating research data from animals to humans. While it might conceivably offer future information relevant to the DES feeding question, the center only now was getting off the ground. *Hearings before the Committee on Nutrition and Human Needs*, 857–59; and *Annual Report of the Food and Drug Administration for 1972*, 150.

30. T. S. Rumsey to T. W. Edminster, Oct. 6, 1972; C. D. Van Houweling to Charles Edwards, Dec. 15, 1972; Van Houweling to E. R. Squibb and Sons, Inc., Dec. 15, 1972; and Gerald F. Meyers, "Progress Report on Diethylstilbestrol (DES) Implants in Cattle," Dec. 19, 1972, FDA Papers, Accession no. 88-82-57, box 27, vol. 259; "DES Implants Clear First Tests," 2, 70; "W. H. Ray—Hess and Clark Research Report, December 12, 1972," FDA Papers, Accession no. 88-82-57, box I, Hero 2 folder; David M. Tennent to Edwards, Dec. 20, 1972, FDA Papers, Accession no. 88-88-07, box I, Hess and Clark folder; "120 Day Withdrawal Period on Stilbestrol Implants Invoked," *Feedstuffs*, 44 (Dec. 25, 1972): I, 38; and "Hess and Clark Reports New DES Test Results," ibid., 45 (Jan. 8, 1973): I, 55.

31. "Drug Firms Seek Repeal of DES Ban, Longer Withdrawal," *Feedstuffs*, 44 (Dec. 4, 1972): 4, 49–50; and "Review of Order Revoking DES in Feeds Sought," ibid., 45 (Jan. 1, 1973): 27. Also see "Firms Lose Appeal on DES Ban," ibid., 44 (Nov. 13, 1972): 63.

32. The quotations are from Victor L. Christie to Charles Edwards, Oct. 31, 1972, FDA Papers, Accession no. 88-82-57, box 27, vol. 355; Betty Ann Skelton to Mr. Currie, Dec. 5, 1972, FDA Papers, Accession no. 88-82-57, box 28, vol. 366; and Ronald F. Riba to Charles H. Percy, Apr. 2, 1973, FDA Papers, Accession no. 88-82-57, box 28, vol. 363. Also see Stephen M. Johnson, "DES Issue 'Exploited': FDA Chief," *Des Moines Register*, Oct. 23, 1972, 3; B. E. Fichte, "Ten Years after *Silent Spring*," *Farm Quarterly*, 27 (Oct. 1972): 24–26; Morton Mintz, "Proxmire Assails DES Implants," *Washington Post*, Nov. 16, 1972, A3; William Proxmire to Edwards, Nov. 13, 1972, FDA Papers, Accession no. 88-82-57, box 27, vol. 258; Luanna S. Hutchinson to Edwards, Nov. 30, 1972, FDA Papers, Accession no. 88-82-57, box 27, vol. 360; James S. Turner, "Eating in the Dark," *Washington Post*, Dec. 8, 1972, D14; "Birth Curb Pill Is Termed Risky," *New York Times*, Dec. 12, 1972, 27; Robert Reinhold, "Link of Drug and a Cancer Confirmed," ibid., Dec. 21, 1972, 41; "Animal Health Group Wants to Replace Delaney Clause with Scientific Judgment," *Feedstuffs*, 44 (Dec. 25, 1972): 4, 38; Gerald F. Meyer to Proxmire, Dec. 4 and 15, 1972, FDA Papers, Accession no. 88-82-57, box 27, vol. 258; Milton R. Young, Clifford P. Hansen, Roman Hruska, Carl T. Curtis, John Tower, Wallace F. Bennett, Paul J. Fannin, Gale W. McGee, Peter H. Dominick, and James O. Eastland to Edwards, Dec. 1, 1972, FDA Papers, Accession no. 88-82-57, box 27, vol. 259; *Congressional Record*, 93d Cong., 1st sess., 49, 496–97, 1738–40, 9965–67; "Texan Introduces Bill to Rescind Ban on DES in Feeds," *Feedstuffs*, 45 (Jan. 15, 1973): 5; "Food Additives, Unprovable Risks," *New York Times*, Feb. 6, 1973, 36; Duane Chapman, "An End to Chemical Farming," *Environment*, 15 (Mar. 1973): 12–17; and "Fountain Sees Drugs Being Banned as Carcinogens," *Feedstuffs*, 45 (Mar. 5, 1973): 4, 49.

33. Morton Mintz, "Changes Opposed in Additive Curb," *Washington Post*, Jan. 21, 1973, A7. Also see *Preventive Medicine*, 2 (1973): 123–70, 608–10; Jane E. Brody, "Group of Scientists Warns against Ending Ban on Cancer-Causing Food Additives," *New York Times*, Jan. 21, 1973, 51; "Reprieve for Delaney Clause?" *Chemical Week*, 112 (Jan. 24, 1973): 16; and "Delaney Amendment," *Feed Management*, 24 (Mar. 1973): 4. Other contemporary statements on the Delaney clause include "Tampering with the Delaney Clause," *Washington Post*, Feb. 1, 1973, A16; "Delaney Amendment under Attack Again," *Organic Gardening and Farming*, 20 (Jan. 1973): 1128–29; and Harrison Wellford and Samuel Epstein, "The Conflict over the Delaney Clause," *New York Times*, Jan. 13, 1973, 31.

34. "Agricultural Scientists Propose Independent Group to Counter Criticism of Food Quality," *Feedstuffs*, 45 (Jan. 22, 1973): 6.

35. William Rice, "The Politics of Food: 'It's Patriotic to Eat Fish,'" *Washington Post*, Feb. 22, 1973, D1, D14; "Mysterious DES Residues Found in Three Livers," *Feedstuffs*, 45 (Feb. 19, 1973): 4, 53; "DES Residue Found in Steer Liver Sample," ibid., 45 (Mar. 5, 1973): 2; "Residue Mystery," *Feedlot Management*, 15 (Apr. 1973): 28; Dale Hunter to John Evans, Apr. 17, 1973, FDA Papers, Accession no. 88-82-57, box 27, vol. 362; Randy Bridson, "High Input Costs, Imports, DES Restriction Face Cattle Feeders," *Feedstuffs*, 45 (Jan. 1, 1973): 9, 34; Mary Bralove, "The Meat Revolt," *Wall Street Journal*, Mar. 29, 1973, 1, 30; "The Meat Boycott," ibid., Apr. 4, 1973, 18; "Environmental Impact Requirement May Delay Animal Drug Approvals Six Months," *Feedstuffs*, 45 (Jan. 29, 1973): 10, 76; "Consumer Breakthrough," *Washington Post*, Feb. 10, 1973, C2; "FDA Relents on Consumer Science Consultants," *Science*, 179 (1973): 879; "Report Declares Food Additive Makes Bacon a Cancer Threat," *New York Times*, Apr. 8, 1973, 23; Morton Mintz, "Safety Test Ordered on Animal Feed Drug," *Washington Post*, Apr. 19, 1973, A4; John McClung, "FDA Gives Antibiotics Manufacturers Two Years to Prove Use in Feed Effective, Safe to Humans," *Feedstuffs*, 45 (Apr. 23, 1973): 1, 50; Victor Cohn, "Academy of Sciences Criticized on Advice," *Washington Post*, Apr. 25, 1973, A2; and "Environmental Impact Statements," *Federal Register*, 38 (Mar. 15, 1973): 7001–6.

36. For laments about the FDA and its situation, see, for instance, Nicholas Wade, "Drug Reg-

ulation: FDA Replies to Charges by Economists and Industry," *Science*, 179 (Feb. 23, 1973): 775–77; Lessel L. Ramsey, "The Role of the Scientist in an Era of Consumerism," *JAOAC*, 56 (1973): 239–45; Malcolm W. Jensen and Thomas M. Folkes, "A New Era in Consumer Safety," *FDA Consumer*, 7 (Feb. 1973): 10–12; Morton Mintz, "Should the FDA Be Dismantled?" *Washington Post*, Feb. 4, 1973, E1, E12; idem, "The FDA: Battling and Embattled," ibid., Feb. 5, 1973, B1, B3; Jack C. Thomas and David P. Ducharme, "The Responsibility of FDA to the Livestock Industry and the Consumer," *JAS*, 37 (1973): 213–25; Cornelius B. Kennedy, "The New Vogue in Rulemaking at FDA: A Foreword," *FDCLJ*, 28 (1973): 172–76; Charles F. Hagan, "Remarks on the Regulatory Philosophy of FDA," ibid., 28 (1973): 195–200; Robert E. Montgomery, Jr., "Comments on the Philosophy of Regulation under the Federal Food, Drug, and Cosmetic Act," ibid., 28 (1973): 201–4; Merrill S. Thompson, "FDA—They Mean Well, But . . . ," ibid., 28 (1973): 205–17; Richard J. Leighton, "The Consumer Protection Agency Bill—Ghosts of Consumerists Past, Present, and Future," ibid., 28 (1973):21–50; and Errett Deck, "Regulated Use of Chemicals in Agricultural Production," ibid., 28 (1973): 628–35. Wade questioned Hutt's integrity in "FDA General Counsel Hutt: A Man Trying to Serve Two Masters," *Science*, 177 (Aug. 11, 1972): 498–500.

37. The quotations are from "Diethylstilbestrol," *Federal Register*, 38 (Apr. 27, 1973): 10485–88. Also see "Edwards' Appointment to Health Post Draws Mixed Reaction," *Science and Government Report*, 3 (Apr. 1, 1973): 5; James E. Patrick and Kenneth I. H. Williams, "The Determination of the Nature of 14C-Liver Residues in Steers Implanted with 14C-DES," FDA Papers, Accession no. 88-88-07, box 1, FDC-D-494 folder; Jonathan Spivak, "Further Limit on DES for Cattle Is Likely as Agriculture Study Shows Traces Linger," *Wall Street Journal*, Apr. 23, 1973, 11; and "Secretary, HEW, from Acting Commissioner, FDA, April 24, 1973," FDA Papers, Accession no. 88-82-57, box 27, vol. 362. Studies were published as T. S. Rumsey, R. R. Oltjen, A. S. Kozak, F. L. Daniels, and P. W. Aschbacher, "Fate of Radiocarbon in Beef Steers Implanted with 14C-Diethylstilbestrol," *JAS*, 40 (1973): 555–60; and P. W. Aschbacher, E. J. Thacker, and T. S. Rumsey, "Metabolic Fate of Diethylstilbestrol Implanted in the Ear of Steers," ibid., 40 (1975): 530–38. Richard D. Lyons, "Cattle Fattener Banned by F.D.A.; Beef Price to Rise," *New York Times*, Apr. 26, 1973, 1, 15; Morton Mintz, "U.S. Ends Use of DES as Livestock Stimulant," *Washington Post*, Apr. 26, 1973, A1 and A4; "DES Livestock Implants Are Prohibited by FDA Because of Cancer Link," *Wall Street Journal*, Apr. 26, 1973; 39; and "Implants Are Extracted," *Chemical News*, 112 (May 2, 1973): 15.

38. "Growth Promotant Similar to DES Being Introduced," *Feedstuffs*, 45 (Apr. 30, 1973): 4; Justin Jay Nell, "The DES Ban," *Cattleman*, 60 (July 1973): 72, 79; "Grown for Stimulants," *Chemical Week*, 112 (May 9, 1973): 16; and "Estradiol Performs Equal to DES as Growth Stimulant," *Feedlot Management*, 15 (June 1973): 31–32.

39. The quotations are from Fred Tunks, "Forget the Crutch and Step Forward," *Feedlot Management*, 15 (June 1973): 30–31; and "The Passing of Stilbestrol," *Feedstuffs*, 45 (Apr. 30, 1973): 10. Also see Daniel Zwedling, "The Meat Risks," *Washington Post*, May 15, 1973, C1, C5; Richard D. Lyons, "In the Case of Safety," *New York Times*, Apr. 29, 1973, sec. 4, p. 2; "Technical Advisory Committee," *Proceedings of the Nutrition Council of the American Feed Manufacturers' Association, Nov. 1973*, 10; "Delaney Clause," *JAVMA*, 163 (1973): 1245–46; "Virginia Feed Men Join Ag Group," *Feedstuffs*, 45 (July 23, 1973): 8, 36; "Animal Groups Discuss Uniting," ibid., 45 (Aug. 13, 1973): 14 *Proceedings of the Agricultural Research Institute*, 22 (1973): 11–13; "Feed Additives and FDA Relations," *JAS*, 39 (1974): 266; W. M. Beeson, "After DES, Then What?" *Proceedings of the Nutrition Council of the American Feed Manufacturers' Association, May 1973*; 28; "'Zero Tolerance' for Feed Additive Residues Need Changing: Beeson," *Feedstuffs* 45 (May 14, 1973): 2, 37; and "Change in Interpretation of Delaney Clause Recommended," ibid., 45 (Nov. 19, 1973): 4, 38.

40. The anticlause quotations are from Wayne Anderson, "More Scientific Input Needed in Risk vs. Benefit Decisions," *Feedstuffs*, 45 (May 14, 1973): 2, 42; Government's Role in Future Beef Industry," *Feedlot Management*, 15 (Nov. 1973): 66; and "Policy on Infinitesimals," *New York*

Times, Apr. 30, 1973, 30. Also see Robert G. Zimbelman, "Hormones and the Delaney Clause," *Cattleman*, 160 (Oct. 1973): 158, 160, 162; Fred Girres, "Better Approach to Animal Drug Regulations Called for," *Feedstuffs*, 45 (Dec. 17, 1973): 1, 38; "Drug Industry Looks at Its Problems," ibid., 45 (May 28, 1973): 9, 42; *Congressional Record*, 93d Cong., 1st sess., 16048–49, 27793, 27858; "Animal Drug Legislation," *JAVMA*, 163 (1973): 703; "'58 Law on Additives Faulty, FDA Aide Says," *Washington Post*, May 16, 1973, A30: and National Academy of Sciences, *How Safe Is Safe?* (Washington, D.C.: National Academy of Sciences, 1974).

41. The quotations are from "Compounds Used in Food-Producing Animals," *Federal Register*, 38 (July 19, 1973): 19226–30. Also see *FDA Consumer*, 7 (May 1973): 38; 7 (July–Aug. 1973): 37; 7 (Sept. 1973): 38; (Oct. 1973): 38; "USDA Finds Sixty-Four Residue Violations in First Quarter," *Feedstuffs*, 45 (May 7, 1973): 6; "Drug Traces in Meat Rise," *Washington Post*, Aug. 12, 1973, B8; George Lakata, "Report on Drug Residues in Animal Tissues," *JAOAC*, 56 (1973): 316–17; Alvin L. Donoho, William S. Johnson, Robert F. Sieck, and William L. Sullivan, "Gas Chromatographic Determination of Diethylstilbestrol and Its Glucuronide in Cattle Tissues," ibid., 56 (1973): 785–92; and G. W. Probst to W. Shimoda, July 12, 1973, FDA Papers, Accession no. 88-82-57, box 28, vol. 365.

42. Harold M. Schmeck, Jr., "F.D.A. to Set Rule for Food Testing," *New York Times*, July 19, 1973, 15; Nancy L. Ross, "Testing Meats for Drug Residues," *Washington Post*, July 19, 1973, B10; "FDA Will Revise Tests for Residue of Drugs in Meat," *Wall Street Journal*, July 19, 1973, 18; John McClung, "FDA Proposes Development of New Assay Standards for Drug Detection," *Feedstuffs*, 45 (July 23, 1973): 1, 39; and "FDA Proposes New Assay Techniques," *Feedlot Management*, 15 (Sept. 1973): 32, 97.

43. For an indication of some of the legal questions facing the FDA, see, for example, William J. Lanouette, "Supreme Court Will Rule on Drug Controls," *National Observer*, Apr. 28, 1973, 10; Joan Arehart-Treichel, "Agonizing over Food and Drugs: How Safe?" *Science News*, 103 (June 2, 1973): 362–63; Scott M. Fisher, "Publicity and the FDA," *FDCLJ*, 28 (1973): 436–46; Peter Barton Hutt, "Safety Regulation in the Real World," ibid., 28 (1973): 460–72; and Wallace F. Janssen, "Toward a New Era in Consumer Protection: The Supreme Court Rulings on Drug Effectiveness," *FDA Consumer*, 7 (Oct. 1973): 19–27.

44. The quotations are from "Brief for Vineland Laboratories, Inc., August 20, 1973," in my possession; "Decision—United States Court of Appeals for the District of Columbia Circuit, Cases 73-1581 and 73-1589, September 14, 1973," in my possession; and "Appeals Court Orders Stay on FDA Ban of DES," *Feedstuffs*, 45 (Sept. 24, 1973): 2, 58. Also see Eugene I. Lambert to Sherwin Gardner, Apr. 30, 1973, James L. Kaler and Frederick S. Hird, Jr., to Hearing Clerk, May 3, 1973; and Sam D. Fine to Lambert, May 7, 1973, FDA Papers, Accession no. 88-88-07, box 1, FDC-D-494 folder; "Drug Firms Ask Court to Stay DES Implant Ban," *Feedstuffs*, 45 (June 11, 1973): 1, 47; and "DES Implants Not Dead Yet," *Feedlot Management*, 15 (July 1973): 7.

45. The quotation is from "Diethylstilbestrol," *Federal Register*, 38 (Oct. 25, 1973): 29510–14. Also see Eugene I. Lambert to A. M. Schmidt, Oct. 4, 1973, FDA Papers, Accession no. 88-88-07, box 1, Hess and Clark folder; "Statement of Vineland Laboratories, Inc., on Issues for Hearing, October 4, 1973"; and "Hess and Clark v. Food and Drug Administration—Petition to Set Aside Orders, October 4, 1973," FDA Papers, Accession no. 88-88-07, box 1, FDC-D-494 folder; Frank J. Wolf to D. M. Tennett, Sept. 24, 1973, "W. H. Ray, Research Reports WHR 73:71, September 26, 1973, and WHR 73:73, October 1, 1973"; "E. M. Crains, Research Report, EMC 73:72, October 2, 1973"; and "Affidavit of Irwin Clark, October 3, 1973," FDA Papers, Accession no. 88-88-07, box 1, Hess and Clark folder; and "FDA Denies Drug Firms' Request for DES Implant Hearing," *Feedstuffs*, 45 (Nov. 5, 1973): 4.

46. "Telephone Conversation—Gilbert Goldhammer to Harry C. Mussman, October 12, 1973," FDA Papers, Accession no. 88-82-57, box 28, vol. 365; Morton Mintz, "Use of Drugs in Cattle Feeding Hit," *Washington Post*, Dec. 9, 1973, A3; and *Regulation of Diethylstilbestrol (DES) and Other Drugs Used in Food Producing Animals: Twelfth Report by the Committee on Government Operations Together with Additional Views* (Washington, D.C.: GPO, 1973).

47. *Regulation of Diethylstilbestrol (DES) and Other Drugs*; Morton Mintz, "Taking a Look at Medicated Livestock Feeds and Possible Cancer-Links," *Washington Post*, Dec. 11, 1973, B1, B9; L. H. Fountain to Alexander M. Schmidt, Dec. 17, 1973, FDA Papers, Accession no. 88-82-57, box 28, vol. 367; "Washington Newsletter," *Chemical Week*, 113 (Dec. 19, 1973): 14; and John McClung, "House Committee Report Criticizes FDA," *Feedstuffs*, 45 (Dec. 24, 1973): 1, 27. For Schmidt, see, for example, "Schmidt Takes on FDA," *Science*, 18 (1973): 526; and John McClung, "Alexander Schmidt New FDA Commissioner," *Feedlot Management*, 15 (Nov. 1973): 57, 60, 62.

48. For the response to Fountain, see Thomas H. Jukes to L. H. Fountain, Jan. 9, 1974, FDA Papers, Accession no. 88-82-57, box 28, vol. 367; "Feed Additives and FDA Regulations," *JAS*, 39 (1974): 266–67; and *Efficiency in Animal Feeding with Particular Reference to Nonnutritive Feed Additives: Council for Agricultural Science and Technology Report No. 22*, Jan. 18, 1974.

49. "Hess and Clark, Division of Rhodia, Inc., v. Food and Drug Administration et al., and Vineland Laboratories, Inc., v. Casper W. Weinburger, Secretary, Department of Health, Education, and Welfare et al.," *Food Drug Cosmetic Law Reporter: Developments* (1973–74), 40436–50; and "Chemetron Corporation, Dawe's Laboratories, Inc., and Hess and Clark, Division of Rhodia, Inc., v. the Department of Health, Education, and Welfare et al.," ibid., 1973–74, 40450–54.

50. "Hess and Clark v. Food and Drug Administration"; "Chemetron Corporation v. Department of Health, Education, and Welfare."

CHAPTER 7. THE NEW SYNTHESIS

1. Ida Honorof stood virtually alone when she dismissed the ruling, declaring the FDA ban "not only legal," but an "action that should have been taken years ago." Ida Honorof to Alexander M. Schmidt, Jan. 28, 1974, FDA Papers, Accession no. 88-82-57, box 28, vol. 368; William Proxmire to Schmidt, Mar. 13, 1974, FDA Papers, Accession no. 88-82-57, box 28, vol. 369; Celia M. Nakano to Daniel W. Clink, Apr. 25, 1974, and Fannie E. Post to FDA, May 20, 1974, FDA Papers, Accession no. 88-82-57, box 28, vol. 370; George Getschow, "Uncertain Status of Growth Drug DES Adds to Woes of Cattle-raising Industry," *Wall Street Journal*, June 17, 1974, 20; "DES Is Back!" *Feed Management*, 25 (March 1974): 6; "DES Is Back—Or Is It?" *Successful Farming*, 72 (Mar. 1974): A3; "DES Is Back," *Wallaces Farmer*, 99 (Mar. 23, 1974): 37; Joseph M. Winski, "Beef Glut," *National Observer*, Apr. 6, 1974, 8; and Bruce Wilkinson, "Industry Outlook," *Feedlot Management*, 16 (Apr. 1974): 70–71.

2. "DES Makers Go Slow," *Chemical Week*, 114 (Feb. 6, 1974): 18; "DES Again in Use, but Future Uncertain," *Iowa Agriculturist*, 75 (Spring 1974): 23; "DES Premix Available," *Feedlot Management*, 16 (May 1974): 64; "Confusion over DES," *Cattleman*, 60 (Mar. 1974): 32; "DES," ibid., 60 (Apr. 1974): 16; "FDA Won't Appeal Court's DES Ruling," *Washington Post*, Jan. 29, 1974, A3; "FDA to Promptly Grant Firms Hearing on Plan to Bar Future DES Use," *Wall Street Journal*, Jan. 28, 1974, 13; "FDA and DES," *JAVMA*, 164 (1974): 588; "Stilbestrol Will Be Marketed for Cattle Again; FDA to Grant Hearing," *Feedstuffs*, 46 (Feb. 4, 1974): 1, 7; "A Second Chance for Stilbestrol," ibid., 46 (Feb. 11, 1974): 10; J. D. Kendall, "Makers of DES Premixes Hold Off Production," ibid., 46 (Feb. 25, 1974): 1, 54; and "Inside Washington," ibid., 46 (Mar. 11, 1974): 6, 46 (Mar. 25, 1974): 2. The column "Inside Washington" was written by members of *Feedstuffs'* Washington bureau. Authors varied, but most articles were written by John McClung.

3. The unflattering quotation is found in "FDA Proposes Revocation of DES Detection Method," *Feedstuffs*, 46 (Apr. 1, 1974): 1, 41. Also see Dick Beeler, "The FDA on Trial," *Animal Nutrition and Health*, Mar. 1974, 22; Richard D. James, "Remember Cyclamates?" *Wall Street Journal*, July 2, 1973, 1, 15; "Cyclamate Comeback," *Chemical Week*, 113 (July 11, 1973): 23; "Cyclamates," *Consumer Register*, 3 (Mar. 1, 1974): 2; "FDA Takes First Step to Comply with DES Order," *FDA Consumer*, 8 (May 1974): 34; Walter E. Byerley, "So Are They All—All Honorable Men; A Review of the DES Revocation Cases to Date," *FDCLJ*, 29 (1974): 460–68; Wayne Pines, "Behind FDA's Regulations," *FDA Consumer*, 8 (Nov. 1974): 10–17; Peter B. Hutt, "FDA Court Actions and Recent Legal Developments," *Quarterly Bulletin of the Association of Food and Drug*

Officials, 39 (Jan. 1975): 11–20; and *Federal Register*, 39 (Mar. 27, 1974): 11299–301, 11323–24; 39 (Apr. 25, 1974): 14611.

4. FDA scientists began studies to determine the carcinogenic threshold of DES and other estrogens in mice at the recently opened National Center for Toxicological Research. Advocates of the view that one molecule may cause cancer predictably denounced the research as impractical and the experiments as poorly designed. Stilbestrol and cattle returned momentarily to the headlines in early April 1974, when Canada refused to import U.S. cattle or beef unless the federal government certified that cattlemen had produced the animals without DES. Neither the USDA nor the FDA possessed the staff necessary to provide that guarantee. Canada remained adamant; and importation, which had averaged about ten thousand animals weekly, ceased. American cattlemen claimed anticipated yearly losses in excess of $150 million. E. H. Allen to W. Shimoda, Jan. 30, 1974; and "Memorandum of Conference, February 7, 1974," FDA Papers, Accession no. 88-82-57, box 28, vol. 367; "Minutes of USDA-FDA Meeting on Diethylstilbestrol (DES), February 13, 1974"; Drug Bioanalysis Branch to Fred Kingma, Feb. 26, 1974; K. F. Johnson to Director, Mar. 7, 1974; Harry C. Mussman to C. D. Van Houweling, Mar. 8, 1974; Shimoda to N. H. Booth, Allen, and W. D. Price, Mar. 11, 1974; and Price to H. D. Washburn, Mar. 21, 1974, FDA Papers, Accession no. 88-82-57, box 28, vol. 368; D. G. Dale to Van Houweling, Mar. 1, 1974, and "Memorandum of Conference, March 19, 1974," FDA Papers, Accession no. 88-82-57, box 28, vol. 368; Director, Bureau of Veterinary Medicine, to Commissioner, Apr. 18, 1974, FDA Papers, Accession no. 88-82-57, box 28, vol. 369; Robert Trumbull, "New Canada Rule Barring Hormone in Meat Imports Halts U.S. Trade," *New York Times*, Apr. 10, 1974, 51, 61; and "Canada Bans Imports of Meat, Livestock Treated with DES," *Wall Street Journal*, Apr. 10, 1974, 28.

5. "House Unit Votes Delaney Study," *Feedstuffs*, 45 (June 11, 1973): 2; Thomas E. Shellenberger, "The National Center for Toxicological Research," *Quarterly Bulletin of the Association of Food and Drug Officials*, 38 (Jan. 1974): 15–22; B. L. Oser, "An Assessment of the Delaney Clause after 15 Years," *FDCLJ*, 29 (1974): 201–9; "Ag Appropriations Bill Includes Funds for DES, Delaney Clause Study," *Feedstuffs*, 46 (July 8, 1974): 2, 36; and "Food and Drug Administration—Study of the Delaney Clause and Other Anticancer Clauses," in *Hearings before a Subcommittee of the Committee of Appropriations, Ninety-third Congress, Second Session* (Washington, D.C.: GPO, 1974), especially 16–17, 20–21, 45–48, 50–83.

6. Burroughs comments for himself and Black in Wise Burroughs to William J. Scherle, May 30, 1974, ISURF Papers, Burroughs Uncatalogued folder. Also see Scherle to Burroughs, May 7, 1974, ISURF Papers, Burroughs Uncatalogued folder; "Scientists Give Time to Analyze Public Issues Affecting Future of Agriculture," *Feedstuffs*, 46 (July 15, 1974): 8, 41; CAST Task Force, *Zero Concepts in Air, Water, and Food Quality Legislation*, Report no. 31 (Ames, Iowa: CAST, 1974), especially 3–6, 15–21; Thomas H. Jukes, "Residues in Meat: Consumer Regulations, Producer Problems," *Feedstuffs*, 46 (July 15, 1974): 26–27; idem, "Estrogens in Beefsteak," *JAMA*, 229 (Sept. 30, 1974): 1920–21; idem. "Food Production and Acceptance: Consumer Attitudes and Problems," *Feedstuffs*, 46 (Dec. 23, 1974): 18, 20; and George H. Gass, "A Discussion of Assay Sensitivity Methodology and Carcinogenic Potential," *FDCLI*, 30 (1975): 111–15.

7. George H. Gass, Jean Brown, and Allan B. Okey, "Carcinogenic Effects of Oral Diethylstilbestrol on C_3H Male Mice with and without the Mammary Tumor Virus," *Journal of the National Cancer Institute*, 53 (Nov. 1974): 1369–70.

8. The quotations are from H. H. Cole, G. H. Gass, R. J. Gerrits, H. D. Hafs, W. H. Hale, R. L. Preston, and L. C. Ulberg, "On the Safety of Estrogenic Hormone Residues in Edible Animal Products," *Bioscience*, 25 (1975): 19, 23. Also see "Feed Additives and FDA Relations," *JAS*, 41 (1975): 441–42; Albert C. Kolbye, Jr., "Zero Tolerance Concept as It Relates to Hormonal Residues in Animal Products," ibid., 40 (1975): 1258–62; S. Nandi, "Comparison of the Tumorigenic Effects of Chemical Carcinogens and Hormones," ibid., 40 (1975): 1263–66; and A. B. Okey and George H. Gass, "Continuous versus Cyclic Estrogen Administration: Mammary Carcinoma in CH3 Mice," *Journal of the National Cancer Institute*, 40 (1968): 225–30.

9. "Inside Washington," *Feedstuffs*, 46 (July 15, 1974): 2; 46 (Aug. 5, 1974): 2; 46 (Aug. 19,

1974): 2; 46 (Nov. 11, 1974): 2; 47 (Jan. 6, 1975): 6; Bernard Brenner, "Cattle and Cancer," *Washington Post*, Dec. 28, 1974, C3; and Wallace F. Janssen, "FDA 1974: Scrutiny from Without and Within," *FDA Consumer*, 9 (Apr. 1975): 4–10.

10. "Telephone Memorandum—Dr. Peckham and R. L. Gillespie, July 2, 1974"; Carolyn N. Andres to William Horwitz, Aug. 2, 1974; and Constantine Zervos to C. D. Van Houweling, Aug. 7, 1974, FDA Papers, Accession no. 88-82-57, box 28, vol. 371; Edward H. Allen to W. D. Price, Oct. 7, 1974; Wilbert Shimoda to Stanley F. Cernosek, Oct. 8, 1974; R. L. Preston to Price, Oct. 11, 1974; "Telephone Memorandum—P. D. Cazier and Marlyn Perez et al., October 22, 1974"; Frank L. Joe, Jr., to Philip D. Cazier, Nov. 14, 1974; "Memorandum—R. P. Lehmann, R. Condon, and W. D. Price, November 14, 1974"; Robert K. Munns to Carolyn N. Andres, Nov. 18 and Dec. 3, 1974; and Price and Shimoda to R. P. Lehmann and H. D. Mercer, Dec. 13, 1974, FDA Papers, Accession no. 88-82-57, box 28, vol. 373; R. L. Preston, "Rationale for Using Hormones in Beef Production," FDA Papers, Accession no. 88-82-57, box 1, Diethylstilbestrol folder; William B. Bixler to Van Houweling, Jan. 21, 1975; Shimoda to Cernosek, Jan. 22, 1975; Shimoda to H. Dwight Mercer, Jan. 22, 1975; Shimoda to Reble Heil, Jan. 27, 1975; John H. Turner to Director, Field Sciences Branch, Feb. 7, 1975, and Bixler to Robert Wetherell, Feb. 20, 1975, FDA Papers, Accession no. 88-82-57, box 28, vol. 374; Shimoda to Heil, Mar. 5, 1975; T. Fazio to Ann Holt, Mar. 5, 1975; "Memorandum—Bernhard Larsen and Philip D. Cazier, March 18, 1975, and March 28, 1975"; Cazier to Perez, Mar. 19, 27, and 31, 1975; "Memorandum—Marlyn Perez and K. F. Johnson, April 8, 1975, and April 17, 1975"; Memorandum—Bernard Larsen and Maryln Perez, April 8, 1975, and April 10, 1975"; and "Memorandum—Tom Fazio and K. F. Johnson, April 17, 1975," FDA Papers, Accession no. 88-82-57, box 28, vol. 375; "Memorandum—Tom Fazio and K. F. Johnson, April 25, 1975"; A. C. Kolbye to Associate Commissioner for Compliance, Apr. 25, 1975; "Memorandum—Bernhard Larsen and Philip Cazier, April 29, 1975"; Shimoda to Mercer, May 19, 1975; David A. Kline to Cazier, June 4, 1975; Cazier to Van Houweling, June 6, 1975; and Shimoda to Mercer, June 24, 1975, FDA Papers, Accession no. 88-82-57, box 28, vol. 376; "USDA Orders Fourteen-Day Withdrawal for Stilbestrol," *Feedstuffs*, 47 (Feb. 17, 1975): 1, 52; "Nine DES Violations Found in Quarter," ibid., 47 (June 9, 1975): 5; "Farm Unit to Require DES Users to Certify Removal from Feed," *Wall Street Journal*, Feb. 12, 1975, 20; "DES Issue Alive," *Feedlot Management*, 17 (Feb. 1975): 20–21; "You Must Certify DES Withdrawal," *Successful Farming*, 73 (Apr. 1975): F4; "USDA Finds Two More DES Violations," *Wallaces Farmer*, 100 (Feb. 22, 1975): 52; "DES Gets Limited 'Yes,'" *Chemical Week*, 116 (Feb. 19, 1975): 15; George Lakata, "Report on Drug Residues in Animal Tissues," *JAOAC*, 58 (1975): 253–54; Edgar W. Day, Jr., Lynn E. Vanatta, and Robert F. Sieck, "The Confirmation of Diethylstilbestrol Residues in Beef Liver by Gas Chromatography–Mass Spectrometry," ibid., 58 (1975): 520–24; "Canada Removes Ban on Meat from the U.S.," *New York Times*, Aug. 3, 1974, 31; "Canada Removes Ban on U.S. Livestock after Accord on DES," *Wall Street Journal*, Aug. 5, 1974, 18; and "Rules for Exporting DES-free Livestock to Canada," *JAVMA*, 165 (1974): 603.

11. *Regulation of New Drug R. and D. by the Food and Drug Administration, 1974: Joint Hearings before the Subcommittee on Health of the Committee on Labor and Public Welfare and the Subcommittee on Administrative Practice and Procedure of the Committee on the Judiciary, United States Senate, Ninety-third Congress, Second Session, September 25 and 27, 1974* (Washington, D.C.: GPO, 1975), 435–68; "Fountain: FDA Failing to Meet Its Responsibilities in DES Program," *Feedstuffs*, 47 (Jan. 27, 1975): 10, 93; R. L. Gillespie to A. C. Kolbye, Oct. 25, 1974, FDA Papers, Accession no. 88-82-57, box 28, vol. 373; Alexander M. Schmidt to L. H. Fountain, Jan. 3, 1975, FDA Papers, Accession no. 88-82-57, box 28, vol. 374; Schmidt to Fountain, Mar. 3, 1975, FDA Papers, Accession no. 88-82-57, box 28, vol. 375; Robert C. Wetherell to C. D. Van Houweling, Apr. 1975; and Caspar Weinberger to Harrison Williams, Apr. 22, 1975, FDA Papers, Accession no. 88-82-57, box 28, vol. 376; *Regulation of Diethylstilbestrol (DES), 1975: Joint Hearing before the Subcommittee on Health of the Committee on Labor and Public Welfare and the Subcommittee on Administrative Practice and Procedure of the Committee on the Judiciary, United States Senate,*

Ninety-fourth Congress, First Session, February 27, 1975 (Washington, D.C.: GPO, 1975); and *Congressional Record*, 94th Cong., 1st sess., 5272, 5319–24.

12. The debate is recorded in *Congressional Record*, 94th Cong., 1st sess., 22800, 22963, 27776–77. Also see John McClung, "DES Ban Said Likely to Take Congressional, Not FDA Action," *Feedstuffs*, 47 (Apr. 14, 1975): 1, 47; "H.E.W. Head Opposes Ban on DES as Birth Curb," *New York Times*, Apr. 24, 1975, 20; "HEW Backs DES Ban in Cattle Feed," *Washington Post*, Apr. 24, 1975, A6; "DES Ban Closer in Senate Action," *Feedstuffs*, 47 (Apr. 28, 1975): 4, 76; *Report to Accompany Senate #963: Ninety-fourth Congress, First Session, Senate Report No. 94-264*: and "Amendment Would Soften DES Bill Impact," *Feedstuffs*, 47 (July 21, 1975): 3.

13. *Congressional Record*, 94th Cong., 1st sess., 27851–82, 28155–80, 28337; "Action against DES," *Washington Post*, Sept. 9, 1975, A18; and "Senate Vote Due Soon on Bill Banning DES Use," *Feedstuffs*, 47 (Sept. 8, 1975): 1, 41.

14. The quotations are from "A Victory for 'Stilbestrophobia,' " *Feedstuffs*, 47 (Sept. 15, 1975): 14; and Thomas H. Jukes, "Kennedy's Good Deed," *Nature*, 258 (Nov. 6, 1975): 5–6. Also see John McClung, "Senate Passes Compromise Bill Banning DES Use," *Feedstuffs*, 47 (Sept. 15, 1975): 9, 65; "Senate Votes to Ban DES," *Feedlot Management*, 17 (Oct. 1975): 24–25; "Action against DES," *JAVMA*, 167 (1975): 722; "More Limits for DES," *Chemical Week*, 117 (Sept. 17, 1975): 19; "Senate Curbs Use of a Cattle Drug," *New York Times*, Sept. 10, 1975, 16; Spencer Rich, "Senate Votes to Ban DES," *Washington Post*, Sept. 10, 1975, A6; "Ban on DES Drug for Use on Livestock Is Voted by Senate," *Wall Street Journal*, Sept. 10, 1975, 6; Fredrick J. Stare, "Emerging Food Patterns: Coping with Reality," *Quarterly Bulletin of the Association of Food and Drug Officials*, 39 (Oct. 1975): 227–36; Elizabeth M. Whelan, "Panic over Food Additives," *Washington Post*, Aug. 10, 1975, B6; and "New Book Defends Use of Additives, Chemicals in Food Production," *Feedstuffs*, 47 (Nov. 24, 1975): 2, 45.

15. P. VanderWal, "General Aspects of the Effectiveness of Anabolic Agents in Increasing Protein Production in Farm Animals, in Particular Bull Calves," in *Anabolic Agents in Animal Production* (Stuttgart: Georg Thieme, 1976), 60–78; Allen Trenkle, "The Anabolic Effect of Estrogens on Nitrogen Metabolism of Growing and Finishing Cattle and Sheep," in ibid., 79–88; B. Hoffman and H. Karg, "Metabolic Fate of Anabolic Agents in Treated Animals and Residue Levels in Their Meat," in ibid., 181–91; A. C. Kolbye and M. K. Perez, "Human Safety Considerations from the Use of Anabolic Agents in Foodproducing Animals," in ibid., 212–18; and Francis J. C. Roe, "Carcinogenicity Studies in Animals Relevant to the Use of Anabolic Agents in Animal Production," in ibid., 227–37; Wilbert Shimoda to Director, Office of International Affairs, Apr. 29, 1975, FDA Papers, Accession no. 88-82-57, box 28, vol. 376; Attallah Kappas and Alvito P. Alvares, "How the Liver Metabolizes Foreign Substances," *Scientific American*, June 1975, 22–31; and R. L. Preston, "Biological Responses to Estrogen Additives in Meat Producing Cattle and Lambs," *JAS*, 41 (1975): 1414–30.

16. See, for example, "A. B. Okey, 'Minimal Time of Exposure to Diethylstilbestrol Required for Mammary Tumor Induction in Mice,' May 20, 1975"; R. L. Gillespie to Norris Alderson, June 4, 1975; Richard P. Lehmann to Allan B. Okey, June 4, 1975; and Bernard H. Haberman to Leonard Friedman, June 10, 1975, FDA Papers, Accession no. 88-82-57, box 28, vol. 376; Nathan Mantel, Neeti R. Bohidar, Charles C. Brown, Joseph L. Ciminera, and John W. Tukey, "An Improved Mantel-Bryan Procedure for 'Safety' Testing of Carcinogens," *Cancer Research*, 35 (April 1975): 865–72; "A.S.A.S. Regulatory Agency Committee Annual Report to the Executive Committee, 1975 to 1976" *JAS*, 43 (1976): 733; *Annual Report of the U.S. Food and Drug Administration*, 1975, 91–93; Rosa M. Gryder to Willy Shimoda, Sept. 25, 1975, FDA Papers, Accession no. 88-82-57, box 28, vol. 374; Nicholas E. Weber to Assistant to the Associate Director, Oct. 3, 1975, FDA Papers, Accession no. 88-82-57, box 1, vol. 11; Philip D. Cazier to Fred Holt, Dec. 30, 1975, FDA Papers, Accession no. 88-82-57, box 1, Hero 2; Manfred Metzler, "Metabolic Activation of Carcinogenic Diethylstilbestrol in Rodents and Humans," *Journal of Toxicology and Environmental Health*, 1, suppl. (1976): 21–35, Lewis L. Engel, Josef Weidenfeld, and George R. Merriam, "Me-

tabolism of Diethylstilbestrol by Rat Liver: A Preliminary Report," ibid., 1, suppl. (1976): 37–43; P. W. Aschbacher, "Diethylstilbestrol Metabolism in Food-Producing Animals," ibid., 1, suppl. (1976): 61–76; Howard A. Bern, Lovell A. Jones, Karen T. Mills, Arthur Kohrman, and Takao Mori, "Use of the Neonatal Mouse in Studying Long-Term Effects of Early Exposure to Hormones and Other Agents," ibid., 1, suppl. (1976): 103–16; J. A. McLachlan and R. R. Newbold, "Reproductive Tract Lesions in Male Mice Exposed Prenatally to Diethylstilbestrol," *Science*, 190 (1975): 991–92; Morton Mintz, "DES May Cause Sterility in Sons, Studies Indicate," *Washington Post*, Dec. 10, 1975, A3; and "New DES Charges," *Chemical Week*, 117 (Dec. 17, 1975): 18.

17. "Inside Washington," *Feedstuffs*, 47 (Dec. 15, 1975): 2; John McClung, "FDA Set to Once Again Prohibit Use of DES for Livestock Production," ibid., 47 (Dec. 22, 1975): 2, 33; and Diane Henry, "Some Birth Pills Face Ban by F.D.A.," *New York Times*, Dec. 29, 1975, 1, 41.

18. John B. Jones, Jr., to John Barrett, Jan. 23, 1976, FDA Papers, Accession no. 88-82-57, box 1, vol. 3; "Cancer-Link Agent Rises in Beef Liver," *Washington Post*, Nov. 16, 1975, A27; Stuart Auerbach, "Ban Sought on DES Use in Livestock," ibid., Jan. 10, 1976, A1, A4; "F.D.A. Again Asks for a Ban on DES," *New York Times*, Jan. 10 1976, 23; and "FDA Moves to Ban DES in Animal Feed Due to Cancer Link," *Wall Street Journal*, Jan. 12, 1976, 5.

19. "Diethylstilbestrol; Notice of Opportunity for Hearing on Proposal to Withdraw Approval of New Animal Drug Application," *Federal Register*, 41 (Jan. 12, 1976): 1804–7; and "Inflation Impact Statement of Proposed Rulemaking—Diethylstilbestrol, January 1976," FDA Papers, Accession no. 88-82-57, box 34.

20. The quotations are from "Beef Management," *Successful Farming*, 74 (Mar. 1976): A2; and Thomas H. Jukes, "Diethylstilbestrol in Beef Production: What Is the Risk to Consumers?" *Preventive Medicine*, 5 (1976): 438–53. Also see John McClung, "FDA Proposes DES Use Ban, Sets Thirty-Day Hearing Proposal Limit," *Feedstuffs*, 48 (Jan. 12, 1976): 1, 66; "Inside Washington," ibid., 48 (Jan. 19, 1976): 2; "Washington Report," *Successful Farming*, 74 (Feb. 1976): 4; "Cancellation of DES," *Feed Management*, 27 (Mar. 1976): 6; "Four Companies Contest Ban on DES Use," *Washington Post*, Feb. 13, 1976, A24; "Ban Sought on Use of DES in Food Animals," *FDA Consumer*, 10 (Feb. 1976): 26; "Written Appearance and Request for Hearing of Vineland Laboratories, Inc., February 9, 1976"; "Dawe's Laboratories, Inc.—Proposed Withdrawal of Approval of NADAs for DES, February 11, 1976," FDA Papers, Accession no. 88-82-57, box 1, vol. 2; "Rhodia, Inc.—Written Appearance and Request for Hearing, February 11, 1976" FDA Papers, Accession no. 88-82-57, box 1, Hero 2; Gordon Van Vleck to John R. Glennie, Jan. 14, 1976; and Thomas J. Marlowe to Hearing Clerk, Feb. 10, 1976, FDA Papers, Accession no. 88-82-57, box 34; Fred H. Holt to Jennie C. Peterson, Feb. 18, 1976, FDA Papers, Accession no. 88-82-57, box 1, vol. 2; "A.S.A.S. Regulatory Agency Committee Report, 1975 to 1976," 734; Thomas H. Jukes to Lesley Stahl, Feb. 11, 1976," FDA Papers, Accession no. 88-82-57, box 27, vol. 361; "Scientific Food Production Methods Endangered by Nutrition Fads: Jukes," *Feedstuffs*, 48 (Mar. 22, 1976): 9, 88; Thomas H. Jukes, "Estrogens in Beef Production," *Bioscience*, 26 (1976): 544–47; and "In Defense of Additives," *Feedstuffs*, 48 (Sept. 6, 1976): 10. For the activities of the three nonparty participants, see Marcia J. Cleveland to Petersen, Jan. 27, 1976, FDA Papers, Accession no. 88-82-57, box 34; "DES Strategy Memo—Gail Cooper to Jackie Warren, Joe Highland, Nancy Stroup, and Bill Butler, October 28, 1975"; "Draft—Citizens' Petition"; Joseph H. Highland to Alex Heingartner, Feb. 25, 1976; and Highland and Jacqueline M. Warren to Alexander M. Schmidt, June 17, 1976, in my possession, courtesy of the Environmental Defense Fund; "DES—Report to the Board of Trustees, March 15, 1976"; and "Memorandum—Safety of DES, July 7, 1976," in my possession, courtesy of the Pacific Legal Foundation; Glenn E. Davis and John H. Sharon to Jeannie C. Petersen, May 24, 1976, FDA Papers, Accession no. 88-82-57, box 1, vol. 2; and "Memorandum of Meeting—Food and Drug Administration, Environmental Defense Fund, and Natural Resources Defense Council, May 21, 1976"; and Warren to Richard Merrill, July 1, 1976, FDA Papers, Accession no. 88-82-57, box 1, vol. 3.

21. Ten Companies Sued on Disputed Drug," *New York Times*, Apr. 21, 1976, 17; Nadine Brozan, "Repercussions of a Drug: An Ordeal for Mothers and Their Daughters—And Maybe

Sons, Too," ibid., June 17, 1976, 41; Morton Mintz, "FDA Aides Say Superior Biased toward Industry," *Washington Post*, July 21, 1976, A4; "Inside Washington," *Feedstuffs*, 48 (July 26, 1976): 2; E.G.F. "Passing the Baton at FDA," *Journal of Pharmaceutical Sciences*, 65 (Nov. 1976): I; and "No DES Discovered in Three-Month Testing of Livers of Cattle," *New York Times*, Aug. 21, 1976, 18. The DES issue surfaced in the media only when three Massachusetts women filed a class action suit against drug companies that had manufactured DES for use during pregnancy. The material on the various consumer campaigns is legion. See, for example, Linda Charlton, "Red Dye No. 2 Is Linked to Cancer by U.S. Study," *New York Times*, Jan. 10, 1976, 20; "Side effects Linked to Hormone Drugs Said to Raise 'Serious' Doubts about Use," *Wall Street Journal*, 22, 1976, 12; Gladwin Hill, "U.S. Agency Urges a Drive to Bar Cancer by Screening Chemicals," *New York Times*, Feb. 28, 1976, 52; Marian Burros, "Hunter's Quarry: FDA," *Washington Post*, Feb. 19, 1976, D1, D8; "HEW Drug Panel Produces a Splendid Muddle," *Science and Government Report*, 6 (June 1, 1976): 1–3; "Chemical Carcinogens: Industry Adopts Controversial 'Quick' Tests," *Science*, 192 (1976): 1215–17; Marion Burros, "Who Regulates the Regulators?" *Washington Post*, June 24, 1976, E1, E9; "Washington Report," *Feedlot Management*, 18 (June 1976): 24; *Regulatory Reform—Food and Drug Administration: Hearings before the Subcommittee on Oversight and Investigations of the Committee on Interstate and Foreign Commerce, House of Representatives, March 15 and 16, 1976* (Washington, D.C.: GPO, 1976); *Food and Drug Administration Appropriations for 1977: Hearings before a Subcommittee of the Committee on Appropriations, House of Representatives* (Washington, D.C.: GPO, 1976); and *Drug Safety and Effectiveness: Hearings before the Interstate and Foreign Commerce Subcommittee, House of Representatives* (Washington, D.C.: GPO, 1976).

22. The quotations are from "Inside Washington," *Feedstuffs*, 48 (Nov. 8, 1976): 2; and "Dawe's Laboratories et al.," *Federal Register*, 41 (Nov. 26, 1976): 52105–6. Also see "A New Consumer Protection Agency," *Cattleman*, 63 (Dec. 1976): 106; J. D. Kendall, "FDA Sets Prehearing Date on DES Ban," *Feedstuffs*, 48 (Dec. 6, 1976): 1, 41–43; and "Elanco Products C., et al.," *Federal Register*, 41 (Nov. 26, 1976): 52106–7.

23. Walter E. Byerley to Hearing Clerk, Dec. 20, 1976; Daniel J. Davidson to Participants, Jan. 6, 1977; "Notice of Pacific Legal Foundation, American National Cattlemen's Association, and the National Livestock Feeders' Association to Enlarge Issues, January 17, 1977"; Joseph A. D'Arco, Walter E. Byerley, James C. McKay, and Frederick S. Hird, Jr., "Memorandum in Support of Motion for Partial Summary Decision," Jan. 17, 1977, Richard F. Kingham to Daniel J. Davidson, Jan. 19, 1977; and "Joint Motion of the Manufacturing Parties to Revise and Enlarge Issues for Hearing, January 17, 1977," FDA Papers, Accession no. 88-82-57, box 1, Hero 1.

24. The quotations are from O. D. Butler to Hearing Clerk, Dec. 22, 1976, FDA Papers, Accession no. 88-82-57, box 1, Hero 1; "CAST Panel to Give Hormone Findings at ANCA," *Feedstuffs*, 49 (Jan. 10, 1977): 26; and "Submissions Pursuant to 21 C.F.R. $$ 2.153 (a) (3)–(4) by Pacific Legal Foundation, American National Cattlemen's Association, and the National Livestock Association, February 7, 1977"; and "Submissions Pursuant to 21 C.F.R. $$ 2.153 (a) (3)–(4) by American Society of Animal Science, February 7, 1977," in my possession, courtesy of Pacific Legal Foundation. Also of interest are "The Voice of Agriculture Speaks Out," *Bioscience*, 26 (Aug. 1976): 481–86; B. P. Cardon, "Council for Agricultural Science and Technology," *Proceedings of the Agricultural Research Institute*, 25 (1976): 23–29; and Ron Sterk, "CAST: Telling America about Agriculture," *Iowa Agriculturist*, 78 (Winter 1977): 26–27.

25. The quotations are from *What Is This Thing Called Food: Council for Agricultural Science and Technology Report No. 61, September 15, 1976*; William Robbins, "Cattle Growth Hormone Is Found by Panel to Pose No Cancer Risk," *New York Times*, Jan. 26, 1977, A16; and George Anthan, "Scientists Say Benefits of DES Far Outweigh 'Slim' Cancer Risks," *Des Moines Register*, Jan. 26, 1977, 7-A. Also see "Proposal for Increased Interaction between Agricultural Research Institute (ARI) and Council for Agricultural Science and Technology (CAST)" *Proceedings of the Agricultural Research Institute*, 26 (1977): 160–61; Steven Marcy, "CAST Report Defends Safety of Stilbestrol Use in Cattle Production," *Feedstuffs*, 49 (Jan. 31, 1977): 5; "Cattle Drug: 'No Evidence

of Cancer Hazard,'" *Science News*, III (Feb. 12, 1977): 102–3; Robert H. Brown, "DES Cuts Fat, Hence Cancer: Jukes," *Feedstuffs*, 49 (Feb. 28, 1977): 8, 68; and *Hormonally Active Substances in Foods: A Safety Evaluation: Council for Agricultural Science and Technology Report No. 66*, March 1977.

26. Patricia Bauman and H. David Banta, "The Congress and Policymaking for Prevention," *Preventive Medicine*, 6 (1977): 277–41; John A. Zapp, Jr., "Extrapolation of Animal Studies to the Human Situation," *Journal of Toxicology and Environmental Health*, 2 (1977): 1425–33; Umberto Saffiotti, "Scientific Bases of Environmental Carcinogenesis and Cancer Prevention: Developing an Interdisciplinary Science and Facing Its Ethical Implications," ibid., 2 (1977): 1435–47; Jerome Cornfield, "Carcinogenic Risk Assessment," *Science*, 198 (Nov. 18, 1977): 693–99; "A.S.A.S. Regulatory Agencies Committee—Annual Report to the Executive Committee, 1976 to 1977," *JAS*, 45 (1977): 182; and T. G. Dunn, C. C. Kaltenbach, D. R. Koritnik, D. L. Turner, and G. D. Niswender, "Metabolites of Estradiol-17b and Estradiol 17b-3-Benzoate in Bovine Tissues," ibid., 45 (1977): 659–73.

27. Charles J. Barnes, "Report on Drug Residues in Animal Tissues," *JAOAC*, 59 (1976): 361–62; John Joseph Ryan and Jean-Claude Pilon, "Chemical Confirmation of Diethylstilbestrol Residues in Beef Livers," ibid., 59 (1976): 817–20; Karen A. Kohrman and Joseph MacGee, "Simple and Rapid Gas-Liquid Chromatographic Determination of Diethylstilbestrol in Biological Specimens," ibid., 60 (1977): 5–8; C. Zervos and J. Staffa, "Development of DES-assaying Methodology Based on Affinity Column Chromatography," FDA Papers, Accession no. 88-82-57, box 28, vol. 374; Jerry Bruton, Lloyd Davis, and William Huber, "Introduction," *Journal of Toxicology and Environmental Health*, 2 (1977): iii; Maryln K. Perez, "Food and Drug Administration Statement of the Problem," ibid., 2 (1977): 727–30; Helen H. Birkhead, "Industry Statement of Problem," ibid., 2 (1977): 731–33; Lewis W. Dittert, "Pharmacokinetic Prediction of Tissue Residues," ibid., 2 (1977): 735–36; George A. Digenis, "Stable and Short-lived Isotopes in the Study of Tissue Distribution," ibid., 2 (1977): 757–85; H. D. Mercer, J. D. Baggot, and R. A. Sams, "Application of Pharmacokinetic Methods to the Drug Residue Profile," ibid., 2 (1977): 787–801; Charles Rosenblum, "Non-Drug-related Residues in Tracer Studies," ibid., 2 (1977): 803–14; Prem S. Jaglan, M. Weldon Glenn, and A. William Neff, "Experiences in Dealing with Drug-related Bound Residues," ibid., 2 (1977): 815–26; Hugo E. Gallo-Torres, "Methodology for the Determination of Bioavailability of Labeled Residues," ibid., 2 (1977): 849–45; James R. Gillete and Lance R. Pohl, "A Prospective on Covalent Binding and Toxicology," ibid., 2 (1977): 849–71; A. L. Donoho, "Summary Comments on Drug Metabolism and Residue Chemistry Requirements for the Use of Drugs in Food-producing Animals," ibid., 2 (1977): 909–12; and W. G. Huber, "Summarization: Assessment of Current Status of Available Methodologies to Utilize Radioactive Drug Metabolism Studies for Determining Safety of Animals," ibid., 2 (1977): 913–15; Heinrich Karg and Karl Vogt, "Control of Hormone Treatment in Animals and Residues in Meat—Regulatory Aspects and Approaches in Methodology," *JAOAC*, 61 (1978): 1201–8; James A. Sphon, "Use of Mass Spectrometry for Confirmation of Drug Residues," ibid., 61 (1978): 1247–52; Malcolm C. Bowman, "Trace Analysis: A Requirement for Toxicological Research with Carcinogens and Hazardous Substances," ibid., 61 (1978): 1253–62; Bernd Hoffmann, "Use of Radioimmunoassay for Monitoring Hormonal Residues in Edible Animal Products," ibid., 61 (1978): 1263–73; John Joseph Ryan and Bernd Hoffmann, "Trenbolone Acetate: Experiences with Bound Residue in Cattle Tissues," ibid., 61 (1978): 1274–79; Donald M. Henircks and Allan K. Torrence, "Endogenous Estradiol-17b in Bovine Tissues," ibid., 61 (1978): 1280–83; and H. D. Hafs, G. D. Niswender, P. V. Malven, C. C. Kaltenbach, R. G. Zimbelman, and R. J. Condon, "Guidelines for Hormone Radioimmunoassays," *JAS*, 45 (1977): 927–28. Also see D. M. Tennent, R. F. Kouba, W. H. Ray, W. J. A. VanenHeuvel, and F. L. Wolf, "Impurities in Labeled Diethylstilbestrol: Identification of Pseudodiethylstilbestrol," *Science*, 194 (Dec. 13, 1976): 1059–60; and G. F. Bories, R. Farrando, J. Wirhaye, J. C. Peleran, and J. P. Valette, "Fate of Tritium in Calves Subcutaneously Implanted with $_3$H-Diethylstilbestrol," *JAS*, 44 (1977): 680–86. Also see, for example, M. Metzler, W. Muller, and W. C. Hobson, "Biotransformation of Diethylstilbestrol in the Rhesus Monkey and

the Chimpanzee," *Journal of Toxicology and Environmental Health*, 2 (1977): 439–50; ibid., 2 (1977): 527–37; and Robert W. Miller, "Relationship between Human Teratogens and Carcinogens," *Journal of the National Cancer Institute*, 58 (Mar. 1977): 471–74.

28. The published material on these points is voluminous. See, for example, "Public's Concern about Food Safety Increasing: Consultant," *Feedstuffs*, 49 (Jan. 3, 1977): 26; "Study Finds FDA Harassed Critical Employees," *Science and Government Report*, 7 (May 1, 1977): 3–4; "Seeking Reform," *Chemical Week*, 120 (June 8, 1977): 26; "FDA Head Plans Faster Action on Drugs, Greater Safeguards against Hazards," *Wall Street Journal*, Aug. 1, 1977, 8; "Antibiotics in Animal Feed: Bans Planned," *Science News*, 112 (Sept. 3, 1977): 151; "FDA Publishes Proposed Ban," *Feedlot Management*, 19 (Oct. 1977): 26; R. Jeffrey Smith, "FDA Reform: An Idea Whose Time Has Come," *Science*, 198 (Oct. 21, 1977): 272–73; Edward G. Feldmann, "Prediction: New Drug Law Coming," *Journal of Pharmaceutical Sciences*, 66 (Dec. 1977): i; "Public Interest Group Petitions for Ban of Nitrate and Nitrite in Meat," *Consumer News*, 7 (Dec. 1, 1977): 3; "Regulation of Food Additives—Saccharin," *JAVMA*, 170 (1977): 1063; Jerry Voorhis, "The Consumer Movement and the Hope of Human Survival," *Journal of Consumer Affairs*, 11 (1977): 1–16; *Food Additives: Competitive, Regulatory, and Safety Problems. Hearings before the Select Committee on Small Business, United States Senate, January 13 and 14, 1977* (Washington, D.C.: GPO, 1977): 1–17, 447–573, 724–49, 869–927; *Impact of Chemical and Related Drug Products and Federal Regulatory Processes: Hearings before the Subcommittee on Dairy and Poultry of the Committee on Agriculture, House of Representatives, May 23–25, 1977* (Washington, D.C.: GPO, 1977): 3–29, 76–101; and *Regulation of Chemicals in Food and Agriculture: Hearings before the Subcommittee on Agricultural Research and General Legislation of the Committee on Agriculture, Nutrition, and Forestry, United States Senate, June 30 and July 19, 1977* (Washington, D.C.: GPO, 1977): 6–19, 48–71, 88–92, 133–42.

29. The quotations are from George E. Moore and William N. Palmer, "Money Causes Cancer: Ban It," *JAMA*, 238 (Aug. 1, 1977): 397. Also see "Filthy Lucre," *Wall Street Journal*, Oct. 26, 1977, 22; and "Money Causes Cancer: Ban It," *Feedstuffs*, 49 (Nov. 7, 1977): 4. The Federation of American Scientists chastised *JAMA* for publishing the study and the Denver researchers for conducting it. To that group it constituted "a political misuse of science." See Jeremy J. Stone, "A Political Misuse of Science," *Bioscience*, 28 (Feb. 1978): 83.

30. "Chicago U. and Eli Lilly Co. Sued by Patsy Mink and Two over Drug," *New York Times*, Apr. 26, 1977, 23; "DES Cover-up Alleged," *Chemical Week*, 120 (May 5, 1977): 14; "Eli Lilly Denies Role in 1950s DES Study," *Washington Post*, Apr. 29, 1977, B16; Helen Dewar, "OSHA Fines employer $34,100 for Workers Harmed by DES," *Washington Post*, June 18, 1977, A2; "DES Maker Fined," *Chemical Week*, 120 (June 29, 1977): 21; Richard D. Lyons, "Nader Group Demands U.S. Ban on Artificial Hormone as Cancer Risk," *New York Times*, Dec. 13, 1977, 24; "DES Blamed for Mothers' Cancers," *Science News*, 112 (Dec. 30, 1977): 422; "Disagree on DES Data," *Chemical Week*, 122 (Jan. 18, 1978): 16; Morton Mintz, "Cancer Alert on DES Urged," *Washington Post*, Jan. 31, 1978, A6; and "Editorial Viewpoints," *JAVMA*, 172 (1978): 1399–1400. In contrast, Herbst's declarations several months earlier that he had found that less than three thousand of the nearly three million daughters whose mothers were identified as having taken DES during pregnancy showed any sign of the characteristic cancer went nearly uncovered. Herbst called the disease "extremely rare among the DES-exposed grop." See Arthur J. Snider, "Doctor: DES Less Risky than Assumed," *Des Moines Tribune*, May 2, 1977, 1; and "DES-Cancer Research," *Washington Post*, May 3, 1977, A20.

31. The quotation is from John McClung, "Carol Foreman: The Consumer Voice inside the Ag Department," *Feedstuffs*, 49 (Apr. 11, 1977): 2, 44–45. Also see Barbara J. Culliton, "Donald Kennedy to Head FDA," *Science*, 195 (Mar. 25, 1977): 1307; "FDA Choice Widely Supported," *Science and Government Report*, 7 (Mar. 15, 1977): 7; "He Will Streamline FDA," *Chemical Week*, 120 (Mar. 16, 1977): 25; "Stanford Professor New FDA Head," *Feedstuffs*, 49 (Mar. 7, 1977): 5; and "Top FDA Job Filled by Biologist," *Bioscience*, 27 (May 1977): 368.

32. The quotation is from John McClung, "FDA Document May Mean Tighter Controls on Use

of Carcinogens," ibid., 49 (Feb. 28, 1977): 1, 69. Also see "Chemical Compounds in Food-Producing Animals," *Federal Register*, 42 (Feb. 22, 1977): 10412–37; "Inside Washington," *Feedstuffs*, 49 (Feb. 7, 1977): 2; and "FDA on Cancer-Linked Drugs," *Feedlot Management*, 19 (Apr. 1977): 23.

33. John McClung, "Petitioners Seek to Rescind or Stay Sensitivity of Method Order," *Feedstuffs*, 49 (Apr. 4, 1977): 2, 38; Steven Marcy, "AHI Files Suit to Halt Implementing Residue Method," ibid., 49 (May 16, 1977): 2, 72–73; "FDA Controversy," *JAVMA*, 171 (1977): 242; and "A.S.A.S. Regulatory Agencies Committee, 1976 to 1977," 183.

34. "Washington," *Feed Management*, 28 (May 1977): 11; Steven Marcy, "FDA Begins DES Hearings; AHI Says It Will Press Sensitivity of Method Suit," *Feedstuffs*, 49 (May 23, 1977): 6, 119; Thomas H. Jukes, "The DES Hearings: What Was Learned," ibid., 49 (Nov. 21, 1977): 22, 23, 30; "Inside Washington," ibid., 50 (Jan. 30, 1978): 2; "Affidavit of Dr. Robert F. Sieck, February 6, 1976"; "Affidavit of Daniel T. Teitelbaum, February 6, 1976"; "Affidavit of H. H. Cole, February 6, 1976"; and "Affidavit of Thomas H. Jukes, February 10, 1976" FDA Papers, Accession no. 88-82-57, box 1, vol. 2; "Environmental Impact Analysis Report and Assessment for Use of Diethylstilbestrol (DES) in Animals, October 13, 1976," FDA Papers, Accession no. 88-82-57, box 1, vol. 3; "Brief for Bureaus of Foods and Veterinary Medicine, Docket No. 76N-0002," FDA Papers, Accession no. 88-82-57, box 34, vol. 445; "Brief for American Home Products Corporation, Dawe's Laboratories, Rhodia, Inc., and Vineland Laboratories, Inc.—Manufacturing Parties, Docket No. 76N-0002," FDA Papers, Accession no. 88-82-57, box 34, vol. 446; and "Brief of Non-Party Participants, March 30, 1978," courtesy of Pacific Legal Foundation, in my possession.

35. "Court to Hear AHI, FDA Arguments on Handling of Sensitivity Proposal," *Feedstuffs*, 49 (Dec. 19, 1977): 2; John McClung, "Judge Rules in Favor of AHI in Sensitivity of Methods Argument," ibid., 49 (Dec. 26, 1977): 4, 51; "Animal Health Institute v. Food and Drug Administration et al.," *Food Drug Cosmetic Law Reporter: Developments*, 1977–78, 38557–65; "Inside Washington," *Feedstuffs*, 50 (Feb. 27, 1978): 2, 50 (Apr. 24, 1978): 2; "Official Transcript of Hearing—Animal Health Institute v. Food and Drug Administration, et al., April 19, 1978"; and "Memorandum—Joseph D'Arco et al., May 8, 1978," FDA Papers, Accession no. 88-82-57, box 34, vol. 448; and "Chemical Compounds in Food-Producing Animals," *Federal Register*, 43 (May 26, 1978): 22675.

36. The quotations are from "DES Daughters and Cancer," *Science News*, 113 (Feb. 18, 1978): 100–1; and "Eating Habits in U.S. Are Linked to Cancer," *New York Times*, June 30, 1978; A8. Also see "Cancer Risk from DES Is Reassessed," *Washington Post*, Feb. 11, 1978, A4; Morton Mintz, "Panel Concerned on Link of DES to Cancer Cases," ibid., June 2, 1978, A8; and "DES Task Force Report," *Science News*, 114 (Oct. 14, 1978): 261. For other information, see "DES Be Not Proud," *New Times*, 10 (June 26, 1978): 24.

37. The quotations are found in John McClung, "New Study Evaluates Risk, Value of Drugs in Feed," *Feedstuffs*, 50 (Apr. 17, 1978): 1, 51. Also see "Inside Washington," ibid., 50 (May 1, 1978): 2; 50 (May 29, 1978): 2; and John McClung, "OTA Suggests Seven Feed Additive Legislative Paths," ibid., 50 (July 24, 1978): 1, 46–47.

38. The quotations in this paragraph and the five subsequent paragraphs are from Daniel J. Davidson, "Initial Decision on Proposal to Withdraw Approval of New Animal Drug Applications for Diethylstilbestrol," FDA Papers, Accession no. 88-82-57, box 34, vol. 448. Also see "Inside Washington," *Feedstuffs*, 50 (Sept. 18, 1978): 2; John McClung, "DES Use Should Be Banned, Administrative Law Judge Decides," ibid., 50 (Sept. 25, 1978): 1, 58; "FDA Seeks to End DES Use for Animal Growth," *Chemical Week*, 123 (Oct. 4, 1978): 16; and "FDA Moves Closer to Banning DES Used in Animals for Growth," *Wall Street Journal*, Sept. 25, 1978, 3.

39. Davidson, "Initial Decision," 49–52. For Davidson's legitimation of cost-benefit considerations, see, for example, "Congress Directs Antibiotics-in-Feed Delay," *JAVMA*, 173 (1978): 1294; "Nitrite Solos as Cancer Hazard," *Science News*, 114 (Aug. 19, 1978): 119; "Washington Newsletter," *Chemical Week*, 123 (Aug. 16, 1978): 14; "New Nitrite Fears," ibid., 123 (Aug. 23, 1978): 16; "Inside Washington," *Feedstuffs*, 50 (Sept. 4, 1978): 2; "Nitrites in Meat," *Cattleman*,

65 (Oct. 1978): 154, 156; "Congress Stops FDA from Restricting Use of Antibiotics," *Bioscience*, 28 (Sept. 1978): 557–59; and "Nitrite Report Blasted by CAST," ibid., 28 (Dec. 1978): 755–56.

40. The quotations are from "Exceptions of the Non-Party Participants to the Initial Decision Dated September 21, 1978, October 23, 1978," FDA Papers, Accession no. 88-82-57, box 34, vol. 448. Also see "The DES Decision," *Feedstuffs*, 50 (Oct. 9, 1978): 8; "Bureaus of Foods and Veterinary Medicine's Exceptions to Initial Decision of the Administrative Law Judge, October 22, 1978"; and "Reply of the Manufacturing Parties to Bureaus' Exceptions, November 13, 1978," FDA Papers, Accession no. 88-82-57, box 34, vol. 449; and "Bureaus' Reply to Manufacturing Parties' Exceptions to the Initial Decision of the Administrative Law Judge, November 13, 1978," FDA Papers, Accession no. 88-82-57, box 34, vol. 450. Toxicologists were experiencing difficulty in securing recognition within the federal bureaucracy. See, for example, "Toxicologists Struggling for Federal Identity," *Science*, 203 (Jan. 12, 1979): 152–53.

41. Among the handful of comments were Clara M. Szego to Donald Kennedy, Oct. 5, 1978; W. C. Foxley to Timothy E. Wirth, Oct. 20, 1978; Peter D. Solomon to Kennedy, Nov. 1, 1978; and Gertrude C. Merzbach to Kennedy, Nov. 1, 1978, FDA Papers, Accession no. 88-82-57, box 34, vol. 451. For the commissioner's action, see "ASAS Regulatory Agencies Committee—Annual Report to the Executive Committee, 1978 to 1979," *JAS*, 49 (1979): 274.

42. For Martin's views, see John McClung, "Martin Will Reintroduce His Bill Amending Delaney Clause in New Congress," *Feedstuffs*, 50 (Dec. 11, 1978): 5, 51; and James G. Martin, "The Delaney Clause and Zero Risk Tolerance," *FDCLJ*, 34 (1979): 43–49. Also see "FDA Informs Syntex of Intention to End Approval of Two Items," *Wall Street Journal*, Jan. 8, 1979, 20; "FDA Proposes Policy on Animal Drug Residues," *JAVMA*, 174 (1979): 554; "Synovex Implants Latest FDA Target," *Beef*, 15 (Mar. 1979): 14; "Bills Seek Delaney Amendment Repeal, Extended Saccharin Ban," *Chemical Week*, 124 (Jan. 31, 1979): 23; "House Ag Committee Chairman Seeks Repeal of Delaney Clause," *Cattleman*, 65 (Mar. 1979): 116; "Proposed Food Safety Legislation," *JAVMA*, 174 (1979): 235; "Views on Food Safety," ibid., 174 (1979): 332; "Zero Tolerance and Carcinogens," ibid., 174 (1979): 470; "Risk-Benefit," ibid., 174 (1979): 691; and "Delaney Clause," *Cattleman*, 65 (Feb. 1979): 126.

43. The quotations are from Rich Jaroslovsky, "FDA's Push to Change Food-Safety Laws May Include Acceptable-Risk Level Plan," *Wall Street Journal*, Feb. 13, 1979, 10; and R. Jeffrey Smith, "Institute of Medicine Report Recommends Complete Overhaul of Food Safety Laws," *Science*, 203 (Mar. 23, 1979): 1221–24. Also see "FDA's Acceptable Risk," *Science News*, 115 (Feb. 24, 1979): 119; "For Saccharin Decision, Revamp Food Laws," ibid., 115 (Mar. 17, 1979): 150; "Cutting the Risk," *Wall Street Journal*, Mar. 13, 1979, 22; John McClung, "Industry Fears FDA New SOM Proposal Is 'Too Conservative,'" *Feedstuffs*, 51 (Mar. 26, 1979): 1, 51, 52; "Chemical Compounds in Food-Producing Animals: Criteria and Procedures for Evaluating Assays for Carcinogenic Residues," *Federal Register*, 44 (Mar. 20, 1979): 17070–114; "Carcinogenic Residues in Food-Producing Animals; Public Hearing on Criteria and Procedures for Evaluating Assays," ibid., 44 (Apr. 20, 1979): 23538–39; and "Carcinogenic Residues in Food-Producing Animals; Public Hearing on Criteria and Procedures for Evaluating Assays; Extension of Time," ibid., 44 (May 8, 1979): 26899–900.

44. Edward G. Feldmann, "Risking Public Credibility," *Journal of Pharmaceutical Sciences*, 68 (Mar. 1979): i; "Speak Out on Delaney," *Industrial Research Development*, 21 (Apr. 1979): 200–01; and "Poll of Scientists Reveals Strong Consensus on Zero Risk," *Bioscience*, 29 (July 1979): 438. For a minority view, see "Cancer Warrior Warns against Weakening Law," *Washington Post*, Mar. 27, 1979, C5.

45. "Agriculture's Strange Bedfellows: CAST-Industry Tie Raised Credibility Concerns," *Bioscience*, 29 (Jan. 1979): 9–12 59.

46. The quotations are from ibid., 12. Also see Gregg Hillyer, "No CASTing Lots to Inform Congress," *Iowa Agriculturist*, 80 (Fall 1978): 14–15, 19. Shock waves radiated from the *Bioscience* report. *The New York Times* and the *Des Moines Register* each published an unflattering series about CAST. Pressure intensified when six members of a CAST antibiotics task force re-

signed, claiming that Black had downplayed their caveats and overstated economic benefits in editing the task force's report. The *Register* called on Iowa State University to terminate its relationship to CAST—Black was using his agronomy department office as CAST headquarters. Michael Jacobson, executive director of Nader's Center for Science in the Public Interest, lobbied the university's president, the president of the university's governing board, and Senator John Culver (D-Iowa) to have the university sever its connections with CAST. See Lauren Soth, "Influence of Agribusiness on Farm Policy is Growing," *Des Moines Register*, Jan. 15, 1979, 10A; idem, "CAST of Specialists Tells Good Side of Ag Chemicals," ibid., Jan. 16, 1979, 10-A; "CAST under Fire on Funding, 'Facts,'" ibid., Jan. 21, 1979, 1-F, 2-F; "ISU and CAST," ibid., Jan. 24, 1979, 10A; George Anthan, "Contributing Sociologists Charge CAST 'Inherently Biased,'" Feb. 27, 1979, 4A; "CAST Poses Reconciliatory Routes to Six Who Resigned," *Feedstuffs*, 51 (Jan. 29, 1979) 6, 50; FC, "CAST," *Farm Chemicals*, 142 (1979): 26, 28, 30, 33; Eliot Marshall, "Scientists Quit Antibiotics Panel at CAST," *Science*, 203 (Feb. 23, 1979): 732–33; and "CAST Profile Evokes Avid Responses," *Bioscience*, 29 (May 1979): 276–80.

47. The quotations are from R. Jeffrey Smith, "NCI Bioassays Yield a Trail of Blunders," *Science* 204 (June 22, 1979): 1287–92. Also see "GAO Criticizes NCI on Cancer Testing," *Chemical Week*, 124 (May 2, 1979): 23.

48. Don Muhm, "Farm Use of Growth Aid DES Prohibited," *Des Moines Register*, June 30, 1979, 3-B; Victor Cohn, "FDA Bans Most Uses of Controversial Drug," *Washington Post*, June 30, 1979, A3; "FDA Bans DES as Animal-Feed Additive Because It's Been Shown to Cause Cancer," *Wall Street Journal*, July 2, 1979, 20; "FDA Bans DES Growth Promotant for Cattle, Sheep," *Feedstuffs*, 51 (July 2, 1979): 1, 65; "Busy Last Week for FDA Commissioner," *Science News*, 116 (July 7, 1979): 4–5; "Washington Newsletter," *Chemical Week*, 125 (July 11, 1979): 14; "Beef Business," *Wallaces Farmer*, 104 (July 28, 1979): 34; "DES Ban Now in Effect," *JAVMA*, 175 (1979): 334; "No DES for Animal Growth," *JAMA*, 242 (Sept. 7, 1979): 1010; "Roundup," *Feedlot Management*, 21 (Aug. 1979): 34; "Update," *FDA Consumer*, 13 (Sept. 1979): 2–3; "Diethylstilbestrol (DES) in Edible Tissues of Cattle and Sheep; Revocations," *Federal Register*, 44 (July 6, 1979): 39387–89; "Diethylstilbestrol (DES) in Edible Tissues of Cattle and Sheep; Withdrawal of Approval of New Animal Drug Applications," ibid., 44 (July 6, 1979): 39618–19; "Diethylstilbestrol (DES) in Edible Tissues of Cattle and Sheep; Revocations; Partial Stay of Effective Dates," ibid., 44 (July 20, 1979): 42679–80; "Diethylstilbestrol (DES) in Edible Tissues of Cattle and Sheep; Revocation of Test Methods Regulation; Partial Stay of Effective Dates," ibid., 44 (Aug. 3, 1979): 45618, 45764–65; "Diethylstilbestrol; Withdrawal of Approval of New Animal Drug Applications; Commissioner's Decision," ibid., 44 (Sept. 21, 1979): 54852–900; "Cattlemen and Associations Expected to Petition FDA for Extension on DES," *Feedstuffs*, 51 (July 9, 1979): 6; "FDA Extends DES Ban Date for End Users," ibid., 51 (July 23, 1979): 4, 57; "Proposed DES Ban Postponed," *Des Moines Register*, Aug. 2, 1979, 7-A; "FDA Again Postpones Its Ban on Use of DES as Animal-Growth Aid," *Wall Street Journal*, Aug. 2, 1979, 4; "FDA Postpones Ban on Existing Supplies of DES," *Feedstuffs*, 51 (Aug. 6, 1979): 2; "DES Ban Again Postponed by FDA," *Wallaces Farmer*, 104 (Aug. 25, 1979): 34; "Washington Roundup," *Cattleman*, 66 (Aug. 1979): 144; 66 (Sept. 1979): 29.

49. "Hess and Clark Appeals FDA Ban on Use of DES," *Feedstuffs*, 51 (July 9, 1979): 4, 50; "DES Ban May Cost $3.1–$5.6 Billion," ibid., 51 (July 23, 1979): 4, 59; "Editor's Roundup," *Beef*, 16 (Sept. 1979): 6;" "Memorandum and Recommendation—Karlissa B. Krombein to Raymond M. Momboisse, July 24, 1979," courtesy of the Pacific Legal Foundation, in my possession.

50. Cost-benefit considerations figured prominently even in the 1976 Vitamin Amendment, generally viewed as a loss of consumer protection. Its passage was the successful conclusion of an attack on the medical-scientific establishment, which claimed authority to set the appropriate and healthful amount of particular nutritional substances. The amendment prohibited that establishment from exercising legal power. Rather than being able to outlaw particular materials, in excess of some specific standard—a standard per se—medical and scientific authorities would have to

make their case in public forums, where they would meet potent competition from others, including other scientists and physicians. Each individual could then weigh the costs versus benefits.

EPILOGUE

1. "USDA to Test Meat for DES after Nov. 1 Ban," *Feedstuffs*, 51 (Oct. 15, 1979): 5; and "DES Residue Testing Continues after Ban," *Cattleman*, 66 (Nov. 1979): 42.

2. John McClung, "USDA, FDA Discover Illegal Use of DES Implants in Beef Cattle," *Feedstuffs*, 52 (Apr. 7, 1980): 1, 46; "More Beef Cattle Identified with Illegal DES," ibid., 52 (Apr. 14, 1980): 1, 42; "Inside Washington," ibid., 52 (Apr. 14, 1980): 2; 52 (May 5, 1980): 2; 52 (June 30, 1980): 2; "DES List Grows; USDA Reinstates Residue Monitoring," ibid., 52 (Apr. 21, 1980): 1, 44; "Washington Newsletter," *Chemical Week*, 126 (Apr. 16, 1980): 14; Joanne Omang, "U.S. Acts to Halt the Sale of Cattle Fattened with the Aid of Banned DES," *Washington Post*, Apr. 5, 1980, A4; "Estimate of DES Violations Up Sharply," ibid., Apr. 16, 1980, A4; Kent Parker, "Iowa Cattle Didn't Get DES, Officials Say," *Des Moines Register*, Apr. 5, 1980, 3A; Larry Fruhling, "U.S. Says Iowa Feedlot Violated Ban on DES," ibid., Apr. 8, 1980, 7-A; and Kent Parker, "The FDA Ban That Didn't Work," ibid., May 11, 1980; 1-F and 2-F. A few thousand cattle were found to have been fed stilbestrol. The FDA also acted against those violators. See "Conditions for Marketing DES-Fed Cattle Established," *Food Drug Cosmetic Law Reporter: Developments*, 1979–80, 40595–97.

3. The quotations are from "Inside Washington," *Feedstuffs*, 52 (Aug. 18, 1980): 2; 52 (Sept. 8, 1980): 2. Also see John McClung, "NCA Works Out Collection System for Holdover Supplies of DES," ibid., 52 (May 26, 1980): 1, 40; "FDA May Prosecute Feed Makers for Sale of DES Implants, Premixes," ibid., 52 (Aug. 11, 1980): 1, 40; and "FDA Maintains Feed Men Violated Law with DES Sales; AFMA: 'No,'" ibid., 52 (Sept. 29, 1980): 1, 33.

4. "FDA, USDA Try to Stem Disregard of Diethylstilbestrol Bans," *Food Drug Cosmetic Law Reporter: Developments*, 1979–80, 40349; "Ante-Mortem Inspection; DES in Cattle—Certification of Reconditioning," *Federal Register*, 45 (Apr. 22, 1980): 26947–50; "Diethylstilbestrol (DES); Food Use of Cattle Illegally Implanted with DES," ibid., 45 (Apr. 22, 1980): 27014–16; "Diethylstilbestrol (DES); Food Use of Cattle Illegally Implanted with DES," ibid., 45 (May 2, 1980): 29413; and "DES-Implanted Cattle Situation Updated, Withdrawal Periods Revised," *Food Drug Cosmetic Law Reporter: Developments*, 1979–80, 40413–19.

5. The quotations are from John McClung, "DES Implant Issue Shifts to 'Why' Illegal Use Occurred," *Feedstuffs*, 52 (Apr. 28, 1980): 1, 51; "DES: The Gamble That Didn't Pay Off," ibid., 52 (May 5, 1980): 10; "Crawford on DES: 'Industry, Government Have Failed World Public,'" ibid., 52 (May 12, 1980): 1, 41; Steve Aldrich, "The DES Debacle: The Cattlemen's Viewpoint," ibid., 52 (June 30, 1980): 7, 28–29; and Steve Frazier and Steve Weiner, "Many Cattlemen Ignored the Federal Ban on Use of DES to Speed Animals' Growth," *Wall Street Journal*, July 15, 1980, 46. For a popular recounting of this episode, especially cattlemen's motives, see Orville Schell, *Modern Meat* (New York: Random House, 1984), 189–254.

6. The quotations are from Charles G. Olentine, "Sidelines," *Feed Management*, 31 (June 1980): 6; and "Many Cattlemen Ignored"; and "Crawford on DES." Also see "The Continuing DES Saga," *FDA Consumer*, 14 (July–Aug. 1980): 2.

7. On these points, see, for example, Leonard A. Cole, *Politics and the Restraint of Science* (Totowa, N.J.: Rowman and Allanheld, 1983), especially 161–73; and Peter M. Sandman, David B. Sandman, Michael R. Greenberg, and Michael Gochfeld, *Environmental Risk and the Press* (New Brunswick, N.J.: Transaction Books, 1987).

8. The quotation is from Sandy Rovner, "Implant Safety: Who's Right?" *Washington Post*, Nov. 12, 1991, WH1. Also see William Booth, "Women Assail, Praise Silicone at Hearing," ibid., Nov. 13, 1991, A1, 4; William Booth, "FDA Advisory Panel Rejects Two Popular Silicone Implants," ibid., Nov. 14, 1991, A33; William Booth, "Breast Implants Allowed," ibid., Nov. 15, 1991, A1; Malcolm Gladwell, "Hearings Focus on Breast Implant Leaks," ibid., Feb. 19, 1992, A6; idem,

"Breast Implant Maker Offers to Monitor Safety," ibid., Feb. 20, 1992, A1, A4; idem, "Panel Urges Limited Use of Implants," ibid., Feb. 21, 1992, A1; and idem, "Silicone Breast Implants," ibid., Mar. 3, 1992, WH10.

9. The quotations are from Carole Sugarman, "The Alarm over Alar," *Washington Post*, Mar. 8, 1989, E1, E14; idem, "Agencies Say Apples Safe, Chemical Not Imminent Risk," ibid., Mar. 17, 1989, A57; and Hobart Rowen, "Public Must Demand an Immediate Ban on Toxic Products," ibid., May 21, 1989, H1. Also see, for instance, Michael Weisskopf, "EPA Targets Chemical Used on Apples," ibid., Feb. 2, 1989, E1, E14; "Nader Urges Ban on Use of Alar; Industry Says Action Already Taken," ibid., May 16, 1989, A4; Philip J. Hilts, "Conflict-of-Interest Issue Arises in Debate on Alar," ibid., May 26, 1989, A21; Bill McAllister, "EPA Plans Total Ban on Alar Pesticide," ibid., Sept. 2, 1989, A6; and Jack Anderson and Dale Van Atta, "Food Chains Taking EPA Bull by Horns," ibid., Oct. 18, 1989, E19. The premise underlying Streep's Alar testimony—a confusion between scientific and performing expertise—has continued. The actress is now identified as an "environmental crusader." See "People in the News," *Des Moines Register*, Jan. 22, 1993, A2.

10. The U.S. Court of Appeals for the District of Columbia upheld the FDA's stilbestrol ban in late 1980. See "Rhone-Poulenc, Inc., Hess and Clark Division, v. FDA; Vineland Laboratories, Inc., v. FDA, Sherwin Gardner, Acting Commissioner of Food and Drugs," *Food Drug Cosmetic Law Reporter: Developments*, 1980–81, 38381–85.

Bibliographical Essay

PRIMARY SOURCE MATERIALS

Periodicals

Mass periodicals, such as *Time, U.S. News and World Report,* or *Newsweek,* proved less useful for this book than did specialty publications. These latter journals and magazines proliferated in the second half of the twentieth century; they were virtually niche publications because each aimed at only a relatively small segment of the public: individuals aligned in groups defined by skills, licensing, occupation, or points of view. A brief examination of the periodicals cited in this book which sought to disseminate scientific information within the agribusiness nexus helps make that point. For example, farmers generally purchased *Wallaces Farmer, Farm Journal, Farm Quarterly,* and *Successful Farming,* while only beef producers subscribed to *Beef, Cattleman, Charolais Journal, Charolais Way,* and *Shorthorn World. Feedstuffs, Feed Bag, Feed Age, Flour and Feed, Feedlot, Feedlot Management,* and *Feed Management* focused on feed manufacturers and cattle feeders. *Corn Belt Livestock Feeder*'s intended audience was narrower and self-explanatory. Scientists employed by feed manufacturers read the proceedings of the Maryland, Cornell, and American Feed Manufacturers' Association nutrition conferences. State experiment stations reached their rural constituencies with *Agricultural Research Report, Arkansas Farm Research, Iowa Farm Science, New York's Food and Life Sciences Quarterly, North Dakota Farm Research, South Dakota Farm and Home Research,* and *Utah Farm and Home Science.* Scientists identified with these stations published in *Journal of Animal Science, Poultry Science, Agricultural Research,* and *Science* to communicate with scientists outside experiment stations and agricultural colleges. *Kansas Farmer, Nebraska Farmer,* and *Wisconsin Agriculturist* served the farmers of those respective states and provided more than state experiment station or cattle-feeding news. These niche publications needed to present only the views of their constituencies—consumers, consumer advocates, pharmaceutical houses, chemical companies, feeders, feed manufacturers, trade associations, physicians, veterinarians, FDA scientists, other government scientists, scientists not in government, cancer specialists, New Journalists, and a host of others—and that extremely narrow focus is useful for historical research. Views were presented in a generally straightforward, undiluted form, not disguised within the bland products of coalition politics. Indeed, those particularistic perspectives were the reason for each of these magazines' existence and persistence; the magazines reflected their constituencies, each of which defined itself as a specific interest group with neatly circumscribed objectives. Several lists of the various types of periodicals which I used follow.

Journals and Magazines

Among the journals and magazines I consulted were *Acta—International Union against Cancer,* 1953–59; *Acta Unio Internationalis contra Cancrum,* 1955–57; *Agricultural Research,* 1953–68; *Agricultural Research Report* (Virginia), 1946–59; *A.M.A. Archives of Pathology,* 1954–56; *American Journal of Veterinary Research,* 1940–61; *Animal Nutrition and Health,* 1974; *Antibiotics and Chemotherapy,* 1953; *Archives of Biochemistry and Biophysics,* 1950; *Arkansas Farm Research,* 1952–73; *Atlantic Monthly,* 1966–72; *Australian Journal of Experimental Biology,* 1949–51; *Australian Veterinary Journal,* 1946; *Beef,* 1972–81; *Beef Cattle Science Handbook,* 1964–68; *Biochimica et Biophysica Acta,* 1962; *Biometrics,* 1966–74; *Bioscience,* 1951–79; *Cancer,* 1970; *Cancer Journal,* 1951; *Cancer Research,* 1949–70; *Cattleman,* 1953–80; *Charolais Journal,* 1977–82; *Charolais Way,* 1969–76; *Chemical and Engineering News,* 1953–69; *Chemicals,* 1954–55; *Chemical Week,* 1954–80; *Clinical Toxicology,* 1978; *Consumer Bulletin,* 1957–73; *Consumer Bulletin Annual,* 1958–73; *Consumer News,* 1971–80; *Consumer Pamphlet,* 1954–67; *Consumer Register,* 1972–79; *Consumer Research Bulletin,* 1941–57; *Consumers' Research Magazine,* 1952–80; *Corn Belt Livestock Feeder,* 1946–68; *Drovers' Journal,* 1970; *Endocrinology,* 1941–51; *Environment,* 1971–78; *Farm Chemicals and Croplife,* 1960–73; *Farm Chemicals, Croplife, and Ag Chemicals and Chemical Fertilizer,* 1973–80; *Farm Journal,* 1952–79; *Farm Quarterly,* 1946–72; *Farm Research,* 1964–67; *FDA Consumer,* 1973–80; *FDA Papers,* 1967–73; *Feed Age,* 1951–80; *Feed Bag,* 1957–61; *Feed Management,* 1966–80; *Feedlot,* 1966–69; *Feedlot Management,* 1969–79; *Feedstuffs,* 1950–82; *Flour and Feed,* 1953–56; *Food and Cosmetic Toxicology,* 1963; *Food and Drug Review,* 1950–80; *Food Drug Cosmetic Law Journal,* 1950–80; *Food Drug Cosmetic Law Quarterly,* 1947–50; *Food Drug Cosmetic Law Reporter: Developments,* 1963–81; *Food Drug Cosmetic Law Reporter: Food Additive and Pesticide Petitions and Proposals,* 1958–73; *Food Drug Cosmetic Law Reporter: New Drug Reports,* 1962–82; *Food Research,* 1953–55; *I.B.I.A. News,* 1974–80; *IBPA Journal,* 1971–75; *I.C.A. Journal,* 1972–75; *Industrial Research Development,* 1978–80; *Iowa Academy of Science,* 1954–57; *Iowa Agriculturist,* 1952–80; *Iowa Beef Council, Annual Report,* 1975–80; *Iowa Beef Industry Council Newsletter,* 1974–80; *Iowa Cattleman,* 1975–80; *Iowa Farm Science,* 1946–71; *Iowa Grain and Feed Review,* 1974–77; *Journal of Agricultural and Food Chemistry,* 1953–79; *Journal of Agricultural Science,* 1972–80; *Journal of Animal Science,* 1942–72; *Journal of Biological Chemistry,* 1950–63; *Journal of Consumer Affairs,* 1967–80; *Journal of Consumer Research,* 1974–78; *Journal of Dairy Science,* 1950–60; *Journal of Experimental Medicine,* 1947–50; *Journal of Food Science,* 1956–59; *Journal of Pharmacology and Experimental Therapeutics,* 1958; *Journal of Pharmaceutical Sciences,* 1948–56; *Journal of Public Law,* 1952–64; *Journal of Range Management,* 1956–75; *Journal of the American Medical Association,* 1969–79; *Journal of the American Pharmaceutical Association: Practical Edition,* 1957–62; *Journal of the American Pharmaceutical Association: Scientific Edition,* 1959–79; *Journal of the American Veterinary Medical Association,* 1943–79; *Journal of the Association of Official Analytical Chemists,* 1955–79; *Journal of the National Cancer Institute,* 1940–79; *Journal of Toxicology and Environmental Health,* 1975–80; *Kansas Farmer,* 1974–81; *National Journal Reports,* 1973–75; *National Police Gazette,* 1954–66; *Nebraska Farmer,* 1944–62; *New England Journal of Medicine,* 1971–75; *New*

York's Food and Life Sciences Quarterly, 1968–70; *North American Charolais Journal,* 1976–77; *North Dakota Farm Research,* 1952–71; *Official Publication—American Feed Control Officials,* 1978–82; *Organic Gardening and Farming,* 1954–79; *Poultry Science,* 1943–50; *Preventive Medicine,* 1972–77; *Proceedings—Agricultural Research Institute,* 1952–79; *Proceedings—Beef Cattle Field Day,* 1977–80; *Proceedings—Distillers' Feed Conference,* 1947–57; *Proceedings—Maryland Nutrition Conference for Feed Manufacturers,* 1972–80; *Proceedings of the American Society for Animal Production,* 1940; *Proceedings of the Cornell Nutrition Conference for Feed Manufacturers,* 1954–59; *Proceedings of the Nutrition Council of the American Feed Manufacturers' Association,* 1953–78; *Proceedings of the Society of Experimental Biology and Medicine,* 1940–44; *Quarterly Bulletin—Association of Food and Drug Officials of the United States,* 1953–76; *Regulation,* 1977–80; *Reports of the Council for Agricultural Science and Technology,* 1973–80; *Sandhills Feeder Cattle,* 1972; *Saturday Review,* 1953–57; *Science,* 1953–82; *Science and Government Report,* 1971–80; *Science News,* 1973–78; *Science News Letter,* 1954–56; *Scientific American,* 1950–76; *Selected Proceedings of the Council on Consumer Information,* 1957–65; *Shorthorn World,* 1954–73; *South Dakota Farm and Home Research,* 1958–66; *Successful Farming,* 1948–79; *Texas Beef Conference Proceedings,* 1973–80; *Toxicology and Applied Pharmacology,* 1959; *Toxicology,* 1973–80; *Utah Farm and Home Science,* 1955–57; *Wallaces Farmer,* 1948–82; and *Wisconsin Agriculturist,* 1963–76.

Nonagricultural Newspapers

The nonagricultural newspapers I consulted included *Ames Daily Tribune,* 1951–80; *Des Moines Register,* 1953–81; *Des Moines Tribune,* 1958–72; *Iowa State Daily,* 1953–80; *National Observer,* 1969–76; *New York Times,* 1950–80; *Wall Street Journal,* 1963–80; and *Washington Post,* 1971–81.

Experiment Station and Agricultural College Bulletins and Reports

I also consulted numerous experiment station and agricultural college bulletins and reports including *California Agricultural Experiment Station Research Bulletins; Indiana Agricultural Experiment Station Research Bulletins; Indiana Agricultural Experiment Station Annual Reports; Iowa State College Experiment Station Annual Reports; Michigan Agricultural Experiment Station Research Bulletins; Minnesota Agricultural Experiment Station Research Bulletins; Missouri Agricultural Experiment Station Research Bulletins; Nebraska Agricultural Experiment Station Research Bulletins; Nebraska Agricultural Experiment Station Annual Feeders' Day Reports; Nebraska Beef Cattle Report; Ohio Agricultural Experiment Station Research Bulletins; Purdue University Agricultural Experiment Station Annual Reports; Texas Agricultural Experiment Station Research Bulletins;* and *Virginia Agricultural Experiment Station Research Reports.*

Popular Books

In the years after 1960, authors wrote numerous books predicting an imminent cataclysm due to the dangers associated with food, drugs, cosmetics, or environmental problems. This literary explosion was quite unlike the famous "guinea pig" literature of the 1930s. That early literature sought to take power out of the hands of consumers who were misled or manipulated by advertisers and to place it in the hands of pristine scientists. The literature argued for scientized regulation, a thrust that was embodied in the 1938 Food, Drug, and Cosmetic Act. With the passage of that act, the guinea pig literature ceased. The literature of the 1960s and later decades was of a very different character. It complained instead that scientists would not protect the individual. Scientists' debt to various special interests, whether acknowledged or unacknowledged, colored and invalidated their research and pronouncements. Individuals, lumped together as "the people" or "consumers" and served by self-defined, allegedly nonpartisan consumer advocates, needed to seize control. Among the studies of that type which I used in this book are Rachel Carson, *Silent Spring* (Boston: Houghton Mifflin, 1962); Everette E. Dennis and William L. Rivers, *Other Voices: The New Journalism in America* (San Francisco: Canfield, 1974); Ralph Adam Fine, *The Great Drug Deception* (New York: Stein and Day, 1972); Richard Harris, *The Real Voice* (New York: Macmillan, 1964); Jim Hightower, *Hard Tomatoes, Hard Times: A Report of the Agribusiness Accountability Project on the Failure of America's Land Grant College Complex* (Cambridge, Mass.: Schenkman, 1973); William Longgood, *The Poisons in Your Food* (New York: Simon and Schuster, 1960); Morton Mintz, *The Therapeutic Nightmare* (Boston: Houghton Mifflin, 1965); James S. Turner, *The Chemical Feast: The Ralph Nader Study Group Report on Food Protection and the Food and Drug Administration* (New York: Grossman, 1970); Harrison Wellford, *Sowing the Wind: A Report from Ralph Nader's Center for Study of Responsive Law on Food Safety and the Chemical Harvest* (New York: Grossman, 1972); and Tom Wolfe, *The New Journalism* (New York: Harper and Row, 1973).

Manuscript Collections

An essential source of information for this book was the papers of the U.S. Food and Drug Administration relating to DES, which have been moved to the Washington National Records Center, Suitland, Maryland. Still under the control of the FDA, the records at the center are stored in containers resembling Page boxes. These have accession lists but are not catalogued. Finding particular documents is not always possible; accession lists provide only gross information, such as the titles of the several folders within a given box. The relevance of those titles to the contents of the folders depends on the diligence of the clerks who filed the documents. Finding most individual documents requires the identification and searching of whole boxes. The material thus selected must be cleared through the Freedom of Information Office prior to use.

The following manuscript collections used in this book are housed at the Special Collections Department, W. Robert and Ellen Sorge Parks Library, Iowa State University, Ames, Iowa: the Floyd Andre Papers; the Iowa State University Research Foundation Papers; the Minute Books of the Iowa State Board of Regents; the Roswell Garst Papers; and the Wise Burroughs Papers.

Published Government Documents

The published government documents used in writing this book included "Report of the Food and Drug Administration," in *Annual Report of the Federal Security Agency*, 1950–52; "Annual Report of the Food and Drug Administration," in *Annual Report of the United States Department of Health, Education, and Welfare*, 1953–70; *Annual Reports of the United States Food and Drug Administration*, 1971–80; *Congressional Record*, 1953–79; *Federal Register*, 1944–81; *Technical Bulletin of the United States Food and Drug Administration*, 1959–60; *United States Statutes at Large*; *National Academy of Sciences/National Research Council Reports;* Food and Nutrition Board, Food Protection Committee, *Science and Food: Today and Tomorrow* (Washington, D.C.: National Academy of Sciences and National Research Council, 1961); Food and Nutrition Board, Food Protection Committee, *Problems in the Evaluation of Carcinogenic Hazard from Use of Food Additives* (Washington, D.C.: National Academy of Sciences and National Research Council, 1960); Food and Nutrition Board, Food Protection Committee, *Toxicants Occurring Naturally in Foods*, Publication no. 1354 (Washington, D.C.: National Academy of Sciences and National Research Council, 1966); National Research Council, Food Protection Committee, *Use of Chemical Additives in Foods* (Washington, D.C.: Food and Nutrition Board, National Research Council; and National Academy of Sciences, 1951); *Safe Use of Chemical Additives in Foods* (Washington, D.C.: Food and Nutrition Board, National Research Council; and National Academy of Sciences, 1952); Committee on Animal Nutrition, Subcommittee on Hormones, *Hormonal Relationships and Applications in the Production of Meats, Milk, and Eggs* (Washington, D.C.: National Academy of Sciences and National Research Council, 1959); Committee on Animal Nutrition, Subcommittee on Hormones, *Hormonal Relationships and Applications in the Production of Meats, Milk, and Eggs*, Publication no. 1415 (Washington, D.C.: National Academy of Sciences and National Research Council, 1966); and National Academy of Sciences, *The Use of Drugs in Animal Feeds: Proceedings of a Symposium*, Publication no. 1679 (Washington, D.C.: National Academy of Sciences, 1969).

Among the transcripts of congressional hearings which I consulted were *Chemicals and the Future of Man: Hearings before the Subcommittee on Executive Reorganization and Government Research of the Committee on Government Operations, United States Senate, Ninety-second Congress, First Session, April 6 and 7, 1971* (Washington, D.C.: GPO, 1971); *Chemicals in Food Products* (Washington, D.C.: GPO, 1951); *Drug Safety and Effectiveness: Hearings before the Interstate and Foreign Commerce Subcommittee, House of Representatives* (Washington, D.C.: GPO, 1976); *Food Additives: Competitive, Regulatory, and Safety Problems: Hearings before the Select Committee on Small Business, United States Senate, January 13 and 14, 1977* (Washington, D.C.: GPO, 1977); "Food and Drug Administration Appropriations for 1977," in *Hearings before a Subcommittee of the Committee on Appropriations, House of Representatives* (Washington, D.C.: GPO, 1976); *Food and Drug Administration—Study of the Delaney Clause and Other Anticancer Clauses: Hearings before a Subcommittee of the Committee on Appropriations, Ninety-third Congress, Second Session* (Washington, D.C.: GPO, 1974); *Hearing on the Regulatory Policies of the Food and Drug Administration before a Subcommittee of the Committee on Government Operations, House of Representatives, Ninety-first Congress, Second Session, June 9, 1970* (Washington, D.C.: GPO, 1970); *Hearings before the Select Committee on Nutrition and Human Needs of the United States Senate, Ninety-second Congress, Second Session, on Nutrition and Human Needs* (Washington, D.C.: GPO, 1973); *Impact of Chemical and Related Drug Products and Federal Regulatory Processes: Hearings before the Subcommittee on Dairy and Poultry of the Committee on Agriculture, House of Representatives, May 23–25, 1977* (Washington, D.C.: GPO, 1977); *Regulation of Chemicals in Food and Agriculture: Hearings before the Subcommittee on Agricultural Research and General Legislation of the Committee on Agriculture, Nutrition, and Forestry, United States Sen-*

ate, June 30 and July 19, 1977 (Washington, D.C.: GPO, 1977); *Regulation of Diethylstilbestrol (DES) and Other Drugs Used in Food Producing Animals: Twelfth Report by the Committee on Government Operations Together with Additional Views* (Washington, D.C.: GPO, 1973); *Regulation of Diethylstilbestrol (DES): Its Use as a Drug for Humans in Animal Feeds (Part I): Hearings before a Subcommittee of the Committee on Government Operations, House of Representatives, Ninety-second Congress, First Session, November 11, 1971* (Washington, D.C.: GPO, 1972); *Regulation of Diethylstilbestrol (DES): Its Use as a Drug for Humans in Animal Feeds, Part 2: Hearings before a Subcommittee of the Committee on Government Operations, House of Representatives, Ninety-second Congress, First Session, December 13, 1971* (Washington, D.C.: GPO, 1971; *Regulation of Diethylstilbestrol (DES), 1972: A Hearing before the Subcommittee on Health of the Committee on Labor and Public Welfare, United States Senate, Ninety-second Congress, Second Session, on S. 2818, July 20, 1972* (Washington, D.C.: GPO, 1972); *Regulation of Diethylstilbestrol (DES), 1975: Joint Hearing before the Subcommittee on Health of the Committee on Labor and Public Welfare and the Subcommittee on Administrative Practice and Procedure of the Committee on the Judiciary, United States Senate, Ninety-fourth Congress, First Session, February 27, 1975* (Washington, D.C.: GPO, 1975); *Regulation of Food Additives and Medicated Animal Feeds: Hearings before a Subcommittee of the Committee on Government Operations, House of Representatives, Ninety-second Congress, First Session, March 16, 17, 18, 29, and 30, 1971* (Washington, D.C.: GPO, 1971); *Regulation of New Drug R. and D. by the Food and Drug Administration, 1974: Joint Hearings before the Subcommittee on Health of the Committee on Labor and Public Welfare and the Subcommittee on Administrative Practice and Procedure of the Committee on the Judiciary, United States Senate, Ninety-third Congress, Second Session, September 25 and 27, 1974* (Washington, D.C.: GPO, 1975); and *Regulatory Reform—Food and Drug Administration: Hearings before the Subcommittee on Oversight and Investigations of the Committee on Interstate and Foreign Commerce, House of Representatives, March 15 and 16, 1976* (Washington, D.C.: GPO, 1976).

SECONDARY SOURCES

The history of an emotionally charged issue such as the use of DES as a cattle feed additive often converts analysts into partisans. The recentness of the story compounds that tendency. For studies that explore some of the issues surrounding DES beef, see Orville Schell, *Modern Meat* (New York: Random House, 1984); Joseph V. Rodricks, "FDA's Ban of the Use of DES in Meat Production: A Case Study," *Agriculture and Human Values*, 3 (Winter–Spring 1986): 10–25; and Jeremy Rifkin, *Beyond Beef: The Rise and Fall of the Cattle Culture* (New York: Dutton, 1992). Susan E. Bell, "The Synthetic Compound Diethylstilbestrol (DES), 1938–1941: The Social Construction of a Medical Treatment" (Ph.D. diss., Brandeis U, 1980); Cynthia Laitman Orenberg, *DES: The Complete Story* (New York: St. Martin's, 1981); Robert Meyers, *D.E.S.: The Bitter Pill* (New York: Putnam's, 1983); and Richard Gillam and Barton J. Bernstein, "Doing Harm: The DES Tragedy and Modern American Medicine," *Public Historian*, 9 (Winter 1987): 57–82, try to explain why the medical use of DES in pregnant women was unwarranted, while Roberta J. Apfel and Susan M. Fisher, *To Do No Harm: DES and the Dilemmas of Modern Medicine* (New Haven: Yale UP, 1984), seek to justify their medical colleagues' decisions.

Several studies of other questions of chemical regulation in modern America, especially outside the province of the Food and Drug Administration, are well known. They include Thomas R. Dunlap, *DDT: Scientists, Citizens, and Public Policy* (Princeton: Princeton UP, 1981); Hugh D. Crone, *Chemicals and Society: A Guide to the New Chemical Age* (Cambridge: Cambridge UP, 1986); Christopher J. Busso, *Pesticides and Politics: The Life Cycle of a Public Issue* (Pittsburgh: U of Pittsburgh P, 1987); and Samuel P. Hays, *Beauty, Health, and Permanence: Environmental Politics, 1955–85* (Cambridge: Cambridge UP, 1987). John H. Perkins, *Insects, Experts, and the Insecticide Crisis: The Quest for New Pest Management Strategies* (New York: Plenum, 1982), is of equal value and has been overlooked. Leonard A. Cole, *Politics and the Restraint of Science* (Totowa, N.J.: Rowman and Allanheld, 1983), especially 161–73; and Peter M. Sandman, David B. Sandman, Michael R. Greenberg, and Michael Gochfeld, *Environmental Risk and the Press* (New Brunswick, N.J.: Transaction Books, 1987), take a somewhat broader, less historical slant. Of use, also, are Sheila Jasanoff, *Risk Management and Political Culture* (New York: Russell Sage Foundation, 1986); and idem, *The Fifth Branch: Science Advisers as Policymakers* (Cambridge, Mass.: Harvard UP, 1990). Many of the aforementioned chemical concerns stem from a fear of cancer. For cancer in contemporary thought, see both Richard A. Rettig, *Cancer Crusade: The Story of the National Cancer Act of 1971* (Princeton: Princeton UP, 1977); and James T. Patterson, *The Dread Disease: Cancer and Modern American Culture* (Cambridge: Harvard UP, 1987).

The late nineteenth-century rise of the scientific professions and their assumption of authority in many arenas, as well as their persistence into the twentieth century, has been studied by numerous scholars. See particularly Samuel P. Hays, *Conservation and the Gospel of Efficiency: The Progressive Conservation Movement, 1890–1920* (Cambridge: Harvard UP, 1959); Louis Galambos, "The American Economy and the Reorganization of the Sources of Knowledge," in Alexandra Oleson and John Vos, eds., *The Organization of Knowledge in Modern America, 1860–1920* (Baltimore: Johns Hopkins UP, 1979), 269–82; Hugh Hawkins, "University Identity: The Teaching and Research Functions," in ibid., 285–312; Daniel J. Kevles, *The Physicists: The History of a Scientific Community in Modern America* (New York: Knopf, 1977); and Edward H. Beardsley, *The Rise of the American Chemical Profession, 1850–1900* (Gainesville: U of Florida P, 1964). I have also added to that literature. See Alan I Marcus, "Professional Revolution and Reform in the Progressive Era: Cincinnati Physicians and the City Elections of 1897 and 1900," *Journal of Urban History*, 5, no. 2 (Feb. 1979): 183–207; idem, "Back to the Present: Historians' Treatment of a City as a Social System during the Reign of the Idea of Community," in Howard Gillette, ed., *American Urbanism: A Historiographic Review* (Westport, Conn. Greenwood, 1987), 7–26; and Alan I Marcus and Howard P. Segal, *Technology in America: A Brief History* (San Diego: Harcourt Brace Jovanovich, 1989), 165–253. For the case of agricultural science, see Margaret W. Rossiter, "The Organization of the Agricultural Sciences," in Oleson and Voss, eds., *Organization of Knowledge*, 211–48; idem, "Graduate Work in the Agricultural Sciences," *Agricultural History*, 60, no. 2 (1986): 37–57; Wallace E. Huffman, "The Supply of New Agricultural Scientists by U.S. Land Grant Universities: 1920–1979," in Lawrence Busch and William Lacy, eds., *The Agricultural Scientific Enterprise* (Boulder, Colo.: Westview, 1986), 108–28; Edward H. Beardsley, *Harry L. Russell and Agricultural Science in Wisconsin* (Madison: U of Wisconsin P, 1969); Roy V. Scott, *The*

Reluctant Farmer: The Rise of Agricultural Extension to 1914 (Urbana: U of Illinois P, 1971); and Harold Lee, *Roswell Garst: A Biography* (Ames: Iowa State UP, 1984). Also see Alan I Marcus, *Agricultural Science and the Quest for Legitimacy: Farmers, Agricultural Colleges, and Experiment Stations, 1870–1890* (Ames: Iowa State UP, 1985); and idem, "The Wisdom of the Body Politic: The Changing Nature of Publicly Sponsored American Agricultural Research since the 1830s," *Agricultural History*, 62, no. 2 (Spring 1988): 4–26.

The early history of federal regulation of food and drugs is ably explored in James Harvey Young, *Pure Food: Securing the Federal Food and Drug Act of 1906* (Princeton: Princeton UP, 1989); and Oscar E. Anderson, Jr., *The Health of a Nation: Harvey W. Wiley and the Fight for Pure Food* (Chicago: U of Chicago P, 1958). Young has carried the analysis forward in his *The Medical Messiahs: A Social History of Health Quackery in Twentieth Century America* (Princeton: Princeton UP, 1967); idem, "The Agile Role of Food: Some Historical Reflections," in *Nutrition and Drug Interrelations* (New York: Academic, 1978), 1–18); idem, "From Oysters to After-Dinner Mints: The Role of the Early Food and Drug Inspector," *Journal of the History of Medicine and Allied Sciences*, 42 (1987): 30–53; and idem, "Food and Drug Enforcers in the 1920s: Restraining and Educating Business," *Business and Economic History*, 21 (1992): 119–28. He is presently writing a history of the FDA before 1940. Also useful on food and drug regulation within the progressive partnership are Charles O. Jackson, *Food and Drug Legislation in the New Deal* (Princeton: Princeton UP, 1970); Peter Temin, *Taking Your Medicine: Drug Regulation in the United States* (Cambridge: Harvard UP, 1980), especially 27–37; Kenneth Helrich, *The Great Collaboration: The First Hundred Years of the Association of Official Analytical Chemists* (Arlington, Va.: Association of Official Analytical Chemists, 1984); David F. Cavers, "The Evolution of the Contemporary System of Drug Regulation under the 1938 Act," in John B. Blake, ed., *Safeguarding the Public: Historical Aspects of Medicinal Drug Control* (Baltimore: Johns Hopkins P, 1970); 158–70; and Louis Lasagna, "1938–1968: The FDA, the Drug Industry, the Medical Profession, and the Public," in ibid, 171–179. For the beginnings of the dissolution of that partnership, see Phyllis Anderson Meyer, "The Last Per Se: The Delaney Cancer Clause in United States Food Regulation," (Ph.D. diss., U of Wisconsin, 1983); and Suzanne R. White, "Chemicals in Food: The Delaney Committee, 1950–52" (Ph.D. diss., Emory U, in progress).

Theories of social organization that undergird the mid-century progressive partnership are given shape in Ellis W. Hawley, *The Great War and the Search for Modern Order: A History of the American People and Their Institutions, 1917–1933* (New York: St. Martin's, 1979); idem, "Three Facets of Hooverian Associationalism: Lumber, Aviation, and Movies, 1921–1930," in Thomas K. McCraw, ed., *Regulation in Perspective: Historical Essays* (Cambridge: Harvard UP, 1981); and David E. Hamilton, *From New Day to New Deal: American Farm Policy from Hoover to Roosevelt, 1928–1933* (Chapel Hill: U of North Carolina P, 1991). Also see Marcus and Segal, *Technology in America*, 257–309. For the destruction of those long-held organizational principles, see William O'Neill's uproariously funny—and poignant—*Coming Apart: An Informal History of the 1960s* (Chicago: Quadrangle Books, 1971). Also see Morris L. Dickstein, *Gates of Eden: American Culture in the 1960s* (New York: Oxford UP, 1987); Tom Shachtman, *Decade of Shocks: Dallas to Watergate, 1963–1974* (New York: Poseidon, 1983); and

Christopher Lasch, *The Culture of Narcissism: American Life in an Age of Diminishing Expectations* (New York: Norton, 1978).

The fractionalization of the 1960s has yielded consumerism, which has quickly been confused with the emergence of consumer advocates. Paeans to consumer advocates and diatribes against them include Erma Angevine, ed., *Consumer Activists: They Made a Difference* (New York: Consumers' Union Foundation, 1982); Robert N. Mayer, *The Consumer Movement: Guardians of the Marketplace* (Boston: Twayne, 1989); Mark V. Nadel, *The Politics of Consumer Protection* (Indianapolis: Bobbs-Merrill, 1971); Dan M. Burr, *Abuse of Trust: A Report on Ralph Nader's Network* (Chicago: Regency Gateway, 1982); Charles McCarry, *Citizen Nader* (New York: Saturday Review, 1972); and Michael Pertschuk, *Revolt against Regulation: The Rise and Pause of the Consumer Movement* (Berkeley: U of California P, 1982).

Index

Abbott Laboratories, 57, 85
Adams, James S., 39
Agricultural Research Institute, 17, 33, 71, 143
Agricultural Research Service, 79; measurement of DES residues by, 106, 111, 121–22, 124–25, 128–29
Agri-industry. *See* American Feed Manufacturers' Association; Animal Health Institute; Pharmaceutical Manufacturers' Association
Alar, 158–59
Allied Chemical, 154
American Academy of Nutrition, 56
American Association for the Advancement of Science, 33, 41
American Bar Association, 19
American Cancer Society, 39, 146
American Chemical Society, 57
American Feed Manufacturers' Association (AFMA), 31; and animal drug legislation, 38, 63, 65, 75, 79; and Delaney clause, 56, 129; and DES, 49–52; and voluntary compliance, 86
American Home Products, 141, 153
American Medical Association, 76, 80, 85
American National Cattlemen's Association, 141–43, 154–55
American Society of Animal Science, and Delaney clause, 129, 137; and DES regulation, 113, 119, 140–41, 143, 150
American Veterinary Medical Association, 129
Anderson, Jack, 54
Anderson, Wayne, 155
Andrews, Frederick N., 13, 27–30
Angevine, Erma, 91
Animal Drug Act (1968), 2, 82
Animal Health Institute (AHI): and animal

drug legislation, 63, 65, 75, 79, 82; challenges SOM, 145–46; and Delaney clause, 56, 130; and DES, 49–52, 141; urges voluntary compliance, 86
Antibiotics, 94; cattlemen fear loss of, 81–82, 114, 123–24; regulation of, 79–81, 95, 116
Appel, James, 80
Association of Official Analytical Chemists (AOAC), 34, 35, 81, 125, 138
Aureomycin, 113
Ayres, Quincy, 14, 15

Ball, Charles, 155–56
Bayh, Birch, 104
Beeson, William M., 13, 129–30
Bellmon, Henry, 138–39
Benson, Ezra Taft, 54–55
Bergland, Bob, 145
Berglund, Roger, 94
Berry, Jesse E. O., 118
Bioassay, mouse uterine, 60, 74–75, 92; defined, 14; and initial regulation of DES, 21; as official method, 65–66, 80–81, 125; as regulatory problem, 36, 54, 97, 103, 142–43
Bioscience, 137, 152
Black, Charles, 127, 136, 140
Brown, Clarence J., 107
Browning, George, 14
Bryan, W. Ray, 73–74
Burroughs, Wise, 6, 24, 59, 95; and animal science research, 70–71, 136; defense of DES to scientists, 27–31; and DES and human cancer, 32; 52–53, 99, 119; and DES research, 11–15; and FDA approval for DES, 18–22; invents DES as growth promoter, 1; publicizes DES, 16
Butler, O. D., 140
Butz, Earl L., 57, 123

229

Cancer, 38–41, 72, 144–46; and animal scientists, 137–38, 141, 143; from beef, public fear of, 6, 32, 42, 112, 126; from beef, public hearings on, 103–6, 108, 118–19, 121–22, 138–39; and cranberries, 54–55; DES daughters, 99–100; and DES investigations at Iowa State, 52–53; and laboratory animal tests, 18, 35–36, 52–54, 74–75, 85, 136–37. *See also* Delaney cancer clause
Cancer Prevention Committee, 39
Cancer research, 63
Cardon, B. P., 116
Carson, Rachel, 64
Carter, Jimmy, 145
Case, Clifford, 104
Cattlemen, 111, 119, 127, 131, 134; accused of illegal DES use, 77, 80, 95–96, 100, 103–4, 108, 112, 117, 130; certification program of, 97–98; demand Delaney clause modification, 123–24, 129; and DES feedlots, 1, 28–29, 59; explanations of, for ignoring DES ban, 154–56; fear general agrichemical ban, 114–15, 134, 141; learn about DES, 15–17, 22–24; and meat quality, 27
Center for the Study of Responsive Law, 83, 84, 99, 120, 126
Centers for Disease Control, 78
Charles Pfizer Company, 36; and DES feeds, 24, 28; and DES implants, 30; fears of research slowdown by, 57, 130
Chemetron Corporation, 124, 125
Chemical and Engineering News, 57
Chemical Feast, The (Turner), 92–94
Chromatography: gas, 60–61; gas-liquid (GLC), 97–98, 100, 103, 109, 112, 116, 119, 124–25, 143; paper, 77, 81
Cipperly, John, 51
Citizens' Advisory Committee: urges FDA expansion, 34; urges FDA redirection, 63, 66, 68, 70–71
Clark, Dick, 139
Clegg, M. L., 29
Coats, Don, 74
Committee of Sixteen, 83
Committee to Get Drugs Out of Meat (DOOM), 112
Congressional hearings, 56; Delaney committee, 18–19; Fountain committee, 98, 104–11, 131–32; Kefauver hearings, 69; Kennedy subcommittee, 120, 122; Ribicoff

committee, 98; Rogers subcommittee, 139–41; Williams subcommittee, 38–39; 42–43
Congressional Office of Technology Assessment, 146–47
Congressional Record, 39
Consumer and Marketing Service (CMS): caught in deception, 103; creation of, 77; and DES residues, 86, 95, 100, 112, 116–18, 154; seeks improved residue testing method, 80–81, 97–98, 119
Consumer Bulletin, 41
Consumer Consultant Program, 49
Consumer Federation of America, 91, 145
Consumerism, 111–12, 118–19, 125–26; and Alar, 158–59; attempts by FDA to mold, 48–49; and Carter, 145; consumer advocates, 56, 57–58, 77–78, 91–92, 116, 122, 123, 134; and Food Additive Amendment passage, 38–43; genesis of, 5–6, 36–37, 41, 42, 69–70; and governmental reorganization, 82, 83, 85, 116, 128; and Johnson, 76–77; and Kennedy, 62–63, 65, 66; and Nader organization, 83–84, 92–94, 101–3, 120, 126–27, 140, 146; and Natural Resources Defense Council, 103–4; and New Journalism, 75–76, 95–96, 106, 110–11, 120–21; and Nixon, 86; and silicone breast implants, 158–59; and thalidomide and *Silent Spring,* 64–65
Consumer Protection and Environmental Health Service, 82–83, 85
Cost-benefit approach: applied to Delaney clause, 151–53; defined 7–8, 135; as foundation for current regulatory policy, 156–58
Council for Agricultural Science and Technology (CAST): as animal agriculture representative, 141, 143, 150; attacked, 152; and DES, 136–37; formed, 127
Crawford, Lester, 155–56
Crofoot, Jay, 155
Crow, Les, 155
Cunha, Tony, 113, 130
Curtis, Carl T., 112, 122, 138–39
Cyclamate, 85, 151

Dairy Science Association, 129
Danielson, Mrs. Melvin A., 112
Darby, William, 58
Davidson, David J., 142, 146, 148–50, 153
Davis, Glenn E., 143
Dawe's Laboratory, 124, 134, 141, 153

DDT, 4, 5, 64
Dean, James, 47
Delaney cancer clause, 59; arguments for
    repeal of, 123, 128–30, 135–36, 145,
    150–52; applied to DES, 49–50, 52–54,
    95–96, 112, 119–21, 126, 132, 138,
    142–43, 148, 150; considered at symposia,
    56–57; and cranberries, 55; and cycla-
    mates, 85, 94–95; enacted, 43–44; identi-
    fied, 1–2; modified by DES Proviso, 65
Delaney committee, 18–19
Delaney, James, 18, 37, 39–40, 42–43, 65,
    111, 118, 126, 138
DES daughters, 2, 104, 144–46; and DES in
    meat, 99, 102, 113, 118, 122, 136, 139, 149;
    and DES sons, 140, 148
DES Proviso, 148; enacted, 65–66; identified,
    2, 47; and radiotracers, 121, 125; and resi-
    due question, 95–96, 108, 135, 143; and
    SOM, 138–39, 145
Diethylstilbestrol (DES; also known as stil-
    bestrol): animal scientists defend, 112–13,
    137–38, 140, 143–44; attempts at science-
    based regulatory policy toward, 30–31,
    53–54, 59–61, 74–75, 136–37; banned,
    121–22, 124–25, 128–29, 153; ban over-
    turned, 132–33; in beef, 77, 100, 117;
    cattlemen violate FDA ban, 154–57; as
    chemical agriculture representative,
    114–17, 123–26, 129–30; consumer suspi-
    cion of, as human carcinogen, 32, 35–37,
    52–53, 57–58, 62, 77–78, 101–2, 111–12,
    144–45; and cost-benefit approach,
    134–35, 140–41, 146–48; and Delaney
    committee, 18–19; and DES daughters,
    99–100; and DES Proviso passage, 65–66;
    and disputes within FDA, 107–11; as
    embodying contemporary regulation, 5–8;
    FDA approval of, 17–18, 20–22; FDA
    freeze on use of, 49–52; as FDA hearing
    subject, 117–20, 142–43, 146; focus of leg-
    islation on, 1–2; and Food Additive Amend-
    ment, 39–40, 42–44; identified, 1; inade-
    quate testing for, 97–98, 103; increase in
    permitted dosage of, 86; judicial ruling
    against, 148–50; invention as cattle feed
    additive, 13–15; as legal action subject,
    103–4, 145–46; Lilly patent of, 20–21; and
    meat quality, 27; in newspapers, 95–97;
    patent infringements, 28; practical implants
    of, 29–30; publicity surrounding, 16,
    22–24, 59; in Science, 121; as subject of

Congressional action, 98, 104–11, 120,
    122–23, 131–32, 138–40, 150–51
District of Columbia Court of Appeals,
    131–33, 142, 153
Dobbs, Sir Charles, 12
Dubos, Rene J., 64
DuShane, Graham, 44

Eastland, James, 138
Eckardt, Robert E., 63–64
Edwards, Charles C., 112; appointment as
    FDA commissioner, 85; assumption of
    office, 94–95; call for DES hearing by, 117,
    119; and Delaney clause, 126; issue of DES
    ban by, 121–22; leaves FDA, 128; testifies
    before Congress, 98–99, 106–9, 120
Eisenhower, Dwight David, 35, 42, 50,
    55–56, 58–59
Elanco, 86, 95, 97
Eli Lilly and Company: and breast cancer,
    144–45; defense of DES by, 52; and
    Delaney clause, 130; and DES ban, 124;
    and DES manufacturer, 20–24; and dosage
    doubling, 86; and patent infringement,
    27–28; and residue tests, 31, 36; sales,
    57, 59
Eller, Burton, 155
Environmental Defense Fund, 2, 103, 141
Environmental Protection Agency, 99, 158
Epstein, Samuel S., 98, 123, 126

Fannin, Paul, 139
Farm Bureau Federation, 78
Federal Register, 19, 44, 140, 153
Federation of Homemakers, 103
Federation of Women's Clubs, 78
Feed Management, 156
Feedstuffs, 17, 51, 94, 106, 155
Feldmann, Edward G., 152
Finch, Robert, 85
Flemming, Arthur, 50, 54–56, 58–61
Flour and Feed, 29
Fluorometric analysis, 80–81
Foley, Thomas S., 150–51
Folkers, Karl, 12
Food Additive Amendment (1958), 42–43,
    48–49, 51, 54, 57, 59
Food and Drug Administration (FDA), 144;
    and Alar, 158–59; actions against DES
    in beef, 116–18, 121–22, 125, 128–30,
    140–42, 148–50, 153; and Animal Drug
    Amendment, 82–83; approval of DES by,

Food and Drug Administration (FDA) (*cont.*)
17–18, 20–22, 24, 31; and CAST, 137–38,
143, 150; and Congressional action,
98–100, 104–10, 120, 122–23, 131–32,
135–36, 138–39, 150–51; and Food Additive Amendment, 42–44; history of, 1–5,
18–19; internal challenges in, 107–11; and
Kennedy, 62–67; legal challenges to,
124–25, 131–33, 135, 145–46, 153; and
medicated feeds symposium, 33–36; "new"
FDA, 68–71, 75–82, 82–86, 92–93,
101–2, 121, 128, 140; and punishment of
violators, 154–55; and regulatory legislation suggestion, 37–38, 41; reorganization
of under Food Additives Amendment,
48–49, 59–61; and response to DES
Proviso passage, 65–66; SOM document-
ing, 130–31, 145–46, 150–51; as subject of
letter writing campaign, 111–12, 118–19,
125–26; and treatment of DES under DES
Proviso, 95–98 100–101, 103, 112; and
treatment of DES under Food Additive
Amendment, 49–52, 54–55, 61–62
Food, Drug, and Cosmetic Act (1938), 4, 42,
142
Food and Nutrition Board of National
Research Council, 19, 33
Food Drug Cosmetic Law Journal, 33
Food Drug Cosmetic Law Quarterly, 19
Food Law Institute, 19, 44
Food Protection Committee, 19, 37, 58,
72–73
Foreman, Carol Tucker, 145
Fountain, L. H.: as Congressional subcom-
mittee chair, 98; criticism of Edwards by,
117–18, 122, 126; and DES hearings,
104–10, 131–32
Frey, C. N., 33
Friedman, Milton, 128

Galbraith, John Kenneth, 76
Gardner, Sherwin, 128–29, 142
Garst, Roswell, 11–12
Gass, George H., 74, 136–37
Gass study, 92, 113; cited by Davidson,
149–50; cited by Wellford, 101–2;
described, 74–75; and Fountain commit-
tee, 108; modification by Gass, 136
General Accounting Office, 152
Gillespie, R. L., 110–11, 121
Glennon, W. E., 38, 63

Goddard, James, as FDA head, 78–82, 85
Goldhammer, Gilbert S., 107–8
Goldwater, Barry, 139
Graham, Nora, 74
Grain and Feed Dealers' National Association,
57
Gross, M. Adrian, 107, 109

Hale, William H., 30, 113, 137
Hansen, Clifford P., 139
Hansford, Mrs. Joseph T., 119
Hardin, Clifford, 99
Hard Tomatoes, Hard Times (Hightower), 116
Harris, Oren, 65
Hart, Gary, 139
Harvey, John L., 19, 21
Hawkins, David, 103–4, 112, 117
Health, Education and Welfare (HEW), 98,
103; attacked by Ley, 86; attacked by
Turner, 93; and cranberries, 54–55; and
Delaney clause, 43, 58, 123; and DES, 50,
70–71, 125, 135, 138, 146; and DES
Proviso, 65; reorganization of, 82, 85
Henderickson, R. M., 130
Herbst, Arthur L., 99, 102, 104, 113, 136
Hertz, Roy, 105–6
Hess and Clark, 59, 116, 124–25, 131, 141,
153–54
Hightower, Jim, 116
Hilton, James, 20, 52
Hines, D. C., 52
Homeyer, Paul, 22, 27
Honorof, Ida, 112
Hruska, Roman, 138–39
Huddleston, Walter, 138–39
Hueper, William C., 41, 63
Humphrey, Hubert, 2, 63, 65–66, 71
Hutt, Peter Barton, 107–8, 128–29, 135

Individuation: and cost-benefit approach,
134–35, 147, 151, 153, 156–59; in decision
making, 7–8, 91–92, 114–15; indications
of, 47–48, 61–63, 68–69
Industrial Research Development, 152
Institute of Medicine, 151–52
International Union against Cancer, 39–41
Iowa Cattle Feeders' Day, 15
Iowa State College: and cancer and DES, 32,
52–53, 136; as center of DES conspiracy,
95; and DES licensing, 20–23; and DES
research, 1, 11–15, 27–28; and DES re-

search publicizing, 15–16; DES royalties, 59; as home of CAST, 127; rivalry with Purdue, 30–31

Iowa State College Research Foundation (ISCRF): established, 14–15; and Hale, 30; licenses DES, 20; patents stilbestrol, 15–16; royalties, 28

John Scott Award for the Benefit of Man, 31
Johnson, Lyndon, 76
Johnston, Bennett, Jr., 139
Journalism, New, 2, 75, 92
Journal of Agricultural and Food Chemistry, 33
Journal of the American Medical Association, 144
Journal of the National Cancer Institute, 74, 137
Journal of the Pharmaceutical Sciences, 152
Journal of Toxicology and Environmental Health, 140, 143
Jukes, Thomas, 6; and animal nutrition, 12; attacks Kennedy, 139; as DES advocate, 126, 141, 143, 150; urges Delaney clause modification, 130, 136

Kefauver, Estes, 2, 65, 69
Kefauver-Humphrey Drug Amendment (1962), 2, 65
Kelsey, Frances, 64
Kennedy, Donald, 145, 150–51, 153
Kennedy, Edward M., 2, 6, 132; attacks Van Houweling, 142; leads Congressional opposition to DES, 120, 122–23, 126, 138–39, 141
Kennedy, John, 62–63, 65–66
Klosterman, Earle W., 29
Kowalk, A. J., 110–11, 121
Krushchev, Nikita, 11

Lang, John S., 95–96
Larrick, George P., 35, 76–77
Laxalt, Paul, 139
Leavitt, Chuck, 156
Lederle Laboratories, 12
Lee, Armistead M., 107
Ley, Herman, 81, 84–86
Lilly. *See* Eli Lilly and Company
Lipsett, Mortimer B., 105
Lonely Crowd, The (Reisman), 47
Longgood, William, 57–58

Lorenz, Fred W., 12–13
Lyng, Richard, 98, 100

McClung, John, 155
McGee, Gale, 139
McGovern, George, 104
Man in the Grey Flannel Suit, The (Wilson), 47
Mantel, Nathan, 73–74, 109–10, 126
Martin, James, 151
Martin, Ronald, 155
Martin, W. Coda, 56
Meat Cutters' Union, 78
Merck, 20, 124
Midwest Feed Manufacturers' Association, 16
Miller, Arthur L., 43
Mink, Patsy, 144
Mintener, Bradshaw, 50, 66
Mintz, Morton: DES reportage for *Washington Post*, 106, 110, 118; *Therapeutic Nightmare*, 75–76. *See also* Journalism, New
Mosher, Mrs. C., 119
Moss, Frank, 104

Nader, Ralph: attacks DES, 2, 6; attacks meat industry, 94, 127; organization founded by, 83–84, 140, 150
National Academy of Sciences, 151; and antibiotics, 79; attacked as biased, 121, 123, 127–28; as HEW advisor, 58; research reports, 31, 77, 145
National Animal Disease Laboratory, 95
National Cancer Institute, 2; attacked as inept, 152–53; defends DES, 36, 41, 52; funds DES research, 61; as HEW advisor, 58; opposes DES, 105, 118; as research agency, 139, 146; and virtual safety, 107, 109
National Center for Toxicological Research, 140
National Institute of Agriculture, 56
National Institutes of Health, 51, 54, 66, 75, 146
National Livestock Feeders' Association, 77, 142–43
National Police Gazette, 32, 69, 76–77
National Research Council, 31, 37, 77
National Welfare Rights Organization, 103
Natural Food and Farming, 77
Natural Resources Defense Council, 2, 108; and Alar, 158; demands immediate DES ban, 117, 141; sues FDA over DES, 103

Nature, 139
Nelson, Gaylord, 56, 84, 86
New Animal Drug Application (NADA), and
    Animal Drug Act, 82; defined, 22; for
    DES, 28; and DES Proviso, 63, 65–66;
    hearing for holders of, 118, 133–34,
    140–41; refusal of FDA to reinstate for
    DES, 124; revoked, 121, 142, 150
New England Journal of Medicine, 99
New York Academy of Sciences, 126
New York City, 42
New Yorker, 64
New York State, 42
New York State Bar Association, 19
New York Times, 36, 39, 41, 130
New York University Law School, 19
Nixon, Richard, 83, 121, 127
No-effects level, 44, 54, 72; and DES,
    60–61, 66, 102, 149–50

Ohio State Experiment Station, 12
Olentine, Charles, 156
Organic Gardening and Farming, 32
Organization Man, The (Whyte), 47

Pacific Legal Foundation, 141–43, 145–46
Pearson, Drew, 54
Pelly, Thomas, 77–78
Perry, T. W., 13
Pfizer. *See* Charles Pfizer Company
Per se rule, defined, 41; and Food Additive
    Amendment, 43; modified, 42–43; re-
    jected, 65
Pharmaceutical interests, 97, 107, 154; and
    antibiotics, 79–82; appeal of FDA action
    on DES, 124–25, 131–33, 141, 153; com-
    petition among for DES, 28, 30; critiques
    of, 75–76, 78–79, 84; and cyclamates, 85;
    defense of DES by, 52, 115–16, 119–20;
    and Delaney clause, 57, 130; and DES li-
    censing, 20–21; and DES testing, 24–25,
    36; expansion of from DES, 28, 59; and
    FDA approval of DES, 21–22, 86; and
    marketing of DES, 22–24, 95, 134; and
    thalidomide, 64
Pharmaceutical Manufacturers' Association,
    107
Pillsbury Company, 50
Poisons in Your Food (Longgood), 57, 69, 77
Poskanzer, David C., 99
Poultry Science Association, 129
Preventive Medicine, 126, 141

Price, Bob, 112, 122
Progressive Partnership: and Burroughs, 11,
    12; collapse of, 5–6, 61–65; decline of,
    47–48, 53–54, 59–61; defined, 3–5; and
    DES, 23–25, 26, 49–51; and Food Addi-
    tive Amendment, 44, 48–49, 54–59, 61,
    65; and ISCRF, 14, 15; Lilly and, 23; and
    regulation, 33–35; 37–38, 52–53, 59,
    66–67; and scientists, 27–31, 33; and
    transmission of knowledge, 16–17, 19
Proxmire, William, 2, 132; demands immedi-
    ate FDA ban of DES, 99–100, 110, 122,
    126; and DES daughters, 99; Senate testi-
    mony of, 120; sponsors Senate DES ban,
    104
Public Citizens' Health Research Group, 140
Purdue University Agriculture Experiment
    Station, 13, 29–31
Pure Food and Drugs Act (1906), 3, 4

Quaker Oats, 24, 50

Radioimmunoassay, 116, 125, 138, 143–144
Radiotracers: problems with, 143–44; use of
    in detection of DES residues, 54, 110, 111,
    121, 124–25, 128–29
Rankin, Winton, 63
Rath Packing Company, 15
Rauscher, Frank J., Jr., 118, 121
Reid, Ogden, 104
Reisman, David, 47
Ribicoff, Abraham, 98–99, 104, 122
Richardson, Elliot L., 43
Roberts, Kenneth A., 63
Roger, Paul, 139–41
Rotto, Theodore, 156
Rowen, Hobart, 158
Ruckelshaus, William D., 99

Saccharin, 144, 151–53
Safety: 100:1, 71–75, 108; virtual, 73–74,
    107–10, 130, 137, 145–46, 151, 158
Saffiotti, Umberto, 105, 126
Scherle, Bill, 116, 119, 122, 135–36
Schmidt, Alexander M., 131, 140, 142
Schweiker, Richard, 138–39
Science, 33, 44, 57, 120, 121, 129
Scientists, 77, 101; acknowledged DES as car-
    cinogen, 18; and agricultural public, 16–17,
    24–25, 29, 31–32, 55–56; in CAST, 127,
    143; collapse of consensus among, 5–7,
    62, 71–75, 77, 92–94; and consumers,

32–35; and cost-benefit approach, 7–8,
134–35, 140–42, 148–50, 156–59; in government, 3–4, 38, 48, 58, 59–61, 83, 84,
130–31; idea of consensus among, 5; in industry, 4–5, 22, 32–33, 51, 52–54, 116,
140; and interpretation, 63–64, 70–71,
84–85, 94–95, 96–97, 102, 104–11, 113,
119, 127–30, 135–38, 143–47, 150–53; at
Iowa State, 11–14, 23; Iowa State-Purdue
tensions, 27–28, 29–31; at Purdue, 13;
symposia of, 35–36, 79–80
Sensitivity of method (SOM), 145–46,
151–52
Shearer, Phineas, 14
Shearson Loeb Rhoades, 156
Silent Spring (Carson), 64, 69
Silicone breast implants, 158
Sioux City livestock market, 15
"Sixty Minutes," 158
Smith, William E., 71; attack on DES feeds,
35–36; cited by Longgood, 58; influence
on Delaney, 40–42; revives International
Union against Cancer, 38–39
Snuggs, Marian, 118–19
Sowing the Wind (Wellford), 126
Specifide, 20
Spiller, Robert, 152
Stahl, Lesley, 141
Stein, Mary Ann, 103
Story, Charles D., 30
Strawman, Clifford, 52
Streep, Meryl, 158
Successful Farming, 29
Sullivan, Leonor K., 37, 65, 122
Swenson, Mrs. Robert, 32, 34

Texas Cattle Feeders' Association, 155–56
Thalidomide, 64, 66, 69, 75
Therapeutic Nightmare, The (Mintz), 75
Thery, Theodore P., 29
Thomas, Helen, 54
Thurmond, Strom, 138
Tower, John, 112, 139
Trenkle, Allen H., 96–97
Trichter, Jerome B., 42
Turner, James S., 92–94, 126

Ulfelder, Howard, 99
Umberger, Ernest, 60
University of California, 12
Upjohn, 130
U.S. Comptroller General, 122–23
U.S. Congress, 18–19, 38, 40–43, 47,
55–57, 59, 61, 65, 71, 116, 128
USDA Agricultural Research Center, 24
U.S. Department of Agriculture (USDA),
2, 31, 36; analytical methods of, 80–81,
97–98, 100, 124–25, 128–29; Consumer
and Marketing Service established, 77; and
DES residue examination, 77–78, 86,
95–97, 116–18, 138, 140, 154; dispute with
FDA, 54–56, 58; as regulatory agency,
37–38, 102–4, 112, 119–20
Utah, 42

Van Houweling, Cornelius, 155; accusation
of, in DES conspiracy, 95–96, 142;
appointment to FDA, 79; and Congress,
98, 100, 112; and DES daughters, 99; position on antibiotics, 81
Vander Jagt, Guy, 130
Vineland Laboratories, 131, 141, 153–54

Wade, Nicholas, 121
Washington Post: and Alar, 158; and New
Journalism, 75–6, 126; role in DES contro-
versy, 106, 110, 140
Weinberger, Caspar, 138
Wellford, Harrison, 98, 100–101, 116, 126
White, John N. S., 77–78, 121
White Laboratories, 20
Whyte, William H., Jr., 47
William S. Merrill Company, 64
Williams, John Bell, 42
Wilson, Sloan, 47
Wilson Packing Company, 15
Wolfe, Sidney, 140, 145–46, 150

Yantla, Harvey E., 17
Yeutter, Clayton, 98, 106

Zero tolerance, 116; attacked, 136; defined, 51